Digital Picture Processing

Computer Science and Applied Mathematics
A SERIES OF MONOGRAPHS AND TEXTBOOKS

Editor
Werner Rheinboldt
University of Maryland

HANS P. KÜNZI, H. G. TSCHACH, and C. A. ZEHNDER. Numerical Methods of Mathematical Optimization: With ALGOL and FORTRAN Programs, Corrected and Augmented Edition

AZRIEL ROSENFELD. Picture Processing by Computer

JAMES ORTEGA AND WERNER RHEINBOLDT. Iterative Solution of Nonlinear Equations in Several Variables

AZARIA PAZ. Introduction to Probabilistic Automata

DAVID YOUNG. Iterative Solution of Large Linear Systems

ANN YASUHARA. Recursive Function Theory and Logic

JAMES M. ORTEGA. Numerical Analysis: A Second Course

G. W. STEWART. Introduction to Matrix Computations

CHIN-LIANG CHANG AND RICHARD CHAR-TUNG LEE. Symbolic Logic and Mechanical Theorem Proving

C. C. GOTLIEB AND A. BORODIN. Social Issues in Computing

ERWIN ENGELER. Introduction to the Theory of Computation

F. W. J. OLVER. Asymptotics and Special Functions

DIONYSIOS C. TSICHRITZIS AND PHILIP A. BERNSTEIN. Operating Systems

ROBERT R. KORFHAGE. Discrete Computational Structures

PHILIP J. DAVIS AND PHILIP RABINOWITZ. Methods of Numerical Integration

A. T. BERZTISS. Data Structures: Theory and Practice, Second Edition

N. CHRISTOPHIDES. Graph Theory: An Algorithmic Approach

ALBERT NIJENHUIS AND HERBERT S. WILF. Combinatorial Algorithms

AZRIEL ROSENFELD AND AVINASH C. KAK. Digital Picture Processing

In preparation

DENNIS C. TSICHRITZIS AND F. H. LOCHOVSKY. Data Base Management Systems

SAKTI P. GHOSH. Data Base Organization for Data Management

Digital Picture Processing

AZRIEL ROSENFELD

Computer Science Center
University of Maryland
College Park, Maryland

AVINASH C. KAK

School of Electrical Engineering
Purdue University
West Lafayette, Indiana

ACADEMIC PRESS   New York   San Francisco   London   1976
A Subsidiary of Harcourt Brace Jovanovich, Publishers

ACADEMIC PRESS, INC.
111 Fifth Avenue, New York, New York 10003

United Kingdom Edition published by
ACADEMIC PRESS, INC. (LONDON) LTD.
24/28 Oval Road, London NW1

Library of Congress Cataloging in Publication Data

Rosenfeld, Azriel,   (date)
      Digital picture processing.

      (Computer science and applied mathematics series)
      "Supersedes an earlier book . . . Picture processing
by computer."
      Includes bibliographies.
      1.   Optical data processing.   I.   Kak, Avinash C.,
joint author.   II.   Title
TA1630.R67   621.3819'598   75-16874
ISBN 0–12–597360–8

Contents

v

Preface

It is now about 20 years since the first appearance of papers dealing with the processing of pictorial information by computer. The growth of the field has been rapid; it is estimated that currently over 500 papers a year are published on the subject, counting only the English-language literature. A large fraction of these papers are application-oriented; but over the years, a considerable body of basic picture-processing techniques has been developed. This book treats the field of picture processing from a technique-oriented standpoint.

The book supersedes an earlier book by one of us (A.R.), *Picture Processing by Computer* (Academic Press, 1969; Japanese translation: Kyoritsu Shippan, 1971; Russian translation: Mir, 1972). We have included much more extensive treatments of picture digitization, compression, and restoration; this material (Chapters 4, 5, 7) and the preliminary Chapter 2, were written by A.C.K. On the other hand, we do not cover optical (or other analog) processing methods. We have also not treated problems specific to the analysis of three-dimensional scenes, such as the use of stereoscopic or range-finding techniques to determine object distances, or the application of projective geometric methods. Neither do we treat the processing of images that are synthesized by computer; this is the principal concern of *computer graphics*.

Some mathematical background is desirable for parts of this book, notably Chapters 4, 5, and 7. Specific areas in which some background would be helpful are linear systems theory (including transform techniques) and probability theory (random variables, stochastic processes). We have provided brief introductions to these areas, and have generally tried to make the presentation relatively self-contained. We felt this to be necessary in order to make the book usable by students in both electrical engineering and computer science.

The book should be suitable for a one- or two-semester advanced undergraduate or graduate course in picture processing. It contains enough material to permit some selection of topics. A variety of exercises has been provided to supplement the text. Courses based on material in this book have been offered repeatedly in both electrical engineering and computer science departments.

Acknowledgments

The continuing support of A.R. by the Information Systems Branch, Office of Naval Research, under Contract N 00014-0239-0012, is gratefully acknowledged. We would also like to express our appreciation to Judith M. S. Prewitt for helpful discussions during the planning of this book, and for comments on the first draft. A number of our colleagues helped us by providing figures (see the individual figure credits); in addition, many of the figures were produced by Joan S. Weszka, Andrew Pilipchuk, and others at the University of Maryland. Shelly Rowe and Mary Ann Macha, among others, did an excellent job of preparing the text. To these, and to others too numerous to mention, our sincerest thanks.

The authors wish to express their indebtedness to the following individuals and organizations for their permission to reproduce the figures listed below.

Chapter 3: Figures 1, 4–7, and 11, from T. N. Cornsweet, *Visual Perception*, Academic Press, New York, 1970. Figure 3, from J. Beck, *Surface Color Perception*, copyright © 1972 by Cornell University; used by permission of Cornell University Press, Ithaca, New York. Figure 8, from S. Coren, "Subjective contours and apparent depth," *Psychological Review* **79**, 1972, 359–367. Figures 9–10, from M. Luckiesh, *Visual Illusions*, Dover, New York, 1965. Figure 12, from J. Beck, "Similarity grouping and peripheral discriminability under uncertainty," *American Journal of Psychology* **85**, 1972, 1–19. Figures 13–16, from M. Wertheimer, "Principles of perceptual organization," in D. C. Beardslee and M. Wertheimer, eds., *Readings in Perception*, 115–135, © 1958 by Litton Educational Publishing, Inc.; reprinted by permission of Van Nostrand Reinhold Company, Princeton, New Jersey. Figures 17–19, from J. Hochberg, *Perception*, © 1964, by permission of Prentice-Hall, Inc., Englewood Cliffs, New Jersey.

Chapter 4: Figure 8, from R. Legault, "The aliasing problems in two-dimensional sampled imagery," in L. M. Biberman, ed., *Perception of Displayed Information*, 279–312, Plenum Press, New York, 1973. Figure 10, from L. M. Biberman, "A summary," *ibid.*, 313–322. Figure 12, from A. Habibi and P. A. Wintz, "Image coding by linear transformation and block quantization," *IEEE Transactions on Communication Technology* **COM-19**, 1971, 50–62. Figure 15, from T. S. Huang, O. J. Tretiak, B. Prasada, and Y. Yamaguchi, "Design considerations in PCM transmission of low-resolution monochrome still pictures," *Proceedings of the IEEE* **55**, 1967, 331–335. Figures 17–19, from T. S. Huang, "PCM picture transmission," *IEEE Spectrum* **2**, 1965, 57–63. Table 1, from J. Max, "Quantizing for minimum distortion," *IRE Transactions on Information Theory* **IT-6**, 1960, 7–12.

Chapter 5: Figures 1 and 3–9, from P. A. Wintz, "Transform picture coding," *Proceedings of the IEEE* **60**, 1972, 809–820. Figures 12–14 and Tables 2–3, from A. Habibi, "Comparison of *n*th-order DPCM encoder with linear transformations and block quantization techniques," *IEEE Transactions on Communication Technology* **COM-19**, 1971, 948–956.

Chapter 6: Figures 11 and 14, from D. A. O'Handley and W. B. Green, "Recent developments in digital image processing at the Image Processing Laboratory of the Jet Propulsion Laboratory," *Proceedings of the IEEE* **60**, 1972, 821–828. Figure 15, from T. G. Stockham, Jr., "Image processing in the context of a visual model," *ibid.*, 828–842.

Chapter 7: Figure 3, from J. L. Harris, Sr., "Image evaluation and restoration," *Journal of the Optical Society of America* **56**, 1966, 569–574. Figure 4, from B. L. McGlamery, "Restoration of turbulence-degraded images," *ibid.* **57**, 1967, 293–297. Figure 5, from J. L. Harris, Sr., "Potential and limitations of techniques for processing linear motion-degraded imagery," in *Evaluation of Motion-Degraded Images*, 131–138, NASA SP-193, December 1968. Figure 6, from B. R. Hunt, "The application of constrained least-squares estimation to image restoration by digital computer," *IEEE Transactions on Computers* **C-22**, 1973, 805–812.

Chapter 9: Figure 15, from J. F. O'Callaghan and J. Loveday, "Quantitative analysis of soil cracking patterns," *Pattern Recognition* **5**, 1973, 83–98.

Chapter 1

Introduction

1.1 PICTURE PROCESSING

Picture processing or *image processing* by computer encompasses a wide variety of techniques and mathematical tools. Most of these have been developed in response to three major problems:

(a) *Picture digitization and coding:* conversion of pictures from continuous to discrete form (digitization); "compression" of the results so as to conserve storage space or channel capacity;

(b) *Picture enhancement and restoration:* improvement of degraded (blurred, noisy) pictures;

(c) *Picture segmentation and description:* conversion of pictures into simplified "maps"; measurement of properties of pictures or picture parts; classification or description of pictures in terms of parts and properties.

Since many aspects of picture processing are closely related to the perception of pictures by humans, Chapter 3 presents a brief introduction to visual perception. Chapter 4 is devoted to the theory of picture digitization, while Chapter 5 treats picture compression. In Chapters 6 and 7 techniques for picture enhancement and restoration are discussed. Chapter 8 deals with methods of segmenting pictures into significant parts, while Chapter 9 discusses the measurement of geometrical properties of picture parts. Chapter 10 considers nongeometrical picture properties (e.g., texture) and the description of pictures in terms of parts, properties, and relationships.

In the remainder of this chapter, we discuss the relationship between continuous and discrete pictures, and the representation of (discrete) pictures in a digital computer. We also give a brief guide to the picture-processing literature. In the next chapter, we review various useful mathematical tools.

1.2   PICTURES AND THEIR COMPUTER REPRESENTATION

1.2.1   Pictures as Functions

Informally, a *picture* is a flat object whose brightness or color may vary from point to point. This variation can be represented mathematically by a function of two spatial variables. When color is involved, the function should be regarded as vector valued, or several functions should be used; see Section 3.3. In this book, however, we will be dealing almost exclusively with black-and-white pictures in which there can be shades of gray, but no color. Such a picture can be represented by a single real-valued function, say $f(x, y)$. The value of this function at a point will be called the *gray level* or *brightness* of the picture at that point.

It is customary to assume that the functions that represent pictures are analytically well behaved—so that, for example, these functions are integrable, have invertible Fourier transforms, etc. It is usual also to regard these functions as having values that are nonnegative and bounded, i.e., $0 \leqslant f(x, y) \leqslant M$ for all x, y.

1.2.2   Pictures as Arrays

When a picture is *digitized* (Chapter 4), a *sampling* process is used to extract from the picture a discrete set of real numbers ("samples"), and a *quantization* process is applied to these samples to yield numbers having a discrete set of possible values. In most practical situations, the samples are the values of the picture at a discrete, usually regularly spaced, set of points—or, more realistically, averages of the values taken over small neighborhoods of such points. Such a set of samples can be represented, for computer-processing purposes, as a (rectangular) *array* of real numbers. The samples are usually quantized to a set of equally spaced gray level values (see Section 4.3). If the unit of measurement is suitably chosen, these values can be taken to be integers; thus a digitized picture, or *digital picture*, can be regarded as an integer array.[†] The elements of a digital picture array are called *picture elements*, *pixels*, or *pels*, or sometimes just "points."

The most common method of picture sampling is to use a regularly spaced *square* array of points, i.e., points (md, nd) whose coordinates are multiples of some unit distance d. In such an array, every point has two kinds of neighbors—four horizontal and vertical neighbors (above, below, to the left, and to the right), at distance d from the point, and four diagonal neighbors at distance $d\sqrt{2}$. (Some of the complications introduced by the existence of these two types of neighbors will be discussed in Chapter 9.) Of course, at the edges of the array, some of these neighbors will not exist.

Another possibility is to use a regular *hexagonal* array of sample points; here each point has six neighbors, all at the same distance from it. One can obtain a similar kind of neighborhood from a square array by shifting, say, odd-numbered rows $d/2$ to the right (see Fig. 1). In fact, rather than do the

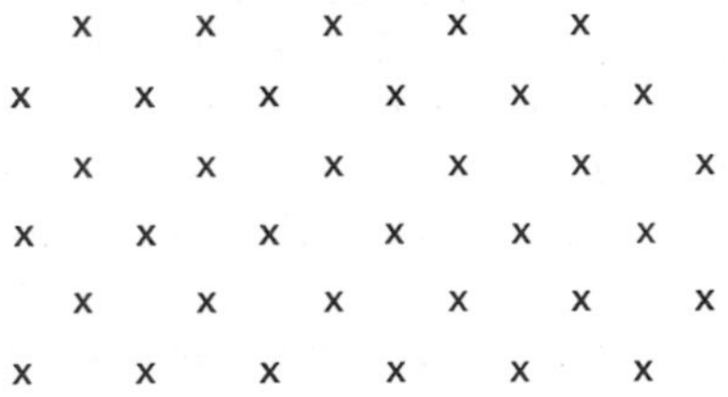

Fig. 1   "Hexagonal" array obtained by shifting odd-numbered rows of a square array.

actual shifting, one can simply stipulate that, for a point p on an odd-numbered row, the neighbors of p are its four horizontal and vertical neighbors together with just *two* of its diagonal neighbors, namely those to the northeast and southeast; and similarly for p on an even-numbered row, but this time allowing only the northwest and southwest diagonal neighbors.

The picture arrays that are processed in practice can be very large. For example, suppose that we want to sample and quantize an ordinary television image finely enough so that it can be redisplayed without noticeable degradation. Then we should use an array of about 500×500 samples, and we should quantize the samples to at least 30–50 discrete gray levels, i.e., a quarter of a million 5- or 6-bit numbers (see Chapter 4). In many cases it is necessary to handle even larger arrays when digitizing high-resolution photographs; and it is often desirable to use finer quantization, to 8 or even 10 bits. (It is customary to use a power of 2 as the number of quantization levels; and one often

---

[†] In particular, if there are just two gray levels, "black" and "white," we usually represent them by 0 and 1, so that the picture becomes a Boolean array.

uses a power that divides the word length of the computer being used, so that several picture elements can be packed in a single word without wasting memory space.)

1.2.3   Curves (or Region Boundaries) as Chains

If a picture consists of relatively few regions, having relatively few gray levels, it is very uneconomical to digitize it into a large array, since the picture is completely determined by specifying the boundaries and gray levels of the regions, which can be done very compactly. Similarly, if a picture is a simple line drawing, consisting of relatively few lines or curves, it can be economically represented by specifying these lines and curves (and perhaps their thicknesses and gray levels, if desired).

It is convenient to approximate boundaries or curves, for digital computer purposes, as polygons composed of short line segments. By taking these to be short enough, this type of digitization permits the curves to be reconstructed with any desired degree of accuracy. The lengths and slopes of the segments can themselves be quantized to discrete sets of values. A widely used scheme of this kind, known as *chain coding*, employs only horizontal and vertical segments of some (small) fixed length d, together with diagonal segments of length $d\sqrt{2}$.

A given (real) curve can be converted into a chain code in a number of ways, one of which we will now briefly describe. Imagine, superimposed on the curve, a regular grid of horizontal and vertical lines spaced d apart, as shown in Fig. 2. For each crossing of one of the grid lines by the curve, the grid point (i.e., the grid line intersection) closest to the crossing point is a point of the digitization of the curve. (If a crossing point is exactly halfway between two grid points, an arbitrary rounding scheme can be used, e.g., always use the grid point below, or to the left of, the crossing point.) Figure 2 shows a curve and the set of its digitization points as defined by this method.

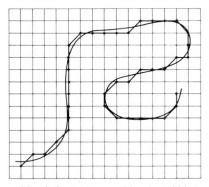

Fig. 2 A curve and its chain, constructed by the grid intersection method.

If the rate of turning of the curve is sufficiently small, and the grid lines are sufficiently closely spaced, then the digitization points defined in this way give rise to a polygonal arc that closely approximates the curve. This polygonal arc, called the *chain* of the curve, is constructed by joining the sequence of digitization points, obtained as one moves along the curve, by line segments. It is evident that if p and q are two successive digitization points, then they must be *neighboring* grid points; in other words, q is one of the eight horizontal, vertical, and diagonal neighbors of p in the grid. Thus the line segment joining p and q is either horizontal or vertical and of length d, or diagonal and of length $d\sqrt{2}$.

A chain can be completely described by specifying the coordinates of its starting point and the sequence of slopes of its segments, or "links." These slopes are multiples of 45° (0°, 45°, 90°, 135°, 180°, 225°, 270°, 315°); hence they can be represented by 3-bit numbers (0, 1, 2, 3, 4, 5, 6, 7) where "k" means "$45k°$," angles being measured, say, counterclockwise starting from the positive x-axis. The sequence of these numbers is called the *chain code* of the curve. For example, the chain code of the curve in Fig. 2 is

$$10112\ 22222\ 21000\ 01000\ 76645\ 44545\ 67000\ 012.$$

(We have broken the code into groups of five for easier reading.)

Exercise 1.   (a)   Draw the curve whose chain code is

$$2222225550000.$$

(b)   Characterize the chain codes of all rectangles whose sides are horizontal and vertical.   ∎

Exercise 2.   Let n_i denote the number of occurrences of i in a chain code, $0 \leqslant i \leqslant 7$. Prove that a chain code represents a *closed* polygon if and only if

$$n_1 + n_2 + n_3 = n_5 + n_6 + n_7 \qquad \text{and} \qquad n_3 + n_4 + n_5 = n_7 + n_0 + n_1. \quad ∎$$

Methods of constructing the chain code of a region boundary in a digital picture will be given in Section 9.1.2. Methods of "thinning" an elongated object in a digital picture into a digital curve (where each point, except the endpoints, has exactly two neighboring points that lie on the curve) will be given in Section 9.4.5; it is straightforward to construct the chain code of such a digital curve.

One can also define a four-direction chain code in which only multiples of 90° are used, so that a code is a string of 2-bit numbers. In fact, any eight-direction chain code can be converted to a four-direction code by simply replacing each 1 in the code by the pair of numbers 0, 2 (or 2, 0); each 3 by 2, 4; each 5 by 4, 6; and each 7 by 6, 0. Note that this converts diagonal lines to "staircases." (To obtain the 2-bit numbers, we simply relabel 0, 2, 4, 6 as 0, 1, 2, 3; that is, we use multiples of 90° rather than even multiples of 45°.)

Exercise 3. Define a six-direction code suitable for use on a hexagonal grid. How would you construct the code of a curve on a real hexagonal grid? A square grid with alternate rows shifted? ▌

BIBLIOGRAPHICAL NOTE

Chain coding was developed by Freeman [1]; for a recent review of the subject see [2].

REFERENCES

1. H. Freeman, On the encoding of arbitrary geometric configurations, *IRE Trans. Electron. Comput.* **EC-10,** 1961, 260–268.
2. H. Freeman, Computer processing of line-drawing images, *Comput. Surv.* **6,** 1974, 57–97.

1.3 A GUIDE TO THE LITERATURE

As of mid-1974, several thousand papers on picture processing had been published, and the literature continues to grow at an accelerating rate. Three textbooks existed (Rosenfeld [19], Andrews [1], Duda and Hart [4]), and many excellent survey papers (too numerous to cite here) had appeared. No attempt will be made in this book to reference this literature comprehensively; the reader is referred to four review papers by Rosenfeld [20–23] which together contain over 1300 references to the English-language literature through 1973.

Picture-processing papers appear in many different journals, particularly in the fields of computing and of optics. (Optical image processing will not be treated in this book. For books and collections of papers on this subject, see [7, 16–17, 25, 27, 28].) Noteworthy are several recent special issues of journals devoted to the subject [2, 9, 10]. There also exist specialized journals which frequently publish papers in the field [14, 24].

A number of collections of papers on picture processing have been published [3, 8, 15].[†] Many of the paper collections in the field of pattern recognition (and, to a lesser extent, artificial intelligence) also often contain papers on picture processing [6, 11, 12, 26, 30–32]. In addition, there are collections on specific applications of pictorial pattern recognition, e.g., to alphanumeric characters or biomedical images [5, 13, 18, 29]. References to more specialized collections will be given in the bibliographical notes at the ends of individual chapters.

[†] Paperbound collections of preprints of papers presented at meetings have not been cited here, since they are usually difficult to obtain.

REFERENCES

1. H. C. Andrews (with contributions by W. K. Pratt and K. Caspari), "Computer Techniques in Image Processing," Academic Press, New York, 1970.
2. H. C. Andrews and L. H. Enloe (eds.), Special issue on digital picture processing, *Proc. IEEE* **60**, 1972, 766–898.
3. G. C. Cheng, R. S. Ledley, D. K. Pollock, and A. Rosenfeld (eds.), "Pictorial Pattern Recognition," Thompson, Washington, D.C., 1968.
4. R. O. Duda and P. E. Hart, "Pattern Classification and Scene Analysis," Wiley, New York, 1973.
5. G. L. Fischer, D. K. Pollock, B. Radack, and M. E. Stevens (eds.), "Optical Character Recognition," Spartan, Baltimore, Maryland, 1962.
6. K. S. Fu (ed.), Special issue on feature extraction and selection in pattern recognition, *IEEE Trans. Comput.* **C-20**, 1971, 965–1120.
7. J. W. Goodman, "Introduction to Fourier Optics," McGraw-Hill, New York, 1968.
8. A. Grasselli (ed.), "Automatic Interpretation and Classification of Images," Academic Press, New York, 1969.
9. E. L. Hall and C. F. George, Jr. (eds.), Special issue on two-dimensional digital signal processing, *IEEE Trans. Comput.* **C-21**, 1972, 633–820.
10. L. D. Harmon (ed.), Special issue on digital pattern recognition, *Proc. IEEE* **60**, 1972, 1117–1233.
11. L. Kanal (ed.), "Pattern Recognition," Thompson, Washington, D. C., 1968.
12. P. A. Kolers and M. Eden (eds.), "Recognizing Patterns: Studies in Living and Automatic Systems," MIT Press, Cambridge, Massachusetts, 1968.
13. V. A. Kovalevsky (ed.), "Character Readers and Pattern Recognition," Spartan, New York, 1968.
14. R. S. Ledley (ed.), *Pattern Recognition* 1968ff.
15. B. S. Lipkin and Λ. Rosenfeld (eds.), "Picture Processing and Psychopictorics," Academic Press, New York, 1970.
16. D. K. Pollock, C. J. Koester, and J. T. Tippett (eds.), "Optical Processing of Information," Spartan, Baltimore, Maryland, 1963.
17. K. Preston, Jr., "Coherent Optical Computers," McGraw-Hill, New York, 1972.
18. D. M. Ramsey (ed.), "Image Processing in Biological Science," Univ. California Press, Berkeley and Los Angeles, 1968.
19. A. Rosenfeld, "Picture Processing by Computer," Academic Press, New York, 1969.
20. A. Rosenfeld, Picture processing by computer, *Comput. Surv.* **1**, 1969, 147–176.
21. A. Rosenfeld, Progress in picture processing: 1969–71, *Comput. Surv.* **5**, 1973, 81–108.
22. A. Rosenfeld, Picture processing: 1972, *Comput. Graph. Image Proc.* **1**, 1972, 394–416.
23. A. Rosenfeld, Picture processing: 1973, *Comput. Graph. Image Proc.* **3**, 1974, 178–194.
24. A. Rosenfeld, H. Freeman, T. S. Huang, and A. van Dam (eds.), *Comput. Graph. Image Proc.* 1972ff.
25. A. R. Shulman, "Optical Data Processing," Wiley, New York, 1970.
26. J. Sklansky (ed.), "Pattern Recognition: Introduction and Foundations," Dowden, Hutchinson, and Ross, Stroudsburg, Pennsylvania, 1973.
27. G. W. Stroke, "An Introduction to Coherent Optics and Holography," Academic Press, New York, 1966 (2nd ed., 1968).
28. J. T. Tippett, D. A. Berkowitz, L. C. Clapp, C. J. Koester, and A. Vanderburgh, Jr., "Optical and Electro-Optical Information Processing," MIT Press, Cambridge, Massachusetts, 1965.
29. W. E. Tolles (ed.), Data extraction and processing of optical images in the medical and biological sciences, *Ann. N.Y. Acad. Sci.* **157**, 1969, 1–530.

30. L. Uhr (ed.), "Pattern Recognition," Wiley, New York, 1966.
31. S. Watanabe (ed.), "Methodologies of Pattern Recognition," Academic Press, New York, 1969.
32. S. Watanabe (ed.), "Frontiers of Pattern Recognition," Academic Press, New York, 1972.

SUPPLEMENTARY REFERENCES

The rapid growth of the picture processing field makes it especially important to keep up with the current literature. It is not practicable to update all of the bibliographies given here in this book, but a few major references are provided here which appeared prior to the end of 1975. (Work on optical information processing is not covered.)

a. H. C. Andrews, ed., Special Issue on Digital Picture Processing, *Computer* **7**(5), 1974, 17–87.
b. A. Rosenfeld, Picture processing: 1974, *Comput. Graph. Image Proc.* **4**, 1975, 133–155.
c. J. K. Aggarwal and R. O. Duda, guest eds., Special Issue on Digital Filtering and Image Processing, *IEEE Trans. Circuits Systems* **CAS-2**, 1975, 161–304.
d. T. S. Huang, ed., *Picture Processing and Digital Filtering*, Springer, New York, 1975.
e. P. H. Winston, ed., *The Psychology of Computer Vision*, McGraw-Hill, New York, 1975.

Chapter 2

Mathematical Preliminaries

2.1 LINEAR OPERATIONS ON PICTURES

2.1.1 Point Sources and Delta Functions

Let $\mathcal{O}$ be an operation that takes pictures into pictures; given the input picture f, the result of applying $\mathcal{O}$ to f is denoted by $\mathcal{O}[f]$. We call $\mathcal{O}$ *linear* if

$$\mathcal{O}[af+bg] = a\mathcal{O}[f] + b\mathcal{O}[g] \tag{1}$$

for all pictures f, g and all constants a, b.

In the analysis of linear operations on pictures, the concept of a *point source* is very convenient. If any arbitrary picture f could be considered to be a sum of point sources, then a knowledge of the operation's output for a point source input could be used to determine the output for f. The output of $\mathcal{O}$ for a point source input is called the *point spread function* of $\mathcal{O}$.

A point source can be regarded as the limit of a sequence of pictures whose nonzero values become more and more concentrated spatially. Note that in order for the total brightness to be the same for each of these pictures, their nonzero values must get larger and larger. As an example of such a sequence of pictures, let

$$\text{rect}(x, y) = \begin{cases} 1 & \text{for} \quad |x| \leqslant \tfrac{1}{2}, \quad |y| \leqslant \tfrac{1}{2} \\ 0 & \text{elsewhere} \end{cases} \tag{2}$$

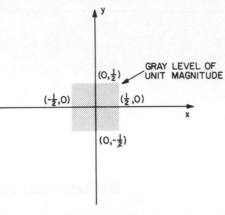

Fig. 1   Rect(x, y).

(see Fig. 1) and let

$$\delta_n(x, y) = n^2 \, \text{rect}(nx, ny), \qquad n = 1, 2, \dots \tag{3}$$

Thus δ_n is zero outside the $1/n \times 1/n$ square described by $|x| \leqslant 1/2n$, $|y| \leqslant 1/2n$ and has constant value n^2 inside that square. It follows that

$$\int\!\!\!\int\limits_{-\infty}^{\infty} \delta_n(x, y) \, dx \, dy = 1 \tag{4}$$

for any n.

As $n \to \infty$, the sequence δ_n does not have a limit in the usual sense, but it is convenient to treat it as though its limit existed. This limit, denoted by δ, is called a *Dirac delta function*. Evidently, we have $\delta(x, y) = 0$ for all (x, y) other than $(0, 0)$ where it is infinite. It follows that $\delta(-x, -y) = \delta(x, y)$.

We can derive a number of important properties of δ as limiting cases of properties of the functions δ_n. Thus, in view of (4), we can write

$$\int\!\!\!\int\limits_{-\infty}^{\infty} \delta(x, y) \, dx \, dy = 1 \tag{5}$$

More generally, consider the integral $\iint_{-\infty}^{\infty} g(x, y) \delta_n(x, y) \, dx \, dy$. This is just the average of $g(x, y)$ over a $1/n \times 1/n$ square centered at the origin. Thus in the limit we retain just the value at the origin itself, so that we can write

$$\int\!\!\!\int\limits_{-\infty}^{\infty} g(x, y) \delta(x, y) \, dx \, dy = g(0, 0)$$

If we shift δ by the amount (α, β), i.e., we use $\delta(x-\alpha, y-\beta)$ instead of $\delta(x, y)$, we similarly obtain the value of g at the point (α, β), i.e.,

$$\int\!\!\int_{-\infty}^{\infty} g(x, y)\,\delta(x-\alpha, y-\beta)\, dx\, dy = g(\alpha, \beta) \tag{6}$$

The same is true for any region of integration containing (α, β). Equation (6) is called the "sifting" property of the δ function.

As a final useful property of δ, we have

$$\int\!\!\int_{-\infty}^{\infty} \exp[-j2\pi(ux+vy)]\, du\, dv = \delta(x, y) \tag{7}$$

where $j = \sqrt{-1}$. For a discussion of this property, see Papoulis [8].

2.1.2   Linear Shift-Invariant Operations

Again let us consider a linear operation (or system) that takes pictures into pictures. The point spread function, which is the output picture for an input point source at the origin of the xy-plane, is denoted by $h(x, y)$.

The linear operation is said to be *shift invariant* (or space invariant, or position invariant) if the response to $\delta(x-\alpha, y-\beta)$, which is a point source located at (α, β) in the xy-plane, is given by $h(x-\alpha, y-\beta)$. In other words, the output is merely shifted by α and β in the x- and y-directions, respectively.

Now let us consider an arbitrary input picture $f(x, y)$. By Eq. (6) this picture can be considered to be a linear sum of point sources. We can write $f(x, y)$ as

$$f(x, y) = \int_{-\infty}^{\infty}\int_{-\infty}^{\infty} f(\alpha, \beta)\,\delta(\alpha-x, \beta-y)\, d\alpha\, d\beta \tag{8}$$

In other words, $f(x, y)$ is a linear sum of point sources located at (α, β) in the xy-plane with α and β ranging from $-\infty$ to $+\infty$. In this sum the point source at a particular value of (α, β) has "strength" $f(\alpha, \beta)$. Let the response of the operation to the input $f(x, y)$ be denoted by $\mathcal{O}[f]$. If we assume the operation to be shift invariant, then by the interpretation just given to the right-hand side of (8), we obtain

$$\mathcal{O}[f(x, y)] = \mathcal{O}\left[\int_{-\infty}^{\infty}\int_{-\infty}^{\infty} f(\alpha, \beta)\,\delta(\alpha-x, \beta-y)\, d\alpha\, d\beta\right]$$

$$= \int\!\!\int f(\alpha, \beta)\,\mathcal{O}[\delta(\alpha-x, \beta-y)]\, d\alpha\, d\beta \tag{9}$$

by the linearity of the operation, which means that the response to a sum of excitations is equal to the sum of responses to each excitation. As stated earlier, the response to $\delta(\alpha-x, \beta-y) = \delta(x-\alpha, y-\beta)$, which is a point source located at (α, β), is given by $h(x-\alpha, y-\beta)$, and if $\mathcal{O}[f]$ is denoted by g, we obtain

$$g(x, y) = \int_{-\infty}^{\infty} \int_{-\infty}^{\infty} f(\alpha, \beta)\, h(x-\alpha, y-\beta)\, d\alpha\, d\beta \qquad (10)$$

The right-hand side is called the *convolution* of f and h, and is often denoted by $f * h$. The integrand is a product of two functions $f(\alpha, \beta)$ and $h(\alpha, \beta)$ with the latter rotated by 180° and shifted by x and y along the x- and y-directions, respectively. This is pictorially shown in Fig. 2. A simple change of variables

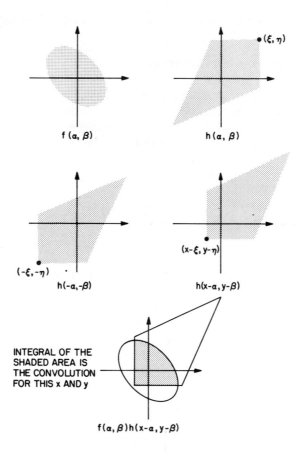

Fig. 2   Convolution.

shows that (10) can also be written as

$$g(x, y) = \int_{-\infty}^{\infty} \int_{-\infty}^{\infty} f(x-\alpha, y-\beta) h(\alpha, \beta) \, d\alpha \, d\beta \qquad (11)$$

so that $f * h = h * f$.

2.1.3 Fourier Analysis

Let $f(x, y)$ be a function of two independent variables x and y; then its *Fourier transform* $F(u, v)$ is defined by

$$F(u, v) = \int_{-\infty}^{\infty} \int_{-\infty}^{\infty} f(x, y) \exp\left[-j2\pi(ux+vy)\right] dx \, dy \qquad (12)$$

In general, F is a complex-valued function of u and v. As an example, let $f(x, y) = \text{rect}(x, y)$. Then it is easily verified that

$$F(u, v) = \frac{\sin \pi u}{\pi u} \frac{\sin \pi v}{\pi v}$$

This last function is usually denoted by $\text{sinc}(u, v)$; it is illustrated in Fig. 3.

Exercise 1. Verify that the Fourier transform of $\text{rect}(x, y)$ is $\text{sinc}(x, y)$ and, more generally, that the Fourier transform of $\text{rect}(nx, ny)$ is

$$(1/n^2) \, \text{sinc}(u/n, v/n). \quad \blacksquare$$

The Fourier transform of f may not exist unless f satisfies certain conditions. The following is a typical set of sufficient conditions for its existence [5]:

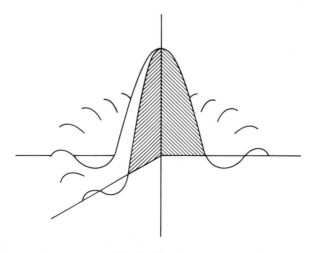

Fig. 3 Sinc(x, y).

(1) $\int_{-\infty}^{\infty}\int_{-\infty}^{\infty}|f(x, y)|\, dx\, dy < \infty$.

(2) $f(x, y)$ must have only a finite number of discontinuities and a finite number of maxima and minima in any finite rectangle.

(3) $f(x, y)$ must have no infinite discontinuities.

Some useful mathematical functions, like the Dirac δ function, do not necessarily obey the preceding conditions. Sometimes it is possible to represent these functions as limits of a sequence of well-behaved functions that do obey these conditions. The Fourier transforms of the members of this sequence will also form a sequence. Now if this sequence of Fourier transforms possesses a limit, then this limit is called the "generalized Fourier transform of the original function." Generalized transforms can be manipulated in the same manner as the conventional transforms, and the distinction between the two is generally ignored; it being understood that when a function fails to satisfy the existence conditions and yet is said to have a transform, then the generalized transform is actually meant [5, 7].

In Section 2.1.1, we introduced the Dirac δ function as a limit of a sequence of the functions $n^2\,\mathrm{rect}\,(nx, ny)$; then by the preceding argument and Exercise 1, the Fourier transform of the Dirac δ function is the limit of the sequence of Fourier transforms $\mathrm{sinc}\,(u/n, v/n)$. In other words, when

$$f(x, y) = \delta(x, y)$$

then

$$F(u, v) = \lim_{n \to \infty} \mathrm{sinc}\,(u/n, v/n) = 1 \tag{13}$$

By multiplying both sides of Eq. (12) by $\exp[j2\pi(u\alpha + v\beta)]$, integrating both sides with respect to u and v, and making use of Eq. (7), it is easily shown that

$$\int_{-\infty}^{\infty}\int_{-\infty}^{\infty} F(u, v)\exp[j2\pi(u\alpha + v\beta)]\, du\, dv = f(\alpha, \beta) \tag{14}$$

or equivalently

$$f(x, y) = \int_{-\infty}^{\infty}\int_{-\infty}^{\infty} F(u, v)\exp[j2\pi(ux + vy)]\, du\, dv \tag{15}$$

which is called the *inverse Fourier transform* of $F(u, v)$. By Eqs. (12) and (15), $f(x, y)$ and $F(u, v)$ form a *Fourier transform pair*.

Equation (15) can be used to give a physical interpretation to $F(u, v)$ and to the coordinates u and v, if x and y represent spatial coordinates. Let us first examine the function

$$\exp[j2\pi(ux + vy)] \tag{16}$$

The real and the imaginary parts of this function are $\cos 2\pi(ux + vy)$ and $\sin 2\pi(ux + vy)$, respectively. In Fig. 4, we have shown the lines corresponding

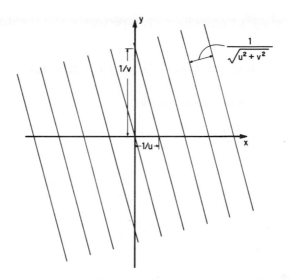

Fig. 4 Lines corresponding to the zeros of $\cos 2\pi(ux+vy)$.

to the zeros of $\cos 2\pi(ux+vy)$. The maxima of this function would be along lines parallel to these but located halfway between. It is clear if one took a section of this two-dimensional pattern parallel to the x-axis, it goes through u cycles per unit distance, while a section parallel to the y-axis goes through v cycles per unit distance. This is the reason why u and v are called the *spatial frequencies* along the x- and y-axes, respectively. Also, from the figure it can be seen that the spatial period of the pattern is $(u^2+v^2)^{-1/2}$. The plot for $\sin 2\pi(ux+vy)$ looks similar to the one in Fig. 4 except that it is displaced by half a period in the direction of maximum rate of change.

From the preceding discussion it is clear that $\exp[j2\pi(ux+vy)]$ is a two-dimensional pattern, the sections of which, parallel to the x- and y-axes, are spatially periodic with frequencies u and v, respectively. The pattern itself has a spatial period of $(u^2+v^2)^{-1/2}$ along a direction that subtends an angle $\tan^{-1}(v/u)$ with the x-axis. By changing u and v, one can generate patterns with spatial periods ranging from 0 to ∞ in any direction in the xy-plane.

Equation (15) can, therefore, be interpreted to mean that $f(x,y)$ can be considered to be a linear combination of elementary periodic patterns of the form $\exp[j2\pi(ux+vy)]$. Evidently, $F(u,v)$ is simply a weighting factor that is a measure of the relative contribution of the elementary pattern, the x- and the y-components of the spatial frequency of which are u and v, respectively, to the total sum. For this reason $F(u,v)$ is called the *frequency spectrum* of $f(x,y)$.

Exercise 2. Show that if $f(x,y)$ is a circularly symmetric function, i.e.,

$f(x, y) = f(\sqrt{x^2 + y^2})$, then its frequency spectrum is also circularly symmetric and is given by

$$F(u, v) = F(\rho) = 2\pi \int_0^\infty rf(r) J_0(2\pi r\rho) \, dr \tag{17}$$

while the inverse relationship is given by

$$f(r) = 2\pi \int_0^\infty \rho F(\rho) J_0(2\pi r\rho) \, d\rho$$

where

$$r = \sqrt{x^2 + y^2}, \qquad \theta = \tan^{-1}(y/x), \qquad \rho = \sqrt{u^2 + v^2}, \qquad \varphi = \tan^{-1}(v/u)$$

and

$$J_0(x) = (1/2\pi) \int_0^{2\pi} \exp[-jx \cos(\theta - \varphi)] \, d\theta$$

where $J_0(x)$ is the zero-order Bessel function of the first kind. The transformation in (17) is also called the *Hankel transform of zero order.* ∎

Exercise 3. Use (7) to show that the Fourier transform of $\cos 2\pi(\alpha x + \beta y)$ is $\frac{1}{2}[\delta(u - \alpha, v - \beta) + \delta(u + \alpha, v + \beta)]$. ∎

Exercise 4. Let us define a comb function as

$$\text{comb}(x, y) = \sum_{m=-\infty}^{\infty} \sum_{n=-\infty}^{\infty} \delta(x - m, y - n) \tag{18}$$

(see Fig. 5). Show that $\text{comb}(x, y)$ and $\text{comb}(u, v)$ are a Fourier transform pair. ∎

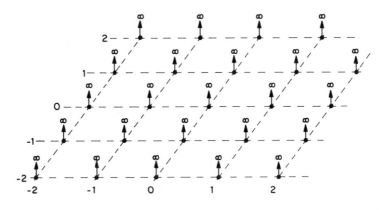

Fig. 5 Comb(x, y).

2.1.4 Properties of Fourier Transforms

Let $\mathscr{F}\{f\}$ denote the Fourier transform of a function $f(x, y)$. Then $\mathscr{F}\{f(x, y)\} = F(u, v)$. We will now present without proof some of the more common properties of Fourier transforms. The proofs are, for the most part, left as exercises for the reader (see the books by Goodman [5] and Papoulis [8]).

(1) *Linearity:*

$$\mathscr{F}\{af_1(x, y) + bf_2(x, y)\} = a\mathscr{F}\{f_1(x, y)\} + b\mathscr{F}\{f_2(x, y)\}$$
$$= aF_1(u, v) + bF_2(u, v) \tag{19}$$

This follows from the linearity of the integration operation.

(2) *Scaling:*

$$\mathscr{F}\{f(\alpha x, \beta y)\} = \frac{1}{|\alpha\beta|} F\left(\frac{u}{\alpha}, \frac{v}{\beta}\right) \tag{20}$$

To see this, introduce the change of variables $x' = \alpha x$, $y' = \beta y$.

(3) *Shift property:*

$$\mathscr{F}\{f(x-\alpha, y-\beta)\} = F(u, v) \exp[-j2\pi(u\alpha + v\beta)] \tag{21a}$$

This, too, follows immediately if we make the change of variables $x' = x - \alpha$, $y' = y - \beta$. The corresponding property for a shift in the frequency domain is

$$\mathscr{F}\{\exp[j2\pi(u_0 x + v_0 y)] f(x, y)\} = F(u - u_0, v - v_0) \tag{21b}$$

(4) *180° rotation:*

$$\mathscr{F}\{\mathscr{F}\{f(x, y)\}\} = f(-x, -y) \tag{22}$$

(5) *Convolution:*

$$\mathscr{F}\left\{\int_{-\infty}^{\infty}\int_{-\infty}^{\infty} f_1(\alpha, \beta) f_2(x-\alpha, y-\beta)\, d\alpha\, d\beta\right\} = \mathscr{F}\{f_1(x, y)\}\, \mathscr{F}\{f_2(x, y)\}$$
$$= F_1(u, v) F_2(u, v) \tag{23a}$$

Note that the convolution of two functions in the space domain is equivalent to the very simple operation of multiplication in the spatial frequency domain. The corresponding property for the convolution in the spatial frequency plane is given by

$$\mathscr{F}\{f_1(x, y) f_2(x, y)\} = \int_{-\infty}^{\infty}\int F_1(u-s, v-t) F_2(s, t)\, ds\, dt \tag{23b}$$

(6)   *Parseval's theorem:*

$$\int_{-\infty}^{\infty} \int_{-\infty}^{\infty} f_1(x, y) f_2^*(x, y)\, dx\, dy = \int_{-\infty}^{\infty} \int_{-\infty}^{\infty} F_1(u, v) F_2^*(u, v)\, du\, dv$$

(24)

where * denotes the complex conjugate. When $f_1(x, y) = f_2(x, y) = f(x, y)$, we have

$$\int_{-\infty}^{\infty} \int_{-\infty}^{\infty} |f(x, y)|^2\, dx\, dy = \int_{-\infty}^{\infty} \int_{-\infty}^{\infty} |F(u, v)|^2\, du\, dv.$$

(25)

In this form, this property is interpretable as a statement of conservation of energy.

We conclude this section by introducing the autocorrelation and cross correlation of functions. The definitions given here are valid only for deterministic functions. These averages are also used to characterize probabilistic functions, and their definitions for that case are given in Section 2.4.2. Under certain conditions to be discussed later, the two definitions are equivalent (see Section 2.4.5).

The *cross correlation* C_{fg} of two functions f and g is defined as

$$C_{fg}(x', y') = \int_{-\infty}^{\infty} \int_{-\infty}^{\infty} f^*(x-x', y-y')\, g(x, y)\, dx\, dy$$

$$= C_{gf}^*(-x', -y')$$

(26)

Note that for real functions the cross correlation differs from the convolution in that before taking the product of the two functions shifted with respect to one another, no rotation is performed.

Exercise 5.   Using (23) show that

$$\mathscr{F}\{C_{fg}\} = \mathscr{F}^*\{f\}\,\mathscr{F}\{g\} = F^*(u, v)\,G(u, v) \quad \blacksquare$$

(27)

The *autocorrelation* of a function $f(x, y)$ is defined as

$$R_f(x', y') = \int_{-\infty}^{\infty} \int f(x, y) f^*(x-x', y-y')\, dx\, dy$$

$$= \int_{-\infty}^{\infty} \int f(x+x', y+y') f^*(x, y)\, dx\, dy$$

(28)

From (27) it is clear that

$$\mathscr{F}\{R_f\} = |F(u, v)|^2$$

(29)

For deterministic functions $f(x, y)$, $|F(u, v)|^2$ is called the *power spectrum*. We see that the power spectrum is the Fourier transform of the autocorrelation function.

2.2 DISCRETE PICTURE TRANSFORMS

Let a matrix $[f]$ represent an $M \times N$ array of numbers:

$$[f] = \begin{bmatrix} f(0,0) & f(0,1) & f(0,N-1) \\ \vdots & \cdots & \vdots & \cdots & \vdots \\ f(M-1,0) & f(M-1,1) & f(M-1,N-1) \end{bmatrix} \quad (30)$$

Note that such an array of numbers could be obtained by sampling a picture $f(x, y)$ on a grid of $M \times N$ points, each number in the array representing the gray level in the picture at the corresponding sampling point.

A general discrete transform of the $M \times N$ matrix $[f]$ is an $M \times N$ matrix $[F]$ defined as the product of the three matrices

$$[F] = [P][f][Q], \quad (31)$$

where $[P]$ and $[Q]$ are nonsingular[†] $M \times M$ and $N \times N$ matrices, respectively. The specific transform is determined by the way the matrices $[P]$ and $[Q]$ are defined. The transform $[F]$, which is an $M \times N$ matrix, can also be written as

$$F(u, v) = \sum_{m=0}^{M-1} \sum_{n=0}^{N-1} P(u, m) f(m, n) Q(n, v) \quad (32)$$

for $u = 0, 1, ..., M-1$, $v = 0, 1, ..., N-1$.

Since the transformation matrices $[P]$ and $[Q]$ are nonsingular, their inverses are uniquely defined. Let $[P]^{-1}$ denote the inverse of $[P]$ and $[Q]^{-1}$ that of $[Q]$. Multiplying both sides of (31) first from the left by $[P]^{-1}$ and then from the right by $[Q]^{-1}$, we obtain

$$[f] = [P]^{-1}[F][Q]^{-1} \quad (33)$$

[†] We will briefly define some of the properties of matrices that we have used in this section.

(1) A square matrix possesses an inverse if its determinant is nonzero, such matrices are called *nonsingular*.

(2) A real square matrix $[T]$ is called *symmetric* if $[T] = [T]^t$ where $[T]^t$ is the transpose of $[T]$. $[T]$ is called *orthogonal* if $[T]^t[T] = [I]$ where $[I]$ is the identity matrix. Therefore, for a real square matrix that is both symmetric and orthogonal, $[T]^{-1} = [T]$.

(3) A complex square matrix $[C]$ is called *Hermitian* if $[C]^{*t} = [C]$, where $[C]^*$ is obtained from $[C]$ by taking the complex conjugate of every element. $[C]$ is called *unitary* if $[C]^{*t}[C] = [I]$. Therefore, for a complex square matrix that is both Hermitian and unitary, $[C]^{-1} = [C]$.

This defines the inverse transform relationship, and we say that $[f]$ is the *inverse transform* of $[F]$.

Before we present special cases of discrete transforms, we would like to mention that if a nonsingular real square matrix is orthogonal and symmetric, then its inverse is equal to itself. Therefore, if the square matrices $[P]$ and $[Q]$ are real, symmetric, and orthogonal, then the transform relationships (31) and (33) become

$$[F] = [P][f][Q], \qquad [f] = [P][F][Q] \tag{34}$$

For the case of complex matrices this is true provided the matrices are Hermitian and unitary.

2.2.1   The Discrete Fourier Transform

Let us define a $J \times J$ transformation matrix $[\Phi_{J,J}]$ whose (m,n)th element is given by

$$(1/J)\,e^{-j(2\pi/J)mn}, \qquad m,n = 0,1,2,\ldots,J-1 \tag{35}$$

With

$$[P] = [\Phi_{M,M}] \qquad \text{and} \qquad [Q] = [\Phi_{N,N}] \tag{36}$$

the *discrete Fourier transform* F of an array f is given by

$$[F] = [P][f][Q]$$

or making use of the expansion in (32), and (35) and (36),

$$F(u,v) = \frac{1}{MN} \sum_{m=0}^{M-1} \sum_{n=0}^{N-1} f(m,n) \exp\left[-j2\pi\left(\frac{mu}{M} + \frac{nv}{N}\right) \right] \tag{37}$$

for $u = 0,1,2,\ldots,M-1$, $v = 0,1,2,\ldots,N-1$. Even though this expression is similar to the discrete sum approximation of the Fourier integral in Eq. (12), note that (37) by itself is not an approximation; $[F]$ as defined in (37) is the exact discrete Fourier transform of the array of numbers $[f]$.

The inverse matrix $[\Phi_{J,J}]^{-1}$ is again $J \times J$, and its (m,n)th element is given by

$$e^{j(2\pi/J)mn} \tag{38}$$

That this is true can be seen by multiplying the two matrices $[\Phi_{J,J}]$ and $[\Phi_{J,J}]^{-1}$ and noting that

$$\sum_{m=0}^{J-1} \exp\left[\frac{-j2\pi}{J} km \right] \exp\left[\frac{j2\pi}{J} mn \right] = \begin{cases} J, & k = n \\ 0, & k \neq n \end{cases} \tag{39}$$

Since $[P] = [\Phi_{M,M}]$ and $[Q] = [\Phi_{N,N}]$, therefore $[P]^{-1} = [\Phi_{M,M}]^{-1}$ and $[Q]^{-1} = [\Phi_{N,N}]^{-1}$. The inverse transform relationship as given by Eq. (33)

is completely defined and in its expanded form is given by using (38):

$$f(m,n) = \sum_{u=0}^{M-1} \sum_{v=0}^{N-1} F(u,v) \exp\left[j2\pi\left(\frac{mu}{M} + \frac{nv}{M}\right)\right] \qquad (40)$$

for $m = 0, 1, 2, \ldots, M-1$, $n = 0, 1, \ldots, N-1$. In Fig. 6a we have shown a picture represented by a 16×16 matrix $[f]$. The real and imaginary parts of its Fourier transform $[F]$ are shown in Figs. 6b and 6c, respectively.

In (37) the discrete Fourier transform $F(u,v)$ is defined for u between 0 and $M-1$ and for v between 0 and $N-1$. If, however, we use the same equation to evaluate $F(\pm u, \pm v)$, we discover that the periodicity properties of the exponential factor imply that

$$F(u, -v) = F(u, N-v)$$
$$F(-u, v) = F(M-u, v) \qquad (41a)$$
$$F(-u, -v) = F(M-u, N-v)$$

Similarly, using (40) we can show that

$$f(-m, n) = f(M-m, n)$$
$$f(m, -n) = f(m, N-n) \qquad (41b)$$
$$f(-m, -n) = f(M-m, N-n)$$

Another related consequence of the periodicity properties of the exponential factors in Eqs. (37) and (40) is that

$$F(aM+u, bN+v) = F(u,v) \qquad \text{and} \qquad f(aM+m, bN+n) = f(m,n) \qquad (42)$$

for $a = 0, \pm 1, \pm 2, \ldots$, $b = 0, \pm 1, \pm 2, \ldots$. Therefore, we have the following conclusion: If a finite array of numbers $[f]$ and its Fourier transform $[F]$ are related by Eqs. (37) and (40), then if it is desired to extend the definition of $f(m,n)$ and $F(u,v)$ beyond the original domain as given by $[0 \leqslant (m \text{ and } u) \leqslant M-1]$ and $[0 \leqslant (n \text{ and } v) \leqslant N-1]$, this extension must be governed by (41) and (42). In other words, the extensions are periodic repetitions of the arrays.

It will now be shown that this periodicity has important consequences when we compute the convolution of two $M \times N$ arrays $[f]$ and $[d]$ by multiplying their discrete Fourier transforms $[F]$ and $[D]$. The convolution of two arrays $[f]$ and $[d]$ is given by

$$g(\alpha, \beta) = \frac{1}{MN} \sum_{m=0}^{M-1} \sum_{n=0}^{N-1} f(m,n)\, d(\alpha-m, \beta-n)$$

$$= \frac{1}{MN} \sum_{m=0}^{M-1} \sum_{n=0}^{N-1} f(\alpha-m, \beta-n)\, d(m,n) \qquad (43)$$

```
0 0 0 0 0 0 0 0 0 0 0 0 0 0 0 0 0 0
0 0 0 0 0 0 0 0 0 0 0 0 0 0 0 0 0 0
0 0 0 0 0 0 0 0 0 0 0 0 0 0 0 0 0 0
0 0 0 0 0 0 0 0 0 0 0 0 0 0 0 0 0 0
0 0 0 0 0 0 0 0 0 0 0 0 0 0 0 0 0 0
0 0 0 0 0 0 0 0 0 0 0 0 0 0 0 0 0 0
0 0 0 0 0 0 1 1 1 0 0 0 0 0 0 0 0 0
0 0 0 0 0 0 1 0 1 0 0 0 0 0 0 0 0 0
0 0 0 0 0 0 1 1 1 0 0 0 0 0 0 0 0 0
0 0 0 0 0 0 0 0 0 0 0 0 0 0 0 0 0 0
0 0 0 0 0 0 0 0 0 0 0 0 0 0 0 0 0 0
0 0 0 0 0 0 0 0 0 0 0 0 0 0 0 0 0 0
0 0 0 0 0 0 0 0 0 0 0 0 0 0 0 0 0 0
0 0 0 0 0 0 0 0 0 0 0 0 0 0 0 0 0 0
0 0 0 0 0 0 0 0 0 0 0 0 0 0 0 0 0 0
0 0 0 0 0 0 0 0 0 0 0 0 0 0 0 0 0 0
```

(a)

Fig. 6a   A picture represented by an array of numbers. The top left hand corner corresponds to $m=0$ and $n=0$.

Fig. 6b   Real part of the Fourier transform matrix $256 \times F(u, v)$ of the picture in Fig. 6a.

Fig. 6c   Imaginary part of the Fourier transform matrix $256 \times F(u, v)$ of the picture in Fig. 6a.

for $\alpha = 0, 1, ..., M-1$, $\beta = 0, 1, ..., N-1$, where we insist that when the values of $f(m, n)$ and $d(m, n)$ are required for indices outside the ranges $0 \leqslant m \leqslant M-1$ and $0 \leqslant n \leqslant N-1$, for which $f(m, n)$ and $d(m, n)$ are defined, then they be obtained by the rules given in Eqs. (41) and (42). With this condition, the convolution previously defined is called a *circular or cyclic convolution*.

We will now show that the discrete Fourier transform of (43) is the product of the discrete Fourier transforms of $[f]$ and $[d]$. By making use of (40), we obtain

$$g(\alpha, \beta) = \frac{1}{MN} \sum_{m=0}^{M-1} \sum_{n=0}^{N-1} f(m, n) \, d(\alpha - m, \beta - n)$$

$$= \frac{1}{MN} \sum_{m=0}^{M-1} \sum_{n=0}^{N-1} \left\{ \sum_{u=0}^{M-1} \sum_{v=0}^{N-1} F(u, v) \exp\left[j2\pi \left(\frac{mu}{M} + \frac{nv}{N} \right) \right] \right\}$$

$$\times \left\{ \sum_{w=0}^{M-1} \sum_{z=0}^{N-1} D(w, z) \exp\left[j2\pi \left(\frac{(\alpha - m) w}{M} + \frac{(\beta - n) z}{N} \right) \right] \right\}$$

Eq. 44 continued on page 24

u →

	-8	-7	-6	-5	-4	-3	-2	-1	0	1	2	3	4	5	6	7
-8	0	.14	-.41	.47	0	-1.06	2.41	-3.55	4.00	-3.55	2.41	-1.06	0	.47	-.41	.14
-7	.14	-.20	.25	-.00	-.71	1.77	-2.81	3.41	-3.27	2.41	-1.17	-.00	.71	-.85	.60	-.28
-6	-.41	.25	0	-.42	1.00	-1.60	2.00	-2.01	1.59	-.83	0	.66	-1.00	1.01	-.83	.60
-5	.47	-.00	-.42	.67	-.71	.59	-.40	.23	-.11	-.00	.17	-.41	.71	-.94	1.01	-.85
-4	0	-.71	1.00	-.71	0	.71	-1.00	.71	0	-.71	1.00	-.71	0	.71	-1.00	.71
-3	-1.06	1.77	-1.60	.59	.71	-1.50	1.25	-.00	-1.64	2.85	-3.01	2.12	-.71	-.41	.66	-.00
-2	2.41	-2.81	2.00	-.40	-1.00	1.25	0	-2.25	4.41	-5.43	4.83	-3.01	1.00	.17	0	-1.17
-1	-3.55	3.41	-2.01	.23	.71	-.00	-2.25	5.03	-6.97	7.11	-5.43	2.85	-.71	-.00	-.83	2.41
0	4.00	-3.27	1.59	-.11	0	-1.64	4.41	-6.97	8.00	-6.97	4.41	-1.64	0	-.11	1.59	-3.27
1	-3.55	2.41	-.83	-.00	-.71	2.85	-5.43	7.11	-6.97	5.03	-2.25	-.00	.71	.23	-2.01	3.41
2	2.41	-1.17	0	.17	1.00	-3.01	4.83	-5.43	4.41	-2.25	0	1.25	-1.00	-.40	2.00	-2.81
3	-1.06	-.00	.66	-.41	-.71	2.12	-3.01	2.85	-1.64	-.00	1.25	-1.50	.71	.59	-1.60	1.77
4	0	.71	-1.00	.71	0	-.71	1.00	-.71	0	.71	-1.00	.71	0	-.71	1.00	-.71
5	.47	-.85	1.01	-.94	.71	-.41	.17	-.00	-.11	.23	-.40	.59	-.71	.67	-.42	-.00
6	-.41	.60	-.83	1.01	-1.00	.66	0	-.83	1.59	-2.01	2.00	-1.60	1.00	-.42	0	.25
7	.14	-.28	.60	-.85	.71	-.00	-1.17	2.41	-3.27	3.41	-2.81	1.77	-.71	-.00	.25	-.20

↓ v

(b)

u →

	-8	-7	-6	-5	-4	-3	-2	-1	0	1	2	3	4	5	6	7
-8	0	.06	-.41	1.14	-2.00	2.55	-2.41	1.47	0	-1.47	2.41	-2.55	2.00	-1.14	.41	-.06
-7	.06	-.20	.60	-1.20	1.71	-1.77	1.17	.00	-1.36	2.41	-2.81	2.50	-1.71	.85	-.25	0
-6	-.41	.60	-.83	1.01	-1.00	.66	0	-.83	1.59	-2.01	2.00	-1.60	1.00	-.42	0	.25
-5	1.14	-1.20	1.01	-.67	.29	.00	-.17	.23	-.27	.33	-.40	.41	-.29	-.00	.42	-.85
-4	-2.00	1.71	-1.00	.29	0	.29	-1.00	1.71	-2.00	1.71	-1.00	.29	0	.29	-1.00	1.71
-3	2.55	-1.77	.66	.00	.29	-1.50	3.01	-4.03	3.97	-2.85	1.25	.00	-.29	-.41	1.60	-2.50
-2	-2.41	1.17	0	-.17	-1.00	3.01	-4.83	5.43	-4.41	2.25	0	-1.25	1.00	.40	-2.00	2.81
-1	1.47	.00	-.83	.23	1.71	-4.03	5.43	-5.03	2.89	0	-2.25	2.85	-1.71	-.33	2.01	-2.41
0	0	-1.36	1.59	-.27	-2.00	3.97	-4.41	2.89	0	-2.89	4.41	-3.97	2.00	.27	-1.59	1.36
1	-1.47	2.41	-2.01	.33	1.71	-2.85	2.25	0	-2.89	5.03	-5.43	4.03	-1.71	-.23	.83	-.00
2	2.41	-2.81	2.00	-.40	-1.00	1.25	0	-2.25	4.41	-5.43	4.83	-3.01	1.00	.17	0	-1.17
3	-2.55	2.50	-1.60	.41	.29	-.00	-1.25	2.85	-3.97	4.03	-3.01	1.50	-.29	-.00	-.66	1.77
4	2.00	-1.71	1.00	-.29	0	-.29	1.00	-1.71	2.00	-1.71	1.00	-.29	0	-.29	1.00	-1.71
5	-1.14	.85	-.42	.00	.29	-.41	.40	-.33	.27	-.23	.17	-.00	-.29	.67	-1.01	1.20
6	.41	-.25	0	.42	-1.00	1.60	-2.00	2.01	-1.59	.83	0	-.66	1.00	-1.01	.83	-.60
7	-.06	0	.25	-.85	1.71	-2.50	2.81	-2.41	1.36	-.00	-1.17	1.77	-1.71	1.20	-.60	.20

↓ v

(c)

$$= \frac{1}{MN} \sum_{u=0}^{M-1} \sum_{v=0}^{N-1} \sum_{w=0}^{M-1} \sum_{z=0}^{N-1} \left\{ F(u,v)\,D(w,z)\, \exp\left[j2\pi \left(\frac{\alpha w}{M} + \frac{\beta z}{N} \right) \right] \right.$$

$$\left. \times \sum_{m=0}^{M-1} \sum_{n=0}^{N-1} \exp\left[j2\pi \frac{m(u-w)}{M} \right] \exp\left[j2\pi \frac{n(v-z)}{N} \right] \right\}$$

$$= \sum_{u=0}^{M-1} \sum_{v=0}^{N-1} f(u,v)\,D(u,v)\, \exp\left[j2\pi \left(\frac{\alpha u}{M} + \frac{\beta v}{N} \right) \right] \qquad (44)$$

where the last equality is obtained by making use of the orthogonality relationship (39). This proves that the discrete Fourier transform of $[g]$ in (43) is $[F] \times [D]$.[†] One can similarly show that the discrete Fourier transform of the "circular" convolution of two $M \times N$ arrays in the frequency domain is equal to the product of the inverse transforms of the two arrays. That is, if we construct the following circular convolution in the frequency domain:

$$G(u,v) = \sum_{w=0}^{M-1} \sum_{z=0}^{N-1} F(w,z)\,D(u-w,v-z) = \sum_{w=0}^{M-1} \sum_{z=0}^{N-1} F(u-w,v-z)\,D(w,z)$$

for $u = 0, 1, 2, ..., M-1$, $v = 0, 1, 2, ..., N-1$, then its inverse discrete Fourier transform is given by

$$g(m,n) = f(m,n)\,d(m,n)$$

for $m = 0, 1, 2, ..., M-1$, $n = 0, 1, 2, ..., N-1$ or

$$[g] = [f] \times [d]. \qquad (45)$$

There is considerable literature on one-dimensional discrete Fourier transforms [12]. They have many other useful properties which can also be easily generalized to the two-dimensional case.

2.2.2 The Hadamard Transform

If the transformation matrices $[P]$ and $[Q]$ in (31) are Hadamard matrices, then $[F]$ is called the *Hadamard transform* of $[f]$. A Hadamard matrix $[H_{J,J}]$ is a symmetric $J \times J$ matrix whose elements are the real numbers $+1$ and -1. The rows (and columns) of a Hadamard matrix are mutually orthogonal. An example of a Hadamard matrix of order 2 is

$$[H_{2,2}] = \begin{bmatrix} 1 & 1 \\ 1 & -1 \end{bmatrix} \qquad (46)$$

[†] Note that this product is *not* the matrix product of $[F]$ and $[D]$ but ordinary multiplication of two functions, which happen to be discrete. If we denote this product by $[F] \times [D]$, then the (m,n)th element of $[F] \times [D]$ is $F(m,n)\,D(m,n)$.

The following theorem is useful in constructing Hadamard matrices of orders that are powers of 2 from $[H_{2,2}]$.

Theorem. If $[H_{J,J}]$ is a Hadamard matrix, then

$$[H_{2J,2J}] = \begin{bmatrix} [H_{J,J}] & [H_{J,J}] \\ [H_{J,J}] & -[H_{J,J}] \end{bmatrix} \tag{47}$$

is also a Hadamard matrix.

Proof: The matrix $[H_{2J,2J}]$ is clearly a symmetric matrix with elements $+1$ and -1. We must now show that the rows of this matrix are orthogonal vectors. Let r_j be the jth row of the matrix $[H_{J,J}]$. The rows of $[H_{2J,2J}]$ can then be denoted by (r_j, r_j) or $(r_j, -r_j)$ depending on whether the string of numbers in r_j is followed by a similar string or by a string of opposite sign. The dot product of the jth and $(J+j)$th rows of $[H_{2J,2J}]$ is

$$(r_j, r_j) \cdot (r_j, -r_j) = (r_j \cdot r_j) + (r_j \cdot (-r_j)) = J - J = 0. \tag{48}$$

For any other combination of rows

$$(r_i, \pm r_i) \cdot (r_j, \pm r_j) = (r_i \cdot r_j) \pm (r_i \cdot r_j) = 0 \pm 0 = 0$$

and thus $[H_{2J,2J}]$ is an orthogonal matrix. ∎

By the Kronecker product $[A] \oplus [B]$ of two matrices $[A_{M,M}]$ and $[B_{N,N}]$ we mean the $MN \times MN$ matrix

$$[A] \oplus [B] = \begin{bmatrix} a_{0,0}[B] & a_{0,1}[B] & \cdots & a_{0,M-1}[B] \\ a_{1,0}[B] & a_{1,1}[B] & \cdots & a_{1,M-1}[B] \\ \vdots & \vdots & \cdots & \vdots \\ a_{M-1,0}[B] & & \cdots & a_{M-1,M-1}[B] \end{bmatrix}$$

Then $[H_{2J,2J}]$ in (47) can be expressed as a Kronecker product of $[H_{2,2}]$ and $[H_{J,J}]$

$$[H_{2J,2J}] = [H_{2,2}] \oplus [H_{J,J}]. \tag{49}$$

By the preceding theorem $[H_{2J,2J}]$ can be further expressed as

$$[H_{2J,2J}] = [H_{2,2}] \oplus [H_{2,2}] \oplus [H_{J/2,J/2}]$$
$$= [H_{2,2}] \oplus [H_{2,2}] \oplus [H_{2,2}] \oplus \cdots \oplus [H_{2,2}] \tag{50}$$

provided J is a power of 2.

The order of a Hadamard matrix need not be a power of 2. It has been conjectured that for J greater than 2, $[H_{J,J}]$ exists for all J equal to a multiple of 4, but this has neither been proved nor disproved. In the applications of Hadamard matrices considered in this book, we have restricted ourselves to those Hadamard matrices with orders equal to powers of 2.

It is easily shown that the inverse of $[H_{J,J}]$ is given by

$$[H_{J,J}]^{-1} = (1/J)[H_{J,J}] \tag{51}$$

Therefore, the discrete Hadamard transform relations take the form

$$[F] = [H_{M,M}][f][H_{N,N}] \quad \text{and} \quad [f] = (1/MN)[H_{M,M}][F][H_{N,N}] \tag{52}$$

In Fig. 7 we show the Hadamard transform of the pattern in Fig. 6a.

By definition, the determination of discrete transforms involves matrix multiplications. Let the matrix $[f]$ in (31) be an $N \times N$ matrix. Then $[P]$ and

```
  8  -2  -8   2   6  -2  -8   2   2   4  -2  -4   2   4  -2  -4
  2   0  -2   0   2   0  -2   0   0   2   0  -2   0   2   0  -2
 -8   2   8  -2  -8   2   8  -2  -2  -4   2   4  -2  -4   2   4
 -2   0   2   0  -2   0   2   0   0  -2   0   2   0  -2   0   2
  8  -2  -8   2   6  -2  -8   2   2   4  -2  -4   2   4  -2  -4
  2   0  -2   0   2   0  -2   0   0   2   0  -2   0   2   0  -2
 -8   2   8  -2  -8   2   8  -2  -2  -4   2   4  -2  -4   2   4
 -2   0   2   0  -2   0   2   0   0  -2   0   2   0  -2   0   2
  2   0  -2   0   2   0  -2   0   0   2   0  -2   0   2   0  -2
 -4   2   4  -2  -4   2   4  -2  -2   0   2   0  -2   0   2   0
 -2   0   2   0  -2   0   2   0   0  -2   0   2   0  -2   0   2
  4  -2  -4   2   4  -2  -4   2   2   0  -2   0   2   0  -2   0
  2   0  -2   0   2   0  -2   0   0   2   0  -2   0   2   0  -2
 -4   2   4  -2  -4   2   4  -2  -2   0   2   0  -2   0   2   0
 -2   0   2   0  -2   0   2   0   0   2   0  -2   0   2   0   2
  4  -2  -4   2   4  -2  -4   2   2   0  -2   0   2   0  -2   0
```

Fig. 7 Hadamard transform of the picture in Fig. 6a.

$[Q]$ are both $N \times N$ matrices. The product $[f][Q]$ is again an $N \times N$ matrix. Given $[f]$ and $[Q]$ it would require N operations to calculate each element of the product, where each operation consists of one multiplication and one addition. Therefore, the total number of operations to compute $[f][Q]$ is, in general, equal to N^3. It follows that determination of the transform $[F]$ ($=[P][f][Q]$) would take $2N^3$ operations.

Good [4] demonstrated that if a matrix can be expressed as a product of p sparse matrices where p is proportional to $\log N$, then a product of the matrix with a vector can be determined using a number of operations proportional to $N \log N$ rather than N^2. It follows that if $[P]$ and $[Q]$ are such matrices, then the product $[f][Q]$ would require a number of operations proportional to $N^2 \log N$. The total number of operations to compute $[P][f][Q]$ and hence $[F]$ from $[f]$ (or vice versa) would then be proportional to $2N^2 \log N$ rather than $2N^3$. In fact, this forms the basis of "fast" Fourier and Hadamard transforms [2, 12]. Another matrix that lends itself to very efficient factorization is the Haar matrix [6, 10] which comprises ones, minus ones, and zeros [1].

2.3   RANDOM VARIABLES

2.3.1   Outcomes, Events, and Probabilities

Let us consider a statistical experiment. The experiment could be rolling a die or selecting a picture from a collection. The experimental outcomes for the case of rolling a die consist of particular faces of the die turning up. In the second case, the outcomes are the selections of particular pictures. If the set of outcomes in an experiment is $\Omega = \{\omega_1, \omega_2, \omega_3, \ldots\}$, then we can assign a probability p_i to each outcome ω_i in Ω. These probabilities satisfy

$$p_i \geqslant 0 \quad \text{and} \quad \sum p_i = 1$$

which correspond to the facts that no outcome can have probability less than zero, and on every performance of the experiment one of the outcomes must occur. Every subset A of Ω in this case also has a well-defined probability which is equal to the sum of probabilities of the outcomes contained in A.

In practice, it turns out that it is necessary to consider probabilities on the subsets of Ω. The subsets of Ω are called *events*. For example, for the case of rolling a die, $\Omega = \{\omega_1, \omega_2, \omega_3, \omega_4, \omega_5, \omega_6\}$ where the outcome ω_i corresponds to the face numbered i showing up. Some possible events for this experiment are

$$A_1 = \{\omega_1, \omega_2\}, \quad A_2 = \{\omega_3, \omega_4\}, \quad A_3 = \{\omega_1, \omega_2, \omega_5, \omega_6\}$$

$$A_{\text{even}} = \{\text{event that an even number will show up}\} = \{\omega_2, \omega_4, \omega_6\}$$

$$A_{\text{odd}} = \{\text{event that an odd number will show up}\} = \{\omega_1, \omega_3, \omega_5\}$$

Note that an event occurs whenever any of the outcomes that constitute that event occur. For example, if face 1 shows up, then events A_1, A_3, and A_{odd} have occurred. It is clear that the collection of all the outcomes Ω is also an event. It is called the *certain event*.

If in an experiment A_i is an event, then the complement of A_i, denoted by $\bar{A}_i$, is also an event. The event $\bar{A}_i$ is the nonoccurrence of A_i. Also, if A_i and A_j are events, then the results of set operations on these events would also be events. That is, if A_i and A_j are any two events, then $A_i \cup A_j$, $A_i \cap A_j$ and $A_i - A_j$ are also events. In the probabilistic description of any experiment, we need one other event—the *null event*. The null event is an empty set, that is, it contains no outcomes and is denoted by $\varnothing$. Two events, A_i and A_j, are called *mutually exclusive* if their intersection equals the null event, or, symbolically, if

$$A_i \cap A_j = \varnothing$$

One can now introduce the axiomatic definition of probability. The probability of an event A is a number $\mathscr{P}(A)$ assigned to this event in such a fashion so as to obey the following conditions:

    I. $\mathscr{P}(A) \geqslant 0$;

    II. $\mathscr{P}(\Omega) = 1$;

    III.   Whenever $A_1, A_2, ..., A_n, ...$ is a sequence of mutually exclusive events, $\mathscr{P}(A_1 \cup A_2 \cup \cdots \cup A_n \cup \cdots) = \sum_{i=1}^{\infty} \mathscr{P}(A_i)$.

Suppose that in an experiment we assign a number to every outcome. That is, if ω_i is an outcome of the experiment, then to this outcome we assign a number $\mathbf{f}(\omega_i)$, which is then a function over the set $\Omega = \{\omega_1, \omega_2, ...\}$ and is called a *random variable*.

Note that the relationship between the outcomes ω_i and the numbers $\mathbf{f}(\omega_i)$ assigned to each of them is deterministic. In other words, once an outcome has occurred, then the value that the random variable takes is a completely determined quantity. What is random is the occurrence of the outcome itself. As an example, consider the experiment of rolling a die. The set of outcomes is $\Omega = \{\omega_1, \omega_2, \omega_3, \omega_4, \omega_5, \omega_6\}$. To each outcome we assign a number as follows:

$$\mathbf{f}(\omega_1) = 5, \quad \mathbf{f}(\omega_2) = 10, \quad ..., \quad \mathbf{f}(\omega_6) = 30$$

thereby defining the random variable $\mathbf{f}(\omega_i)$. Suppose we roll the die and the face ω_2 shows; then the random variable takes the value 10. The random variable $\mathbf{f}(\omega_i) = 5\omega_i$ constructed here could be our gain if ω_i shows in a dice game.

We will use boldface lower case letters to denote random variables. The symbol $\mathbf{f}(\omega_i)$ will indicate the number assigned to the specific event ω_i. Also, we will use the notation $\{\mathbf{f} \leqslant z\}$ to represent a set consisting of all the outcomes ω_i such that $\mathbf{f}(\omega_i) \leqslant z$. Note that z is an arbitrary but fixed number and for any z the set $\{\mathbf{f} \leqslant z\}$ is an event. Similarly, given two arbitrary but fixed numbers z_1 and z_2

$$\{z_1 \leqslant \mathbf{f} \leqslant z_2\}$$

is also an event. The set $\{\mathbf{f} = z\}$ is an event and is a collection of all the outcomes ω_i for which $\mathbf{f}(\omega_i) = z$. In particular, the sets $\{\mathbf{f} = +\infty\}$ and $\{\mathbf{f} = -\infty\}$ are also events, i.e., in general, we may allow the random variable $\mathbf{f}$ to equal $+\infty$ or $-\infty$ for some outcomes ω_i. We shall insist, however, that the set of such outcomes have zero probability.

2.3.2   Distributions and Densities

The probability of $\{\mathbf{f} \leqslant z\}$ is a number depending on z, that is, it is a function of z. This function will be denoted by $P_f(z)$ and will be called the *distribution function* of the random variable $\mathbf{f}$. Thus

$$P_f(z) = \mathscr{P}\{\mathbf{f} \leqslant z\} \tag{53}$$

where $\mathscr{P}(A)$ denotes the probability of event A. We will assume here that $P_f(z)$ is differentiable.

A distribution function has the following properties:

(a) $P_f(-\infty) = \mathscr{P}\{\mathbf{f} \leqslant -\infty\} = 0$ and $P_f(\infty) = \mathscr{P}\{\mathbf{f} \leqslant +\infty\} = 1$.

The first property follows directly from the specifications for a random variable in the preceding section. As for the second, the event $\{\mathbf{f} \leqslant \infty\}$ must be the set of all the outcomes and, hence, is the certain event whose probability is 1.

(b) It is a nondecreasing function of z:

$$P_f(z_1) \leqslant P_f(z_2) \qquad \text{for} \quad z_1 \leqslant z_2$$

This can be proved by showing that for $z_1 \leqslant z_2$, $\{\mathbf{f} \leqslant z_1\}$ is always a subset of $\{\mathbf{f} \leqslant z_2\}$ and further by showing that if event A is included in event B in the set-theoretic sense, then $\mathscr{P}(A) \leqslant \mathscr{P}(B)$.

(c) $\mathscr{P}\{z_1 < \mathbf{f} \leqslant z_2\} = P_f(z_2) - P_f(z_1)$. (54)

Property (c) is equivalent to saying the probability of the random variable taking a value between z_1 and z_2 is equal to the difference between the values of the distribution function at these two points.

The derivative

$$p_f(z) = dP_f(z)/dz \tag{55}$$

of the distribution function $P_f(z)$ is called the *density function* of the random variable $\mathbf{f}$. By property (c) and Eq. (55), we have

$$\mathscr{P}\{z_1 \leqslant \mathbf{f} \leqslant z_2\} = \int_{z_1}^{z_2} p_f(z)\, dz \tag{56}$$

so that $P_f(z) = \int_{-\infty}^{z} p_f(z)\, dz$. Also for Δz sufficiently small

$$\mathscr{P}\{z \leqslant \mathbf{f} \leqslant z + \Delta z\} \simeq p_f(z)\, \Delta z$$

The density function can thus be defined as a limit

$$p_f(z) = \lim_{\Delta z \to 0} \mathscr{P}\{z \leqslant \mathbf{f} \leqslant z + \Delta z\}/\Delta z \tag{57}$$

This equation points to a direct method for determining the density function of a continuous random variable. The steps involved are: Repeat the experiment a large number of times, say N. Count the number of times the random variable takes values between z and $z + \Delta z$ with Δz being small compared to the range of values taken by the random variable. Let this number be $N_{\Delta z}$. Clearly, the probability that the random variable takes a value between z and $z + \Delta z$ is approximately $N_{\Delta z}/N$. Therefore, the value of $p_f(z)$ at z is approximately

$$N_{\Delta z}/N \, \Delta z. \tag{58}$$

Note that by Eq. (56) the area under the density function between z_1 and z_2 equals the probability of $\mathbf{f}$ taking a value between these two limits.

The *expected value* or the *mean value* $E\{\mathbf{f}\}$ of a random variable $\mathbf{f}$ is defined as

$$E\{\mathbf{f}\} = \int_{-\infty}^{\infty} z p_f(z) \, dz \tag{59}$$

That this definition agrees with the meaning we commonly associate with averages can be shown as follows:

$$E\{\mathbf{f}\} \simeq \sum_{i=-\infty}^{\infty} z_i \, p_f(z_i) \, \Delta z$$

where the integration has been approximated by a discrete sum. By using Eq. (57), we obtain

$$E\{\mathbf{f}\} \simeq \sum_{i=-\infty}^{\infty} z_i \mathscr{P}\{z_i \leqslant \mathbf{f} \leqslant z_{i+1}\}$$

By the arguments leading to (58), this equation can be written as

$$E\{f\} \simeq \sum_{i=-\infty}^{\infty} z_i \frac{[\text{number of times } \mathbf{f} \text{ falls in the interval } (z_i, z_{i+1})]}{[\text{total number of times the experiment is performed}]}$$

which is the usual definition of an average.

Let $\mathbf{g} = \mathscr{L}(\mathbf{f})$ be a random variable that is a function of the random variable $\mathbf{f}$; then it can be shown that

$$E\{\mathbf{g}\} = E\{\mathscr{L}(\mathbf{f})\} = \int_{-\infty}^{\infty} \mathscr{L}(z) \, p_f(z) \, dz \tag{60}$$

Let μ_f be the mean of the random variable $\mathbf{f}$. Then the *variance* σ_f^2 of $\mathbf{f}$ is defined as

$$\sigma_f^2 = E\{(\mathbf{f} - \mu)^2\} = \int_{-\infty}^{\infty} (z - \mu)^2 p_f(z) \, dz \tag{61}$$

where the last equality follows from (60). The square root of the variance is called the *standard deviation*.

As an example of a density function commonly encountered, we say a random variable $\mathbf{f}$ is *normally distributed* if its density function is a Gaussian curve given by

$$p_f(z) = \frac{1}{\sigma_f \sqrt{2\pi}} \exp \frac{-(z - \mu_f)^2}{2\sigma_f^2}$$

where it can be shown that μ_f and σ_f^2 are, respectively, the mean and the variance of $\mathbf{f}$.

2.3.3 Conditional Densities

Consider two events A and B. As stated earlier, their intersection $A \cap B$ is also an event. The *conditional probability* of A, given that B has occurred, is defined by

$$\mathscr{P}(A \mid B) = \mathscr{P}(A \cap B)/\mathscr{P}(B) \tag{62}$$

The *conditional distribution* $P_f(z \mid B)$ of a random variable $\mathbf{f}$, assuming B has occurred, is defined as the conditional probability of the event $\{\mathbf{f} \leqslant z\}$; that is,

$$P_f(z \mid B) = \mathscr{P}\{\mathbf{f} \leqslant z \mid B\} = \mathscr{P}(\{\mathbf{f} \leqslant z\} \cap B)/\mathscr{P}(B) \tag{63}$$

where the numerator is the event consisting of all outcomes ω_i such that

$$\mathbf{f}(\omega_i) \leqslant z \quad \text{and} \quad \omega_i \in B \tag{64}$$

At the points of continuity of $P_f(z \mid B)$, the *conditional density* is defined as

$$p_f(z \mid B) = dP_f(z \mid B)/dz \tag{65}$$

$$= \lim_{\Delta z \to 0} \mathscr{P}\{z \leqslant \mathbf{f} \leqslant z + \Delta z \mid B\}/\Delta z \tag{66}$$

The last equality follows by arguments similar to those leading to (57).

The *conditional expected value* of the random variable $\mathbf{f}$ is given by integral (59), where $p_f(z)$ is replaced by $p_f(z \mid B)$:

$$E\{\mathbf{f} \mid B\} = \int_{\infty}^{\infty} z p_f(z \mid B) \, dz \tag{67}$$

2.3.4 Joint, Marginal, and Conditional Densities

A random variable $\mathbf{f}$ was defined to be a function $\mathbf{f}(\omega_i)$ over the set of all outcomes $\Omega = (\omega_1, \omega_2, \ldots)$ of an experiment. Let us define n functions

$$\mathbf{f}_1(\omega_i), \mathbf{f}_2(\omega_i), \ldots, \mathbf{f}_n(\omega_i)$$

over this set of outcomes. From our earlier discussion, the sets

$$\{\mathbf{f}_1 \leqslant z_1\}, \{\mathbf{f}_2 \leqslant z_2\}, \ldots, \{\mathbf{f}_n \leqslant z_n\}$$

are events with respective probabilities

$$\mathscr{P}\{\mathbf{f}_1 \leqslant z_1\} = P_{f_1}(z_1)$$

$$\mathscr{P}\{\mathbf{f}_2 \leqslant z_2\} = P_{f_2}(z_2)$$

$$\vdots$$

$$\mathscr{P}\{\mathbf{f}_n \leqslant z_n\} = P_{f_n}(z_n)$$

where $P_{f_1}(z_1), P_{f_2}(z_2), ..., P_{f_n}(z_n)$ are the distribution functions of the random variables $\mathbf{f}_1, \mathbf{f}_2, ..., \mathbf{f}_n$.

The intersection

$$\{\mathbf{f}_1 \leqslant z_1\} \cap \{\mathbf{f}_2 \leqslant z_2\} \cap \cdots \cap \{\mathbf{f}_n \leqslant z_n\} = \{\mathbf{f}_1 \leqslant z_1, \mathbf{f}_2 \leqslant z_2, ..., \mathbf{f}_n \leqslant z_n\}$$

of these n sets is also an event consisting of all the outcomes ω_i such that $\mathbf{f}_1(\omega_i) \leqslant z_1, \mathbf{f}_2(\omega_i) \leqslant z_2, ..., \mathbf{f}_n(\omega_i) \leqslant z_n$. The *joint distribution function*

$$P_{f_1, ..., f_n}(z_1, ..., z_n)$$

is the probability of this event. That is

$$P_{f_1, ..., f_n}(z_1, z_2, ..., z_n) = \mathscr{P}\{\mathbf{f}_1 \leqslant z_1, \mathbf{f}_2 \leqslant z_2, ..., \mathbf{f}_n \leqslant z_n\} \tag{68}$$

The nth order partial derivative

$$p_{f_1, ..., f_n}(z_1, z_2, ..., z_n) = \frac{\partial^n P_{f_1, ..., f_n}(z_1, z_2, ..., z_n)}{\partial z_1 \, \partial z_2 \, \cdots \, \partial z_n} \tag{69}$$

wherever it exists, is known as the *joint density function* of the random variables $\mathbf{f}_1, \mathbf{f}_2, ..., \mathbf{f}_n$.

Since $\mathbf{f}_1, \mathbf{f}_2, ..., \mathbf{f}_n$ are defined over the same set Ω, and since the set $\{\mathbf{f}_k \leqslant \infty\}$ for any k must be the collection of all possible outcomes, then by the property of intersections, we have for $i \leqslant n$

$$\{\mathbf{f}_1 \leqslant z_1\} \cap \{\mathbf{f}_2 \leqslant z_2\} \cap \cdots \cap \{\mathbf{f}_i \leqslant z_i\}$$

$$= \{\mathbf{f}_1 \leqslant z_1\} \cap \{\mathbf{f}_2 \leqslant z_2\} \cap \cdots \cap \{\mathbf{f}_i \leqslant z_i\} \cap \{\mathbf{f}_{i+1} \leqslant \infty\} \cap \cdots \cap \{\mathbf{f}_n \leqslant \infty\}$$

Therefore, it follows for $i \leqslant n$ that

$$P_{f_1, ..., f_i}(z_1, z_2, ..., z_i) = P_{f_1, ..., f_n}(z_1, z_2, ..., z_i, \infty, ..., \infty) \tag{70}$$

where there are $(n-i)$ ∞'s.

It can be shown from (69) that

$$P_{f_1, ..., f_n}(z_1, z_2, ..., z_n)$$

$$= \int_{-\infty}^{z_1} \int_{-\infty}^{z_2} \cdots \int_{-\infty}^{z_n} p_{f_1, ..., f_n}(\xi_1, \xi_2, ..., \xi_n) \, d\xi_1 \, d\xi_2 \cdots d\xi_n \tag{71}$$

By making use of (70) and (71), we obtain for $i \leqslant n$

$$p_{f_1, ..., f_i}(z_1, z_2, ..., z_i)$$

$$= \frac{\partial^i P_{f_1, ..., f_i}(z_1, z_2, ..., z_i)}{\partial z_1 \, \partial z_2 \, \cdots \, \partial z_i}$$

$$= \int_{-\infty}^{\infty} \int_{-\infty}^{\infty} \cdots \int_{-\infty}^{\infty} p_{f_1, ..., f_n}(z_1, z_2, ..., z_n) \, dz_{i+1} \, dz_{i+2} \cdots dz_n \tag{72}$$

So we see that if we integrate $p_{f_1, \ldots, f_n}(z_1, z_2, \ldots, z_n)$ with respect to some of the variables, we obtain the joint density function over the rest of the variables. Joint density functions thus obtained are called *marginal density functions*.

By making use of Eqs. (63)–(67), one can show that

$$p_{f_1, \ldots, f_k}(z_1, z_2, \ldots, z_k \,|\, \mathbf{f}_{k+1} = z_{k+1}, \mathbf{f}_{k+2} = z_{k+2}, \ldots, \mathbf{f}_n = z_n)$$

$$= \frac{p_{f_1, \ldots, f_n}(z_1, z_2, \ldots, z_k, z_{k+1}, \ldots, z_n)}{p_{f_1, \ldots, f_n}(z_{k+1}, z_{k+2}, \ldots, z_n)}$$

where the left-hand side denotes the conditional density of $\mathbf{f}_1, \mathbf{f}_2, \ldots, \mathbf{f}_k$ given that the random variables $\mathbf{f}_{k+1}, \mathbf{f}_{k+2}, \ldots, \mathbf{f}_n$ have taken the specific values $z_{k+1}, z_{k+2}, \ldots, z_n$, respectively. The denominator on the right-hand side is the marginal density over $\mathbf{f}_{k+1}, \mathbf{f}_{k+2}, \ldots, \mathbf{f}_n$.

Before we end this subsection, we would like to give the expression for the expectation of a function of several random variables. Let a random variable $\mathbf{g}$ be a function of the random variables $\mathbf{f}_1, \mathbf{f}_2, \ldots, \mathbf{f}_n$ as follows:

$$\mathbf{g} = \mathscr{L}(\mathbf{f}_1, \mathbf{f}_2, \ldots, \mathbf{f}_n)$$

By making use of definition (59), it can be shown that

$$E\{\mathbf{g}\} = \int\int \cdots \int \mathscr{L}(z_1, z_2, \ldots, z_n)\, p(z_1, z_2, \ldots, z_n)\, dz_1\, dz_2 \cdots dz_n \qquad (73)$$

2.3.5 Independent, Uncorrelated, and Orthogonal Random Variables

The random variables $\mathbf{f}_1, \mathbf{f}_2, \ldots, \mathbf{f}_n$ are called *independent* if

$$p_{f_1, \ldots, f_n}(z_1, z_2, \ldots, z_n) = p_{f_1}(z_1)\, p_{f_2}(z_2) \cdots p_{f_n}(z_n) \qquad (74a)$$

where $p_{f_i}(z_i)$ is the density function of the random variable $\mathbf{f}_i$. Similarly, $\mathbf{f}_1, \mathbf{f}_2, \ldots, \mathbf{f}_r$ are called "independent" of $\mathbf{f}_{r+1}, \mathbf{f}_{r+2}, \ldots, \mathbf{f}_n$ if

$$p_{f_1, \ldots, f_n}(z_1, \ldots, z_n) = p_{f_1, \ldots, f_r}(z_1, \ldots, z_r)$$

$$\times p_{f_{r+1}, \ldots, f_n}(z_{r+1}, \ldots, z_n) \qquad (74b)$$

The random variables $\mathbf{f}_1, \mathbf{f}_2, \ldots, \mathbf{f}_n$ are called *uncorrelated* if

$$E\{\mathbf{f}_i \mathbf{f}_j\} = E\{\mathbf{f}_i\}\, E\{\mathbf{f}_j\} \qquad \text{for all } i, j \text{ with } i \neq j \qquad (75)$$

If μ_{f_i} denotes the expected value of the random variable $\mathbf{f}_i$, then the *covariance* of two random variables $\mathbf{f}_i$ and $\mathbf{f}_j$ is defined by

$$C_{ij} = E\{(\mathbf{f}_i - \mu_{f_i})(\mathbf{f}_j - \mu_{f_j})\} \qquad (76)$$

By expanding this it can easily be shown that if $\mathbf{f}_i$ and $\mathbf{f}_j$ are uncorrelated, then their covariance equals zero.

The *correlation coefficient* r_{ij} of two random variables $\mathbf{f}_i$ and $\mathbf{f}_j$ is defined as

$$r_{ij} = C_{ij}/\sigma_{f_i}\sigma_{f_j} \tag{77}$$

where the σ's were defined in (61).

Exercise 6. Given n random variables $\mathbf{f}_1, \mathbf{f}_2, ..., \mathbf{f}_n$ with $\sigma_{f_i}^2 = \sigma^2$, $i = 1, 2, ..., n$, let us define a random variable $\mathbf{f}$ as

$$\mathbf{f} = (\mathbf{f}_1 + \mathbf{f}_2 + \cdots + \mathbf{f}_n)/n$$

If $\mathbf{f}_1, \mathbf{f}_2, ..., \mathbf{f}_n$ are uncorrelated, show that the variance σ_f^2 of $\mathbf{f}$ is given by

$$\sigma_f^2 = \sigma^2/n \quad \blacksquare \tag{78}$$

The n random variables $\mathbf{f}_1, \mathbf{f}_2, ..., \mathbf{f}_n$ are called *orthogonal* if

$$E\{\mathbf{f}_i \mathbf{f}_j\} = 0 \qquad \text{for} \quad \text{all } i, j \quad \text{with } i \neq j \tag{79}$$

Note that uncorrelated random variables that have zero mean are orthogonal.

2.3.6 Functions of Random Variables

Let n random variables $\mathbf{g}_1, \mathbf{g}_2, ..., \mathbf{g}_n$ be functions of the random variables $\mathbf{f}_1, \mathbf{f}_2, ..., \mathbf{f}_n$ as

$$\mathbf{g}_1 = \mathscr{L}_1(\mathbf{f}_1, \mathbf{f}_2, ..., \mathbf{f}_n)$$
$$\vdots \tag{80}$$
$$\mathbf{g}_n = \mathscr{L}_n(\mathbf{f}_1, \mathbf{f}_2, ..., \mathbf{f}_n)$$

Let us denote the joint density function of the random variables $\mathbf{g}_1, \mathbf{g}_2, ..., \mathbf{g}_n$ by $p_{g_1, ..., g_n}(y_1, y_2, ..., y_n)$; that is,

$$p_{g_1, ..., g_n}(y_1, y_2, ..., y_n) = \frac{\partial^n \mathscr{P}\{\mathbf{g}_1 \leqslant y_1, \mathbf{g}_2 \leqslant y_2, ..., \mathbf{g}_n \leqslant y_n\}}{\partial y_1\, \partial y_2 \cdots \partial y_n}.$$

Our aim is to express the relationship between the joint density function

$$p_{g_1, ..., g_n}(y_1, y_2, ..., y_n) \qquad \text{and} \qquad p_{f_1, ..., f_n}(z_1, z_2, ..., z_n).$$

To determine $p_{g_1, ..., g_n}(y_1, y_2, ..., y_n)$ for a given set of numbers $y_1, y_2, ..., y_n$, we solve the following set of simultaneous equations:

$$\mathscr{L}_1(z_1, z_2, ..., z_n) = y_1$$
$$\mathscr{L}_2(z_1, z_2, ..., z_n) = y_2$$
$$\vdots \tag{81}$$
$$\mathscr{L}_n(z_1, z_2, ..., z_n) = y_n$$

Let this system have a single real solution $z_1, z_2, \ldots, z_n$; then it can be shown that

$$p_{g_1, \ldots, g_n}(y_1, y_2, \ldots, y_n) = \frac{p_{f_1, \ldots, f_n}(z_1, z_2, \ldots, z_n)}{|J(z_1, z_2, \ldots, z_n)|} \qquad (82)$$

where J, called the *Jacobian* of transformation (80), is given by

$$J(z_1, z_2, \ldots, z_n) = \begin{vmatrix} \dfrac{\partial \mathscr{L}_1}{\partial z_1} & \dfrac{\partial \mathscr{L}_1}{\partial z_2} & \cdots & \dfrac{\partial \mathscr{L}_1}{\partial z_n} \\ \vdots & \vdots & \vdots & \vdots \\ \dfrac{\partial \mathscr{L}_n}{\partial z_1} & \dfrac{\partial \mathscr{L}_n}{\partial z_2} & \cdots & \dfrac{\partial \mathscr{L}_n}{\partial z_n} \end{vmatrix} \qquad (83)$$

If Eqs. (81) have more than one solution, then we add in the right-hand side (82) the corresponding expressions resulting from all the solutions. For a proof of (82), see, for example, Papoulis [9].

2.4 RANDOM FIELDS

2.4.1 Definition of a Random Field

In Section 2.3.4, we discussed sets of random variables over the set Ω. We will now generalize this discussion in the following manner. Let us define a one-parameter family of functions $\mathbf{f}_s(\omega_i)$ over the set of all outcomes $\Omega = \{\omega_1, \omega_2, \ldots\}$ of an experiment, where the parameter $s \in I$ and I is an interval on the real axis or a region of a multidimensional Euclidean space. *Note that for each value of s we have a function defined over the set Ω.* Also, note that for each value of ω_i, $\mathbf{f}_s(\omega_i)$ is a function of s over the region I.

When I is one-dimensional and a collection of discrete points, e.g., $I = \{1, 2, \ldots, n\}$, our family of functions is just a set of n random variables as defined in Section 2.3.4.

This discussion indicates that there are two ways of looking at $\mathbf{f}_s(\omega_i)$, $\omega_i \in \Omega$, $s \in I$. They are as follows:

(i) $\mathbf{f}_s(\omega_i)$ is clearly a family of random variables, each member of the family being generated by a value of s.

(ii) $\mathbf{f}_s(\omega_i)$ can also be looked on as a family of functions of s, each member of this family corresponding to an outcome ω_i.

When I is one-dimensional, the family of functions $\mathbf{f}_s(\omega_i)$ is called a *stochastic process*. When the dimensionality of I is two or greater, the family of functions $\mathbf{f}_s(\omega_i)$ is called a *random field*. In what follows, we shall concentrate on random fields with I two-dimensional.

Let I be the xy-plane, so that s is a point in the xy-plane, and s can be represented either by its coordinates or by its position vector $\vec{r}$. The random field $\mathbf{f}_s(\omega_i)$, which can now be expressed as $\mathbf{f}_{\vec{r}}(\omega_i)$, will henceforth be denoted by $\mathbf{f}(\vec{r}, \omega_i)$. For a given value of $\vec{r}$, $\mathbf{f}(\vec{r}, \omega_i)$ is a random variable, while for a given outcome ω_i, $\mathbf{f}(\vec{r}, \omega_i)$ is a function over the xy-plane.

In Fig. 8, we have given an example of a random field $\mathbf{f}(\vec{r}, \omega_i)$ with $\Omega = \{\omega_1, \omega_2, \omega_3\}$. We will remind the reader that the ω_i's are the outcomes of the underlying experiment. The underlying experiment could, for example, be the selection of a picture from a collection. The outcome ω_i corresponds to the selection of the ith picture.

It is seen that with $\vec{r} = OA$ in Fig. 8, $\mathbf{f}(\vec{r}, \omega_i)$ is a function of ω_i. Its value for ω_1 equals 2, for ω_2 its value is 4, and for ω_3 it is 3. So we see that for a fixed $\vec{r}$, $\mathbf{f}(\vec{r}, \omega_i)$ is a function of ω_i and is, therefore, a random variable. For a

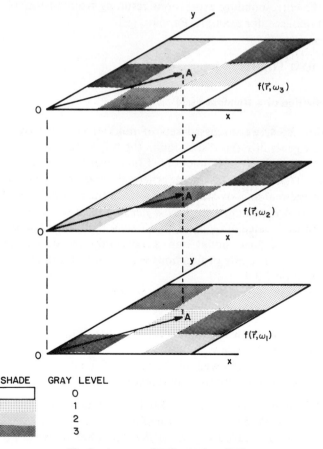

Fig. 8   An example of a random field.

fixed ω_i, however, $\mathbf{f}(\vec{r}, \omega_i)$ is a two-dimensional function (picture) in the xy-plane.

The random field $\mathbf{f}(\vec{r}, \omega_i)$ is a random variable for a specific value of $\vec{r}$. In general, this random variable will not have the same statistical properties for all values of $\vec{r}$. In other words, the distribution and density functions for the random variable $\mathbf{f}(\vec{r}, \omega_i)$ will depend on the value of $\vec{r}$ that is chosen. Let

$$P_f(z; \vec{r}) = \mathscr{P}\{\mathbf{f}(\vec{r}, \omega_i) \leqslant z\} \tag{84}$$

Clearly, $P_f(z; \vec{r})$ is the distribution function for the random variable $\mathbf{f}(\vec{r}, \omega_i)$ at the point $\vec{r}$ in the xy-plane. The density function at the point $\vec{r}$ would then be given by

$$p_f(z; \vec{r}) = \partial P_f(z; \vec{r})/\partial z \tag{85}$$

Henceforth, we will denote $\mathbf{f}(\vec{r}, \omega_i)$ by $\mathbf{f}(\vec{r})$ (or $\mathbf{f}(x, y)$), it being understood that this represents a family of two-dimensional functions, each function corresponding to an outcome ω_i.

Given points $\vec{r}_1, \vec{r}_2, ..., \vec{r}_n$ in the xy-plane, we have n random variables $\mathbf{f}(\vec{r}_1), \mathbf{f}(\vec{r}_2), ..., \mathbf{f}(\vec{r}_n)$. The joint distribution and density functions of these random variables are given by

$$P_f(z_1, z_2, ..., z_n; \vec{r}_1, \vec{r}_2, ..., \vec{r}_n) = \mathscr{P}\{\mathbf{f}(\vec{r}_1) \leqslant z_1, ..., \mathbf{f}(\vec{r}_n) \leqslant z_n\} \tag{86}$$

and

$$p_f(z_1, ..., z_n; \vec{r}_1, ..., \vec{r}_n) = \frac{\partial^n P_f(z_1, ..., z_n; \vec{r}_1, ..., \vec{r}_n)}{\partial z_1 \cdots \partial z_n} \tag{87}$$

$P_f(z_1, ..., z_n; \vec{r}_1, ..., \vec{r}_n)$ and $p_f(z_1, ..., z_n; \vec{r}_1, ..., \vec{r}_n)$ are called the nth order distribution and density functions, respectively, of the random field $\mathbf{f}(\vec{r})$.

2.4.2   Mean, Autocorrelation, Autocovariance[†]

The random field $\mathbf{f}(\vec{r})$ is a random variable for a given $\vec{r}$. Since, in general, the density function of this random variable depends on the value of $\vec{r}$ that is chosen, its expected value must also be a function of $\vec{r}$. If $\mu_f(\vec{r})$ denotes this expectation, then by (59)

$$\mu_f(\vec{r}) = E\{\mathbf{f}(\vec{r})\} = \int_{-\infty}^{\infty} z p_f(z; \vec{r}) \, dz \tag{88}$$

This expectation is called the *mean* of the random field $\mathbf{f}$ at $\vec{r}$.

[†] A random field may be real or complex. Since in this book we will only be concerned with real random fields, the definitions given here and elsewhere apply only for that case, unless otherwise mentioned.

The *autocorrelation* $R(\vec{r}_1, \vec{r}_2)$ of a random field $\mathbf{f}(\vec{r})$ is defined as the expected value of the product of the two random variables $\mathbf{f}(\vec{r}_1)$ and $\mathbf{f}(\vec{r}_2)$ and by using (73) is shown to be given by

$$R_{ff}(\vec{r}_1, \vec{r}_2) = E\{\mathbf{f}(\vec{r}_1)\mathbf{f}(\vec{r}_2)\} = \int\int_{-\infty}^{\infty} z_1 z_2 \, p_f(z_1, z_2; \vec{r}_1, \vec{r}_2) \, dz_1 \, dz_2 \quad (89)$$

The *autocovariance* $C(\vec{r}_1, \vec{r}_2)$ of a random field $\mathbf{f}(\vec{r})$ is defined as

$$C_{ff}(\vec{r}_1, \vec{r}_2) = E\{[\mathbf{f}(\vec{r}_1) - \mu_f(\vec{r}_1)][\mathbf{f}(\vec{r}_2) - \mu_f(\vec{r}_2)]\}$$
$$= R_{ff}(\vec{r}_1, \vec{r}_2) - \mu_f(\vec{r}_1)\mu_f(\vec{r}_2). \quad (90)$$

The *cross correlation* of two real random fields is defined by

$$R_{fg}(\vec{r}_1, \vec{r}_2) = E\{\mathbf{f}(\vec{r}_1)\mathbf{g}(\vec{r}_2)\} \quad (91)$$

and their *cross covariance* by

$$C_{fg}(\vec{r}_1, \vec{r}_2) = E\{[\mathbf{f}(\vec{r}_1) - \mu_f(\vec{r}_1)][\mathbf{g}(\vec{r}_2) - \mu_g(\vec{r}_2)]\}$$
$$= R_{fg}(\vec{r}_1, \vec{r}_2) - \mu_f(\vec{r}_1)\mu_g(\vec{r}_2) \quad (92)$$

Two random fields $\mathbf{f}(\vec{r})$ and $\mathbf{g}(\vec{r})$ are called *uncorrelated* if for any $\vec{r}_1$ and $\vec{r}_2$

$$C_{fg}(\vec{r}_1, \vec{r}_2) = 0$$

which is equivalent to

$$E\{\mathbf{f}(\vec{r}_1)\mathbf{g}(\vec{r}_2)\} = E\{\mathbf{f}(\vec{r}_1)\} E\{\mathbf{g}(\vec{r}_2)\} \quad (93)$$

They are called *orthogonal* if for any $\vec{r}_1$ and $\vec{r}_2$

$$R_{fg}(\vec{r}_1, \vec{r}_2) = 0 \quad (94)$$

We say that the random fields $\mathbf{f}(\vec{r})$ and $\mathbf{g}(\vec{r})$ are *independent* if the set

$$\mathbf{f}(\vec{r}_1), \mathbf{f}(\vec{r}_2), ..., \mathbf{f}(\vec{r}_n)$$

is independent of the set

$$\mathbf{g}(\vec{r}_1'), \mathbf{g}(\vec{r}_2'), ..., \mathbf{g}(\vec{r}_m') \quad (95)$$

for any $\vec{r}_1, \vec{r}_2, ..., \vec{r}_n, \vec{r}_1', \vec{r}_2', ..., \vec{r}_m'$, as defined in (74b).

2.4.3 Homogeneous Random Fields

A random field is called *homogeneous* [11] if its expected value $\mu(\vec{r})$ is independent of position $\vec{r}$, that is, if

$$\mu_f(\vec{r}) = \mu = \text{constant independent of } \vec{r} \quad (96)$$

and if its autocorrelation function is translation invariant, that is, if

$$R_{ff}(\vec{r}_1, \vec{r}_2) = E\{\mathbf{f}(\vec{r}_1)\mathbf{f}(\vec{r}_2)\} = E\{\mathbf{f}(\vec{r}_1 + \vec{r}_0)\mathbf{f}(\vec{r}_2 + \vec{r}_0)\}$$

$$= R_{ff}(\vec{r}_1 + \vec{r}_0, \vec{r}_2 + \vec{r}_0) \qquad (97)$$

for all $\vec{r}_1$, $\vec{r}_2$, and $\vec{r}_0$ in the xy-plane. Setting $\vec{r}_0 = -\vec{r}_2$ in (97), we see that the correlation function

$$R_{ff}(\vec{r}_1, \vec{r}_2) = R_{ff}(\vec{r}_1 - \vec{r}_2, 0) \equiv R_{ff}(\vec{r}_1 - \vec{r}_2) \qquad (98)$$

depends only on the difference vector $(\vec{r}_1 - \vec{r}_2)$. Since $E\{\mathbf{f}(\vec{r}_1)\mathbf{f}(\vec{r}_2)\} = E\{\mathbf{f}(\vec{r}_2)\mathbf{f}(\vec{r}_1)\}$ implies $R_{ff}(\vec{r}_1, \vec{r}_2) = R_{ff}(\vec{r}_2, \vec{r}_1)$, therefore, in the homogeneous case we may conclude that

$$R_{ff}(\vec{r}_1 - \vec{r}_2) = R_{ff}(\vec{r}_2 - \vec{r}_1) \qquad (99)$$

If x_1, y_1 and x_2, y_2 denote the coordinates of the position vectors $\vec{r}_1$ and $\vec{r}_2$, respectively, then (99) can be expressed as

$$R_{ff}(x_1, y_1, x_2, y_2) = R_{ff}(x_1 - x_2, y_1 - y_2) = R_{ff}(x_2 - x_1, y_2 - y_1)$$
$$(100)$$

If we use α and β to denote the difference coordinates $x_1 - x_2$ and $y_1 - y_2$, respectively, (100) can be expressed as

$$R_{ff}(x_1, y_1, x_2, y_2) = R_{ff}(\alpha, \beta) = R_{ff}(-\alpha, -\beta)$$

Clearly, then, for a real homogeneous random field the autocorrelation function is given by

$$R_{ff}(\alpha, \beta) = E\{\mathbf{f}(x + \alpha, y + \beta)\mathbf{f}(x, y)\} \qquad (101a)$$

We see that, in general, the autocorrelation function of a random field is a function of four variables. For homogeneous random fields, however, it is a function of only two variables α and β. Also, it is invariant with respect to 180° rotation. The autocorrelation function may, of course, possess a higher symmetry than this. For example, $R_{ff}(\alpha, \beta)$ may be invariant with respect to all rotations. In this case, the autocorrelation is a function of only one variable, that is, the Euclidean distance $|\vec{r}_1 - \vec{r}_2| = \sqrt{\alpha^2 + \beta^2}$. Random fields with this property are called *homogeneous and isotropic* random fields.

Similar conclusions can be drawn about the cross-correlation function $R_{fg}(x_1, y_1, x_2, y_2)$ defined in (91). Two random fields are called *jointly homogeneous* if this function depends on only $x_1 - x_2$ and $y_1 - y_2$. In other words for two real random fields that are jointly homogeneous, the cross-correlation function is given by

$$R_{fg}(\alpha, \beta) = E\{\mathbf{f}(x + \alpha, y + \beta)\mathbf{g}(x, y)\} \qquad (101b)$$

The concept of homogeneous random fields is a generalization of one-dimensional wide-sense stationary stochastic processes to higher dimensions. In fact homogeneous random fields are also sometimes referred to as "wide-sense stationary random processes or fields."

2.4.4 Spectral Density

The *spectral density* $S_{ff}(u, v)$ of a homogeneous random field $\mathbf{f}(\vec{r})$ is the Fourier transform of its autocorrelation

$$S_{ff}(u, v) = \int\int_{-\infty}^{\infty} R_{ff}(\alpha, \beta) \exp[-j2\pi(\alpha u + \beta v)] \, d\alpha \, d\beta \qquad (102a)$$

Similarly, the *cross spectral density* of two homogeneous random fields $\mathbf{f}$ and $\mathbf{g}$ is the Fourier transform of their cross correlation

$$S_{fg}(u, v) = \int\int_{-\infty}^{\infty} R_{fg}(\alpha, \beta) \exp[-j2\pi(\alpha u + \beta v)] \, d\alpha \, d\beta \qquad (102b)$$

If we use $\vec{w}$ to denote a vector in the uv-plane, and $\vec{r}$ a vector in the $\alpha\beta$-plane,[†] we will sometimes find it more convenient to write (102a) as

$$S_{ff}(\vec{w}) = \int_{\alpha\beta\text{-plane}} R_{ff}(\vec{r}) \exp(-j2\pi\vec{r}\cdot\vec{w}) \, d\vec{r} \qquad (103)$$

From (102a), $R_{ff}(\alpha, \beta)$ is the inverse Fourier transform of $S(u, v)$ and is given by

$$R_{ff}(\alpha, \beta) = \int\int_{-\infty}^{\infty} S_{ff}(u, v) \exp[+j2\pi(\alpha u + \beta v)] \, du \, dv \qquad (104)$$

In the more compact notation, this can be rewritten as

$$R_{ff}(\vec{r}) = \int_{uv\text{-plane}} S_{ff}(\vec{w}) \exp(+j2\pi\vec{r}\cdot\vec{w}) \, d\vec{w} \qquad (105)$$

With $\alpha = \beta = 0$, (104) gives

$$\int_{-\infty}^{\infty} \int_{-\infty}^{\infty} S_{ff}(u, v) \, du \, dv = R_{ff}(0, 0) = E\{[\mathbf{f}(r)]^2\} \geqslant 0 \qquad (106)$$

2.4.5 Linear Operations on Random Fields

In Section 2.1.2, we showed that a linear shift-invariant operation on a picture $f(x, y)$ can be expressed as a convolution of $f(x, y)$ with a point spread

† The context should make it clear whether $\vec{r}$ is a vector in the xy-plane or in the $\alpha\beta$-plane.

function $h(x, y)$. Let us now consider the action of such a linear operation having point spread function $h(x, y)$ on a homogeneous random field $\mathbf{f}(\vec{r})$, which henceforth will be written as $\mathbf{f}(x, y)$. Since a random field is a family of two-dimensional functions defined over the xy-plane, the result of such an operation can be expressed as

$$\mathbf{g}(x, y) = \int\!\!\!\int\limits_{-\infty}^{\infty} \mathbf{f}(x-\alpha, y-\beta) h(\alpha, \beta) \, d\alpha \, d\beta \qquad (107)$$

Clearly, (107) is a family of equations, one for each member in the family of functions represented by $\mathbf{f}(x, y)$. Each member of $\mathbf{f}(x, y)$ on the right-hand side of (107) gives rise to a function of x and y; the family of these functions is the random field $\mathbf{g}(x, y)$. If $\mathbf{f}$ is homogeneous, it can be verified that $\mathbf{g}$ is also.

We will now present a theorem on the relationship between the spectral densities of the "input" $\mathbf{f}(x, y)$ and the "output" $\mathbf{g}(x, y)$ in (107).

Theorem.   Let $S_{ff}(u, v)$ and $S_{gg}(u, v)$ denote the spectral densities of the homogeneous random fields $\mathbf{f}(x, y)$ and $\mathbf{g}(x, y)$, respectively. If $\mathbf{g}$ is derived from $\mathbf{f}$ by (107), then

$$S_{gg}(u, v) = S_{ff}(u, v) |H(u, v)|^2 \qquad (108)$$

where $H(u, v)$ is the Fourier transform of the point spread function $h(x, y)$.

Proof: From (107), we can write

$$\mathbf{g}(x+a, y+b) = \int\!\!\!\int\limits_{-\infty}^{\infty} \mathbf{f}(x+a-\alpha, y+b-\beta) h(\alpha, \beta) \, d\alpha \, d\beta$$

Multiplying both sides by $\mathbf{g}(x, y)$, taking the expectation, interchanging the order of integration and expectation, and using (101a) and (101b), we obtain

$$R_{gg}(a, b) = \int\!\!\!\int\limits_{-\infty}^{\infty} R_{fg}(a-\alpha, b-\beta) h(\alpha, \beta) \, d\alpha \, d\beta$$

Therefore, symbolically,

$$R_{gg}(\alpha, \beta) = R_{fg}(\alpha, \beta) * h(\alpha, \beta) \qquad (109)$$

Similarly, by multiplying both sides of (107) by $\mathbf{f}(x+a, y+b)$ and taking the expectation, one can show that

$$R_{fg}(a, b) = \int\!\!\!\int\limits_{-\infty}^{\infty} R_{ff}(a+\alpha, b+\beta) h(\alpha, \beta) \, d\alpha \, d\beta$$

which by a transformation of variables $\alpha = -\alpha'$ and $\beta = -\beta'$ can easily be shown to be a convolution of $R_{ff}(\alpha, \beta)$ and $h(-\alpha, -\beta)$. Therefore,

$$R_{fg}(\alpha, \beta) = R_{ff}(\alpha, \beta) * h(-\alpha, -\beta) \tag{110}$$

Substituting (110) in (109), we obtain

$$\dot{R}_{gg}(\alpha, \beta) = R_{ff}(\alpha, \beta) * h(-\alpha, -\beta) * h(\alpha, \beta) \tag{111}$$

Since the Fourier transform of $h(-\alpha, -\beta)$ equals $H^*(u, v)$, we obtain (108) from Eq. (111) and the convolution theorem as given by (23a). This completes the proof of the theorem. ∎

In our derivation of (108), $\mathbf{f}$ and $\mathbf{g}$ were assumed to be real random fields. The theorem is easily shown to be true for complex random fields, also. In the complex case, the autocorrelation is defined as $E\{\mathbf{f}(r_1)\mathbf{f}^*(r_2)\}$.

Exercise 7.   The random fields $\mathbf{f}$ and $\mathbf{g}$ are related as

$$\mathbf{g}(x, y) = \int\int \mathbf{f}(x - \alpha, y - \beta)\, h(\alpha, \beta)\, d\alpha\, d\beta + \mathbf{v}(x, y)$$

where the random field $\mathbf{v}(x, y)$ represents additive noise. If $\mathbf{f}$, $\mathbf{g}$ and $\mathbf{v}$ are homogeneous random fields and if $\mathbf{f}$ and $\mathbf{v}$ have zero means and are uncorrelated, show that

$$S_{gg}(u, v) = S_{ff}(u, v)|H(u, v)|^2 + S_{vv}(u, v) \quad∎ \tag{112}$$

2.4.6   Ergodicity

In this subsection, we will discuss spatial and ensemble averages and bring out the connection between the definition of autocorrelation as introduced in Section 2.4.2 and the definition of autocorrelation as introduced in Section 2.1.4 for deterministic functions. *In what follows, we will be concerned only with homogeneous random fields.*

Expressions (88) and (89) for the mean and autocorrelation are called *ensemble averages.* Let us first consider (88). Some thought shows that (88) is equivalent to repeating the underlying experiment a large number of times, and for every performance, observing the outcome ω_i, sampling the corresponding two-dimensional function from the family $\mathbf{f}(\bar{r})$ at the position $\bar{r}$, and averaging the values so obtained.

Let us define a spatial average as

$$\mathbf{E} = \lim_{S \to \infty} \frac{1}{S} \int\int_{\mathscr{S}} \mathbf{f}(x, y)\, dx\, dy \tag{113}$$

where $\mathscr{S}$ is a bounded region of the xy-plane, S is the area of $\mathscr{S}$, and by $\lim_{S \to \infty}$ is meant a limiting process which tends to include the entire xy-plane.

Equation (113) represents a family of equations. Each member of the family of functions represented by the random field $\mathbf{f}(x, y)$ when substituted on the right-hand side in (113) yields a number. The collection of these numbers, denoted by $\mathbf{E}$, is a function over the set Ω of outcomes. Therefore, $\mathbf{E}$ is a random variable.

Suppose the random variable $\mathbf{E}$ is such that its value is the same for all the outcomes ω_i, i.e., that $\mathbf{E}$ is equal to a constant. If this constant is equal to the mean μ (see Eq. (96)), the random field is called *ergodic with respect to the mean*.

Note that if a collection of pictures is a random field that is ergodic with respect to the mean, the mean value of the random field can be obtained simply by spatially averaging any one of the pictures in the collection.

To see that definition (89) for the autocorrelation is an ensemble average, we note that this definition is equivalent to performing the underlying experiment a large number of times; and for every performance, observing the outcome ω_i, sampling the associated two-dimensional function (from the family $\mathbf{f}(x, y)$) at points (x_1, y_1) and (x_2, y_2), determining the product of these two samples, and averaging the result so obtained.

Let us now define the following spatial average for a homogeneous random field:

$$\mathbf{R}(\alpha, \beta) = \lim_{S \to \infty} \frac{1}{S} \int\!\!\int_{\mathscr{S}} \mathbf{f}(x, y)\, \mathbf{f}(x+\alpha, y+\beta)\, dx\, dy \qquad (114)$$

Clearly $\mathbf{R}(\alpha, \beta)$ is a random variable. If this random variable is equal to a constant for each (α, β) and if this constant value is equal to $R_{ff}(\alpha, \beta)$, then the random field is called *ergodic with respect to autocorrelation*.

If a collection of pictures is ergodic with respect to autocorrelation, then the autocorrelation function $R_{ff}(\alpha, \beta)$ can be obtained simply by performing the integration on the right-hand side in (114) for *any* picture in the collection. For a more general discussion of ergodicity, see, for example, [11].

REFERENCES

1. G. Alexists, "Convergence Problems of Orthogonal Series," p. 62, Pergamon, New York, 1961.
2. H. C. Andrews, "Computer Techniques in Image Processing," Academic Press, New York, 1970.
3. R. N. Bracewell, "The Fourier Transform and its Applications," McGraw-Hill, New York, 1968.

4. I. J. Good, The interaction algorithm and practical Fourier series, *J. Roy. Statist. Soc. Ser. B.* **20**, 1958, 361–372; **22**, 1960, 372–375,.
5. J. W. Goodman, "Introduction to Fourier Optics," McGraw-Hill, New York, 1968.
6. A. Haar, Zur theorie der orthogonalen funktionensysteme, *Math. Ann.* **69**, 1910, 331–371.
7. M. J. Lighthill, "Introduction to Fourier Analysis and Generalized Functions," Cambridge Univ. Press, London and New York, 1960.
8. A. Papoulis, "The Fourier Integral and Its Applications," McGraw-Hill, New York, 1962.
9. A. Papoulis, "Probability, Random Variables, and Stochastic Processes," McGraw-Hill, New York, 1965.
10. C. Watari, A generalization of Haar functions, *Tohoku Math. J.* **8**, 1956, 286–290.
11. E. Wong, "Stochastic Processes in Information and Dynamical Systems," McGraw-Hill, New York, 1971.
12. Special issues of *IEEE Trans. Audio Electroacoust.* on Fast Fourier Transforms, **AE-15**, June 1967; **AE-17**, June 1969.

Chapter 3

Visual Perception

Knowledge about the human visual system (VS) can be very useful to the designer and user of picture-processing techniques. In particular, one must know something about subjective picture quality, and about the fidelity of a picture to an original scene, in designing systems for picture digitization or coding, or for image enhancement, when the pictures are intended for viewing by humans. (Picture quality will be discussed in greater detail in Section 6.1.) Similarly, in analyzing the structure of a picture for purposes of picture description, one wants to extract picture parts that correspond to those seen by humans, and to describe them in terms corresponding to those used by humans.

The subject of visual perception is very broad and complex; the present chapter cannot do more than briefly introduce a few basic topics. For more detailed overviews, including quantitative data, see, e.g., Graham [4], Cornsweet [3], and Zusne [8]. Many important topics will not even be touched on here; for example, the ways in which perceptual abilities are acquired (perceptual learning), or the way in which one adapts to perceptual distortions. We shall not discuss the anatomy and physiology of the VS, but rather shall treat it from the "black box" standpoint of perception psychology.

An important point to keep in mind when studying the VS is that it cannot be treated simply as a special type of image digitization and transmission system. To regard it as such is to commit the "homunculus fallacy"; when an image has been transmitted by the eye to the brain, there is no "little man" inside the brain to look at it! The input to the VS may be an image, but the output which it furnishes to the higher brain centers must be something quite different. This fact has many implications as regards the perception of even the simplest stimuli, as we shall see in the course of this chapter.

3.1 BRIGHTNESS AND CONTRAST

In this and the next section we discuss the visual perception of simple stimuli such as spots, edges, and bars of light. Perception of color, texture, form, etc., will be discussed in later sections.

The ability to detect a bright spot or flash depends not only on such properties as the brightness, size, and duration of the spot, but also on the brightness of the background against which the spot appears. There is, of course, an absolute threshold—at the quantum level—below which detection is impossible (see also Section 3.7); but more generally, there is a contrast threshold—a "*just noticeable difference*" between spot and background.

In general, detection thresholds depend on the previous pattern of illumination (in space and time); this dependence is called *adaptation*. When there has been no previous illumination for a long period (of the order of an hour or more), there is complete "dark adaptation," and thresholds are at their lowest. Vision under conditions of dark adaptation is called *scotopic*, and is characterized by reduced ability to perceive colors; under light adaptation, vision is called *photopic*. Detection thresholds also depend on position of the stimulus relative to the visual axis (i.e., the direction in which the eye is turned), being generally lower at the periphery. There are adaptation-like effects that depend on *subsequent* illumination (up to about 0.1 sec); the contrast threshold rises even before perception of a light flash. Such time-dependent phenomena are known as *metacontrast* effects.

If the background illumination I is uniform and extensive (in space and time), the threshold "just noticeable difference" ΔI in illumination is approximately proportional to I over a wide range; this relationship is known as *Weber's law*. Thus the higher I, the higher must ΔI be for detectability. One can say that the VS has a logarithmic response to brightness, since detectability depends on the ratio, rather than the difference, between I and $I + \Delta I$.

The *apparent brightness* of a stimulus also depends on adaptation; in addition, it is a nonlinear function of the intensity of the stimulus. In general, the apparent magnitude of a perceived stimulus can be approximated by a power function of intensity. It should also be pointed out that the VS is much less accurate at judging the magnitude of a single stimulus than it is at determining which of two stimuli is greater in magnitude; there is much less accuracy for absolute judgments than for relative judgments.

Fig. 1   Simultaneous contrast. (From Cornsweet [3], p. 279, Fig. 11.7.)

Apparent brightness depends strongly on the *local* background intensity. In Fig. 1, the small squares all have equal intensities, but they appear quite different in brightness, because their backgrounds have widely different intensities. This phenomenon is called *simultaneous contrast*. If the contrast ratio of an object to its background remains constant, the apparent brightness of the object can remain constant over a wide range of luminances (i.e., actual intensities) of the object and its background; this phenomenon is called *brightness constancy*.

It must be emphasized that the definition of "background" in this connection is not always straightforward. Thus in Fig. 2, making a small "cut"

Fig. 2   The Benussi ring.

in the gray ring causes its two halves to be seen as "belonging to" the white and black backgrounds, respectively, so that they have different apparent brightnesses. In Fig. 3, the upper gray triangle is seen as belonging to the black cross, but the lower one is not; hence the upper one is contrasted with the cross, and so looks lighter than the lower one (which is contrasted with the white background), even though the latter has more black in its vicinity.

Fig. 3 The Benary cross. (From J. Beck, "Surface Color Perception," p. 44, Plate 2. Cornell University Press, Ithaca, New York, 1972.)

3.2 ACUITY AND CONTOUR

When visual stimuli have complex space or time patterns, it becomes more difficult to predict their detectability. Such predictions would be relatively easy if the VS were linear, i.e., if the effect of a composite stimulus were the sum of the effects of the parts. There would then, for example, be a simple tradeoff between the intensity, size, and duration of a stimulus, e.g., one could compensate for a decrease in intensity by increasing area or duration. (Within restricted ranges of size and time, such tradeoffs do exist.) It is frequently assumed that the VS is in fact linear once an initial logarithmic transformation (Section 3.1) has been performed.

If linearity is assumed, it becomes straightforward to predict the response of the VS to various types of patterned stimuli, provided that we know its responses to certain simple patterns such as points, lines, edges, or sinusoids (i.e., its point, line, or edge spread function, or its modulation transfer function; see Section 6.1.1). The response to sinusoids, in particular, has been extensively measured. For a wide range of background luminances, this response has a bandpass characteristic, being greatest in the vicinity of 5 to 10 cycles per degree (of angle subtended by the stimulus at the eye), and falling off at both lower and higher frequencies. This characteristic is demonstrated in Fig. 4, where the spatial frequency increases to the right while the contrast increases upward; the curve along which the pattern is just visible represents the modulation transfer function of the VS.

The ability of the VS to detect fine spatial detail is known as *acuity*. It should be pointed out that acuity not only depends on background luminance, etc., as just discussed, but also varies greatly with the position of the stimulus. It is greatest within about a degree of the visual axis, but falls off very rapidly

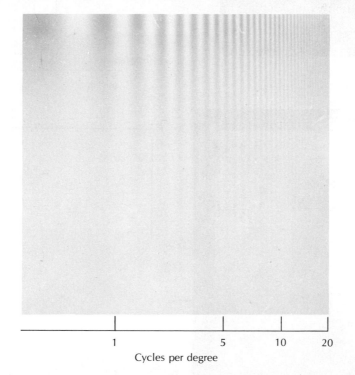

Cycles per degree

Fig. 4   Sinusoidally modulated pattern. (From Cornsweet [3], p. 343, Fig. 12.21; photograph made by F. W. Campbell and J. G. Robson.)

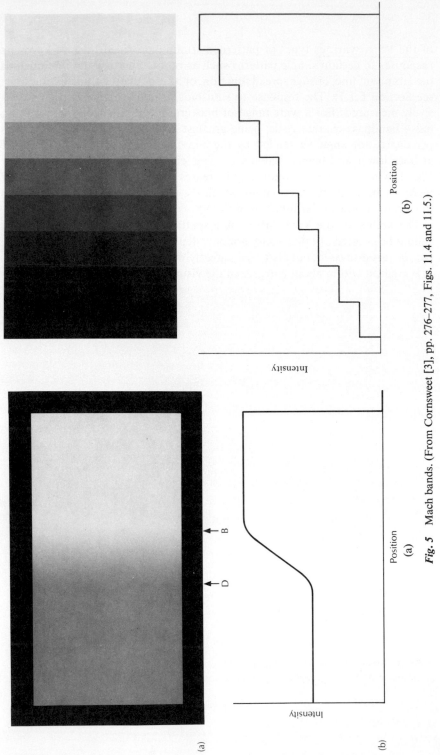

Fig. 5 Mach bands. (From Cornsweet [3], pp. 276–277, Figs. 11.4 and 11.5.)

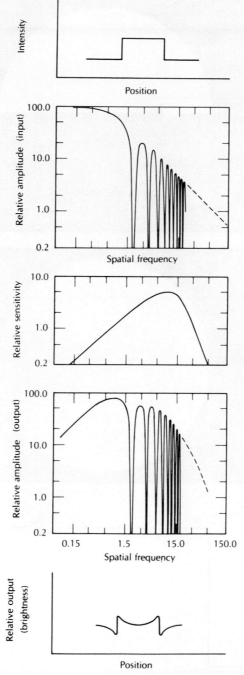

Fig. 6 Response of the visual system to a steplike pattern. (From Cornsweet [3], p. 346, Fig. 12.24.)

51

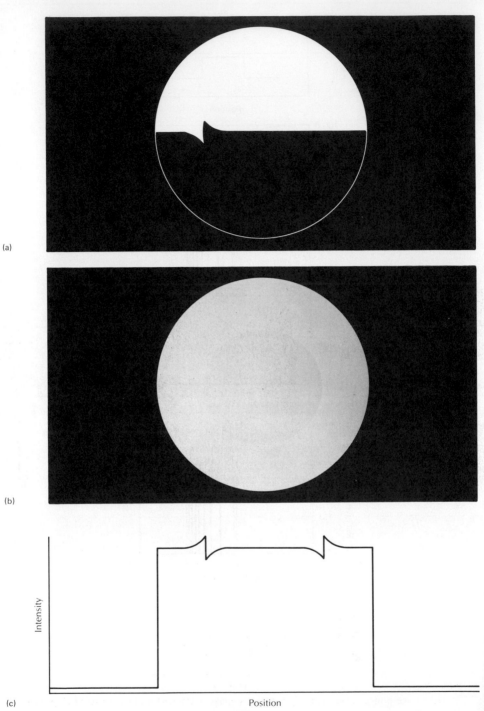

(a)

(b)

(c)

Intensity

Position

52

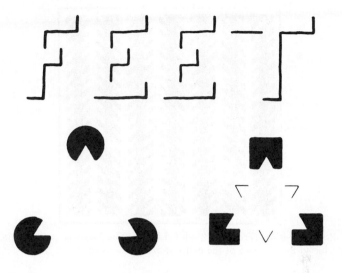

Fig. 8   "Filling in" from edges having gaps. From S. Coren, Subjective contours and apparent depth, *Psych. Rev.* **79**, 1972, 359–367, Figs. 2A, 3, 8A.

toward the periphery of the visual field. This can be demonstrated by fixing one's eyes on the middle of a row of books, and trying to read the titles of the books some distance away from the point of fixation, without moving the eyes.

The response of the VS to abrupt changes in luminance ("edges" or "contours") displays "overshoots" known as *Mach bands* (Fig. 5), which have the effect of enhancing or deblurring the edges. Such overshoot responses are, in fact, predicted if one multiplies the spatial frequency spectrum of a steplike function by the modulation transfer function of the VS (Fig. 6). The VS also seems to "fill in" the apparent brightness in the interior of a region based on that at the region's edges (perhaps at the frequency for which the response is strongest). This is demonstrated in Fig. 7, where the two regions have the same luminance except near their common edge, but each of them appears to have a different uniform brightness. This phenomenon can take place even when there are gaps in the edges (Fig. 8); the contours appear to be inter-

---

Fig. 7   "Filling in" of regions from their edges. (From Cornsweet [3], p. 273, Fig. 11.2.) When the disk shown in part (a) is spun rapidly about its center, the brightness cross section through the center is as shown in part (c), but what one sees is shown in the photograph in part (b).

Fig. 9 Zollner illusion. (From Luckiesh [7], p. 77, Fig. 37.) The vertical bars appear to be slanted, due to the presence of the many short oblique crossbars.

polated across the gaps. The insensitivity of the VS to low spatial frequency information when edges are present has important consequences for picture compression and enhancement (see Chapters 5 and 6).

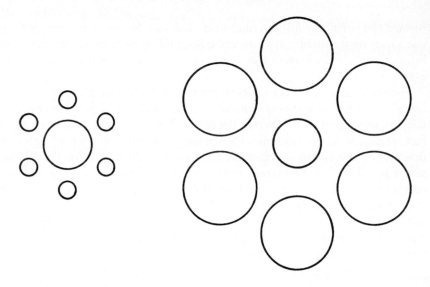

Fig. 10 Ebbinghaus illusion. (From Luckiesh [7], p. 56, Fig. 17.) The inner circle on the left looks larger than the one on the right.

Contours and simple patterns such as spots, lines, or bars produce adaptation and contrast effects on the perception of other contours, including "backward masking" effects (analogous to metacontrast), in which the perceptibility of a visual stimulus is reduced if a second stimulus is presented immediately afterwards. Many of these effects are dependent on the *slant* or *size* of the contours involved. For example, adaptation to bars of a particular orientation θ raises the detection threshold for bars oriented at θ but does not affect the threshold for bars in very different orientations, e.g., those perpendicular to θ. As another example, a background of (or previous exposure to) many bars in a given orientation θ tends to cause a single bar to be perceived as more nearly perpendicular to θ than it actually is (Fig. 9). Analogous phenomena hold for bar width or spot size (Fig. 10). For another approach to explaining these "illusions" of size and slant, see Section 3.5. A large collection of illusions is given by Luckiesh [7].

3.3   COLOR

Up to now we have ignored the *wavelength* of the light which produces the visual stimulus. As this wavelength varies, the perceived *color* (or, more precisely, *hue*) of the light changes, from red (corresponding to the longest visible wavelength) through orange, yellow, green, and blue to violet. Mixtures of wavelengths also give rise to perceived hues, some of which (e.g., purple) cannot be obtained from any single wavelength. "White" (or gray) light can also be obtained by mixing colored lights. In fact, any perceivable hue can be obtained by mixing three colors (e.g., a red, a green, and a blue) in appropriate proportions. Visual detection thresholds are color dependent; the VS is more sensitive to green than to red or blue.

The appearance of a colored spot of light can be described in terms of its brightness, its hue, and its *saturation*. This last can be thought of as related to the degree to which a hue is "undiluted" by white light; for example, pink is an unsaturated red.

Many of the adaptation, contrast, and constancy phenomena described in Sections 3.1 and 3.2 also hold for color. Adaptation to, or contrast with, a particular hue distorts the perception of other hues; for example, gray seen after green, or seen against a green background, looks purplish. Some of these effects are slant specific, e.g., after viewing vertical green bars, vertical gray bars look purplish, but horizontal ones do not. There seem to be important differences, however, between hue and brightness; for example, abrupt changes in hue do not always produce sharp contours when brightness is constant.

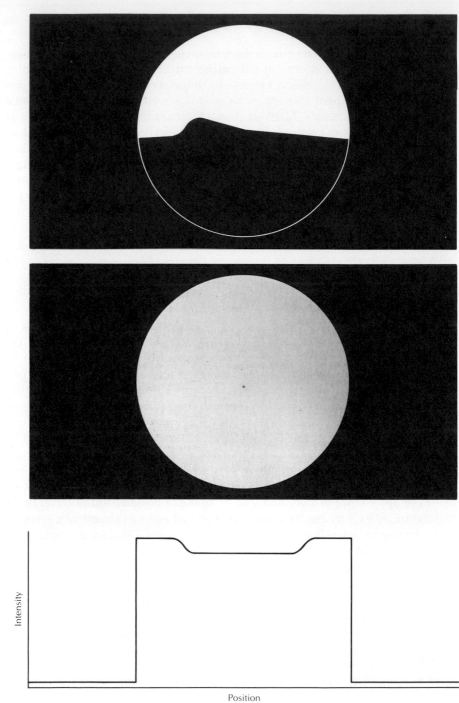

(a)

(b)

(c)

Intensity

Position

56

3.4   PATTERN AND TEXTURE

The visual field is not perceived as an array of independent picture points; it is usually seen as consisting of a relatively small number of regions, e.g., objects on a background. Only a few of the possible subsets of the visual field can be perceived as regions. For example, a region must be, at least partially, surrrounded by edges (i.e., abrupt changes in luminance); if it is not, one cannot see it as an individual entity (Fig. 11). Moreover, not all possible combinations of edges can be perceived as defining regions or objects. The *Gestalt laws of organization* describe those groupings or patterns which are most naturally seen as units.

The *law of similarity* states that similar entities tend to group together (Fig. 12). Here "similar" seems to be definable in terms of a small set of entity properties such as brightness, color, slope, and size. According to the *law of proximity*, closely clustered entities tend to group (Fig. 13). By the *law of good continuation*, when curves cross or branch, parts that smoothly continue one another are seen as belonging together (Fig. 14); while by the *law of closure*, closed figures tend to be seen as units (Fig. 15). Groupings which violate the laws are difficult to see as units, even if they constitute familiar objects; one can realize intellectually that Fig. 16 contains an "M" and a "W," but one does not readily "see" them there.

When a visual field is seen as composed of regions, not all of these regions

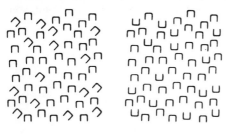 

Fig. 12   Similarity. (From J. Beck, Similarity grouping and peripheral discriminability under uncertainty, *Amer. J. Psych.* **85**, 1972, 1–19, Fig. 1.) When slope differences are present, it is easier to separate the mixture of U's.

Fig. 13   Proximity. (From M. Wertheimer, Principles of perceptual organization, *in* "Readings in Perception," D. C. Beardslee and, M. Wertheimer (eds.), p. 122. Van Nostrand-Reinhold, Princeton, New Jersey, 1958.) We see the dots grouped in 3's, not (e.g.) in horizontal rows.

Fig. 11   A region cannot be seen as an entity if it has *no* edges. (From Cornsweet [3], p. 275, Fig. 11.3.) Compare Fig. 7.

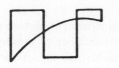

Fig. 14 Good continuation. (From M. Wert-
heimer, Principles of perceptual organization, *in*
"Readings in Perception," D. C. Beardslee and
M. Wertheimer (eds.), p. 130, Fig. 23. Van
Nostrand-Reinhold, Princeton, New Jersey, 1958.)
We see a rectangular wave and a smooth curve,
not three closed regions.

Fig. 15 Closure. (From M. Wertheimer, Principles of perceptual organization, *in* "Readings in
Perception," D. C. Beardslee and M. Wertheimer
(eds.), p. 130, Fig. 25. Van Nostrand-Reinhold,
Princeton, New Jersey, 1958.) We see two touching closed regions, not a curve that crosses itself.

Fig. 16 Hidden figures. (After M. Wertheimer,
Principles of perceptual organization, *in* "Readings in
Perception," D. C. Beardslee and M. Wertheimer
(eds.), p. 134, Fig. 39. Van Nostrand-Reinhold,
Princeton, New Jersey, 1958.) This is not easily seen
as a "W" on top of an "M".

Fig. 17 Rubin vase. (From Hochberg [6], p. 84,
Fig. 5-8.) This can be seen either as a vase or as two
faces.

can be seen as "objects" at the same time. For example, in Fig. 17, one can see either the two profiles or the vase, but not both at once. In general, only the region on one side of an edge is seen, at a given time, as a "figure" that has a shape; the region on the other side is then seen as a shapeless "ground," which usually appears to extend behind the figure. If one of the two regions is brighter, or smaller in size, or more symmetrical, or bounded (i.e., "closed"), it is generally easier to see as the figure, but one can also see it as the ground by making a conscious effort.

The properties that are important in producing similarity grouping are also important in the perception of *visual textures*. These are complex visual patterns composed of entities, or subpatterns, that have characteristic brightnesses, colors, slopes, sizes, etc. Thus a texture can be regarded as a similarity grouping. The local subpattern properties give rise to the perceived lightness, directionality, coarseness, etc., of the texture as a whole. For a large collection of examples of textures, see Brodatz [1].

3.5 SHAPE AND SPACE

In determining which arrangements of parts in a scene are seen as figures, *simplicity* seems to play an important role [6]. In Fig. 14, for example, the preferred grouping is the simpler one (a smooth curve and a rectangular wave, not three irregular closed figures); similar remarks apply to Figs. 15 and 16. One can attempt to formulate this criterion of simplicity in terms of *information theory*: The grouping that can be specified using the least information is the preferred figure. Here the specification might be formulated in terms of the numbers of (unequal) lines, angles, line crossings, etc. in the grouping. For a review of this and other applications of information theory to pattern perception see Corcoran [2]. It should be emphasized, however, that a single information measure cannot capture all aspects of shape perception. The shape of a figure is a highly multidimensional concept—shapes can be compact, elongated, or dispersed; smooth or jagged; etc.

The image formed by the eye is two-dimensional; many different arrangements of objects in three-dimensional space could give rise to the same image. The VS uses a variety of "cues" to reduce or eliminate this ambiguity. Here again, many of these cues can be regarded as expressing principles of simplicity. Thus the larger of two similar objects may be seen as nearer—perhaps because, by judging the objects to be at different distances, one can assume them to be congruent, which simplifies the description of the scene. More generally, if texture coarseness decreases in a given direction, one can see the textured objects as uniformly textured but receding in that direction. If object A appears partly to hide object B, object A is seen as nearer; this too may be because the visible portion of object B has a relatively complex shape, which can be simplified by imagining a suitable "completion" of B in back of A. Converging lines can be seen as receding, perhaps because this allows one to regard them as parallel; and in general, objects drawn in perspective are easy to see as three-dimensional, because this makes their shapes simpler or more familiar (e.g., Fig. 18).

When the VS treats size changes as distance changes, or shape changes as perspective changes, it is said to be displaying *size and shape constancy*. Many of the illusions involving slant and size (see Section 3.2) can also be interpreted along these lines. Similar interpretations can be given for the cues that involve blur (interpreted as due to atmospheric effects on the image of a distant object) and shadow (interpreted in terms of the scene lit from above and behind, as by sunlight).

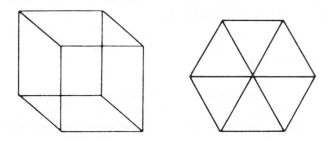

Fig. 18   Kopfermann cubes. (From Hochberg [6], p. 87, Fig. 5-10.) When we see the left-hand figure as three-dimensional, its description becomes simpler, since its lines now can be regarded as all being of equal length.

The following are some other important cues to three-dimensional depth. When the eyes or head are moved, the relative motions of parts of a scene give clues to their relative distances (if one makes the simplifying assumption that the parts themselves have not moved!); this cue is called *motion parallax*. Comparison of the images seen by the two eyes also gives a distance cue, called *binocular parallax*; the distance to an object is related to the amount of shift required to bring its two images into registration. Here too, of course, one is making a simplifying assumption, namely, that one has correctly identified the two images as images of the same object. Finally, for a relatively close object, the eyes may have to focus ("accommodate") or to turn inward ("converge") in order to see it sharply, and the degrees to which these actions are necessary also provide depth measures.

When the VS has chosen a particular three-dimensional interpretation of a scene, it tends to ignore any cues that are inconsistent with the chosen interpretation, or to reinterpret these cues in other ways even when this is inconsistent with normal experience. For example, if a room is built in a trapezoidal shape, it will appear rectangular when seen (with one eye) from the proper viewpoint, and people moving around in it will appear to change size.

3.6   DURATION AND MOTION

As indicated in Section 3.1, the response of the VS to a stimulus depends on temporal, as well as spatial factors. In this connection, it is useful to study the response of the VS to simple temporal patterns, e.g., to impulses ("flashes") or to modulated ("flickering") illumination. As was the case with spatial patterns, there is a tendency for this response to fall off at both the high and low ends of the frequency spectrum: totally unchanging patterns (provided they are immobile relative to the eye) tend to fade, while patterns that change too rapidly "fuse" and appear to be unchanging. It should be stressed that the spatial and temporal characteristics of the VS are interdependent; the response to a flickering pattern depends in a complex way on its spatial and temporal frequencies.

When part of the visual image is displaced, e.g., if a light is turned off, and another light is turned on after a short enough time, one perceives a single light moving continuously from the first position to the second. As in Section 3.5, one can regard this perception as a "simple" interpretation of the stimulus, in which the succession of distinct images can be accounted for by assuming that a single object is in motion. In this connection, if a collection of parts move in unison, one tends to see them as a single figure; this is the Gestalt *law of common fate*. If an object changes size or shape as well as position, and it is possible to account for this in terms of translational and rotational motions in three dimensions, one tends to perceive such motions.

Some of the adaptation effects described in Sections 3.1 and 3.2 are motion-specific. For example, after prolonged viewing of a downward-moving pattern, stationary patterns appear to move upward (the "waterfall effect"); and the detection threshold for downward-moving patterns is raised, but that for upward-moving patterns is not.

3.7   DETECTION AND RECOGNITION

The processes of detecting, classifying, and recognizing a visual (or any other) stimulus can be analyzed from the standpoint of *statistical decision theory*. The observer can be thought of as estimating the prior and posterior probabilities that the stimulus was actually present, and then using some criterion (e.g., thresholding a likelihood ratio) to make his decision. His choice of this criterion determines the probabilities of his missing an actual stimulus (false dismissal), and of reporting a stimulus when none was actually present (false alarm). The tradeoff between these probabilities, as a function of the criterion used, is called the observer's "receiver operating characteristic" (ROC). From the decision-theoretic standpoint, one should speak not only of a detection "threshold," but also of detection criteria. For a review of this topic see Green and Swets [5].

Perceptual decision criteria are influenced by many factors besides the physical nature of the stimulus. Expectation or "perceptual set" plays an important role; in particular, it affects the estimation of prior probabilities. It also affects how ambiguous figures are perceived; for example, prior viewing of the young woman in Fig. 19a makes it less likely that the old woman will be seen in Fig. 19b, and conversely. "Response biases," e.g., estimates of the

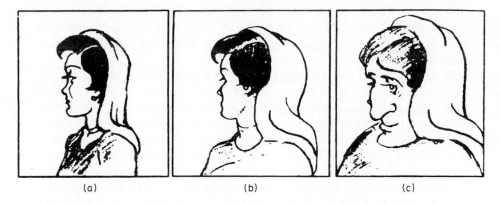

(a)   (b)   (c)

Fig. 19   Ambiguous face. (From Hochberg [6], p. 70, Fig. 4-28.) The center face can be seen as resembling either the one on the left or the one on the right.

values of hits and the costs of misses, also enter into the formulation of decision criteria.

Visual detection and recognition tasks, particularly the latter, involve not only the VS, but memory processes as well. The stored information that might be involved in recognition, say, of a shape or pattern, could be of a variety of types, ranging from a "template" of the pattern (or perhaps of an idealized prototype), to a set of features or measurements that characterize the pattern, or a structural description based on such features. As we shall see in Chapter 10, the use of features makes it possible to formulate invariant descriptions of the pattern, i.e., descriptions that do not change when the pattern varies—at least, to limited degrees—in brightness, color, position, orientation, or size.

The transformations, however, under which a pattern is still easily recognized are hard to characterize. For example, recognition is impaired by change of the three-dimensional reference frame, or by reversal of figure and ground, even though the pattern itself remains exactly the same.

Little is known about the nature of the features that the VS may use for recognition, although some clues can be obtained from the neurophysiology of the VS, as well as from studies of the movements that the eye makes when it examines patterns. One can also raise many interesting questions about the manner in which the information used for recognition is extracted from the observed pattern, and the manner in which the stored information is retrieved (e.g., are these processes serial or parallel?); see Corcoran [2] for a review of this topic.

REFERENCES

1. P. Brodatz, "Textures." Dover, New York, 1966.
2. D. W. J. Corcoran, "Pattern Recognition." Penguin, Baltimore, Maryland, 1971.
3. T. N. Cornsweet, "Visual Perception." Academic Press, New York, 1970.
4. C. H. Graham *et al.* (eds.), "Vision and Visual Perception." Wiley, New York, 1965.
5. D. M. Green and J. A. Swets, "Signal Detection Theory and Psychophysics." Wiley, New York, 1966.
6. J. Hochberg, "Perception." Prentice-Hall, Englewood Cliffs, New Jersey, 1964.
7. M. Luckiesh, "Visual Illusions." Dover, New York, 1965.
8. L. Zusne, "Visual Perception of Form." Academic Press, New York, 1970.

Chapter 4

Digitization

Pictures can, in general, be considered continuous functions, the gray level being a continuous function of position in the picture. Before such a picture can be processed by a digital computer or, in many cases, transmitted over a channel, it needs to be digitized.

In its ordinary sense, digitization consists of sampling the gray level in the picture at an $M \times N$ array of points. Since the gray level at these points may take any value in a continuous range, for digital processing the gray level needs to be quantized. By this we mean that we divide the range of gray level into K intervals, and require the gray level at any point to take on only one of these values. In order for the picture reproduced from these numbers to be a "good" reproduction of the original, M, N, and K have to be large. Ordinarily, the finer the sampling and quantization, the better the reproduced picture. Nothing is gained, however, by increasing M, N, and K beyond the spatial and gray scale resolution capabilities of the receiver.

In a more general sense, the aim of sampling is to represent a continuous picture by a finite string or array of numbers, called *samples*. The only constraint on these numbers is that it should be possible to reconstruct a picture from them. There may be small errors involved in the reconstruction provided that these errors do not cause any impairment in the representation of the scene with respect to the spatial and gray scale resolution capabilities of the receiver.

In the next two sections, we will discuss two different approaches to picture sampling. In the first approach, the samples are the values of the picture function (gray levels) at a discrete array of points, and we discuss the constraints that a picture must satisfy for this type of sampling to be adequate. In the second approach, the picture function is expanded in terms of orthonormal functions, and the coefficients of expansion are the samples.

4.1 SAMPLING USING AN ARRAY OF POINTS

We will now present the theory of a sampling technique in which the samples are the gray levels of the picture at an array of points. Before we expose the reader to such sampling of pictures, we will first consider the case of a one-dimensional signal. This is to facilitate the presentation of the well-known Whittaker–Kotelnikov–Shannon theorem. A two-dimensional analog of the theorem will then be presented.

4.1.1 One-Dimensional Functions

Let us consider a one-dimensional function $f(t)$, which we seek to represent by samples $f(kT)$, where k takes on integer values from $-\infty$ to $+\infty$ and T is the sampling period. Evidently the validity of the representation depends on whether it is possible to reconstruct the function from its samples.

One way to reconstruct the original $f(t)$ from the samples $f(kT)$ would be to interpolate suitably between the samples. Following the classic paper of Peterson and Middleton [13], this can be done using an interpolation function $g(t)$ as follows:

$$f(t) = \sum_{k=-\infty}^{\infty} f(kT)\, g(t-kT) \tag{1}$$

In other words, the contribution that the sample $f(kT)$ makes to the reconstruction at time t is weighted by the factor $g(t-kT)$ which is the interpolation function $g(t)$ displaced by time kT along the t-axis. We will now assume that f and g are Fourier transformable.

By Eq. (6) of Chapter 2, we can write

$$f(kT)\, g(t-kT) = \int_{-\infty}^{\infty} f(\tau)\, g(t-\tau)\, \delta(\tau-kT)\, d\tau \tag{2}$$

Substituting in (1) and inverting the order of summation and integration, we obtain

$$f(t) = \int_{-\infty}^{\infty} f(\tau)\, g(t-\tau) \left(\sum_{k=-\infty}^{\infty} \delta(\tau-kT) \right) d\tau \tag{3}$$

The function in the large parentheses is periodic with period T, and can, therefore, be represented by a Fourier series of the form

$$\sum_{k=-\infty}^{\infty} \delta(\tau-kT) = \sum_{n=-\infty}^{\infty} a_n \exp\left(+j\frac{2\pi n}{T}\tau\right) \qquad (4)$$

where the coefficients a_n of the Fourier expansion are given by

$$a_n = \frac{1}{T}\int_{-T/2}^{T/2}\left(\sum_{k=-\infty}^{\infty}\delta(\tau-kT)\right)\exp\left(-j\frac{2\pi n}{T}\tau\right)d\tau$$

$$= \frac{1}{T}\int_{-T/2}^{T/2}\delta(\tau)\exp\left(-j\frac{2\pi n}{T}\tau\right)d\tau$$

since only the $k = 0$ term of the sum is nonzero in the range of integration. By Eq. (6) of Chapter 2, it follows that

$$a_n = 1/T \qquad \text{for} \quad \text{all } n \qquad (5)$$

Substituting (4) and (5) in (3), we can write

$$f(t) = \sum_{n=-\infty}^{\infty}\int_{-\infty}^{\infty}\left[f(\tau)\exp\left(+j\frac{2\pi n}{T}\tau\right)\right]\left(\frac{g(t-\tau)}{T}\right)d\tau \qquad (6)$$

Each term in the summation is a convolution of the functions

$$f(t)\exp\left(+j\frac{2\pi nt}{T}\right) \qquad \text{and} \qquad \frac{g(t)}{T} \qquad (7)$$

By the convolution property of Fourier transforms as given by Eq. (23) of Chapter 2, the Fourier transform of each term in the summation in (6) is the product of the Fourier transforms of these two functions. Let $F(\omega)$ and $G(\omega)$ denote the Fourier transforms of $f(t)$ and $g(t)$, respectively. Then using the one-dimensional analog of property (21b) in Chapter 2, the Fourier transforms of the two functions in (7) are $F(\omega-(2\pi n/T))$ and $G(\omega)/T$, respectively. Taking the Fourier transform of both sides of (6), we obtain

$$F(\omega) = \frac{G(\omega)}{T}\sum_{n=-\infty}^{\infty}F\left(\omega-\frac{2\pi n}{T}\right) \qquad (8)$$

Since the steps taken in deriving (8) from (1) are all reversible, (8) is a *necessary and sufficient* condition for the exact reconstructability of $f(t)$ from its samples $f(kT)$ using (1). We shall now formulate sufficient conditions on G and T in order for reconstructability to hold.

Now $F(\omega-(2\pi n/T))$ is just $F(\omega)$ shifted by $2\pi n/T$. Suppose that $f(t)$ is bandlimited, i.e., that $F(\omega) = 0$ for $|\omega| \geqslant 2\pi f_c$, say, as in Fig. 1a. If $T \leqslant 1/2f_c$, then the shifted copies of $F(\omega)$ are far enough apart that no two of them can have nonzero values at any one point, as shown in Fig. 1b. For such an F,

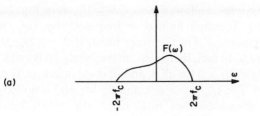

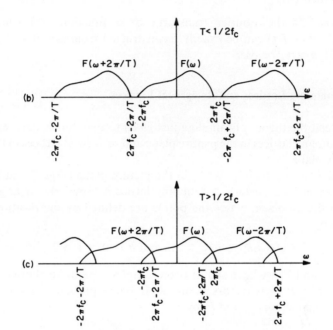

Fig. 1 (a) $F(\omega)$ for a one-dimensional bandlimited function. (b) Nonoverlapping copies of $F(\omega)$. (c) Overlapping copies of $F(\omega)$.

if we take

$$G(\omega) = \begin{cases} T, & |\omega| < 2\pi f_c \\ 0, & \text{otherwise} \end{cases}$$

then (8) is immediately satisfied, since on the right-hand side the unshifted copy of $F(\omega)$ has weight 1, while all the other copies have weight 0. Since this G is a rect function, its inverse Fourier transform g is a sinc function; in fact, we have

$$g(t) = \frac{1}{2\pi} \int_{-2\pi f_c}^{2\pi f_c} T e^{j\omega t}\, d\omega = \frac{\sin 2\pi f_c t}{\pi t/T} \tag{9}$$

Thus if $T = 1/2f_c$, we have $g(t) = \text{sinc}(2\pi f_c t)$.

If overlap of the copies is allowed, i.e., $T > 1/2f_c$, as in Fig. 1c, and f is still permitted to be any function limited in frequency by $2\pi f_c$, no $G(\omega)$ will satisfy (8) for all possible F. On the other hand, if $T < 1/2f_c$, we have some flexibility in defining G. In fact, there exist intervals (e.g., AB and CD in Fig. 1b) for which $F(\omega - (2\pi n/T)) = 0$ for all n, and in these intervals, G can be chosen arbitrarily. Thus for $T < 1/2f_c$, our solution (9) for $g(t)$ is no longer unique.

The preceding discussion has established the Whittaker–Kotelnikov–Shannon theorem:

Theorem. If the Fourier transform of a function $f(t)$ vanishes for $|\omega| \geqslant 2\pi f_c$, then $f(t)$ can be exactly reconstructed from samples of its values taken $1/2f_c$ apart or closer.

4.1.2 Sampling Lattices and Reciprocal Lattices

To extend the theory of sampling just given to two dimensions, we need to define sampling lattices in the picture plane and reciprocal lattices in the spatial frequency plane.

Consider two vectors $\vec{r}_1$ and $\vec{r}_2$ in the picture plane (Figs. 2a and 2b). We will call them *basis vectors*. A sampling lattice is a periodic arrangement of points in the xy-plane, where the points are defined by the position vectors $\vec{r}_{mn}$ given by

$$\vec{r}_{mn} = m\vec{r}_1 + n\vec{r}_2, \qquad m \text{ and } n = 0, \pm 1, \pm 2, \ldots \tag{10}$$

In Figs. 2a and 2b we have shown two examples of sampling lattices.

The reciprocal basis vectors $\vec{w}_1$ and $\vec{w}_2$ in the uv-plane are uniquely derived from the basis vectors $\vec{r}_1$ and $\vec{r}_2$ by

$$\vec{r}_i \cdot \vec{w}_j = \begin{cases} 0, & i \neq j \\ 1, & i = j \end{cases} \tag{11}$$

By the previous definition, the vector $\vec{w}_1$ is perpendicular to $\vec{r}_2$. If θ is the angle between $\vec{r}_1$ and $\vec{r}_2$, it is easy to show that the magnitude of $\vec{w}_1$ is $1/|\vec{r}_1| \sin \theta$. Similarly, $\vec{w}_2$ is perpendicular to $\vec{r}_1$ and its magnitude is equal to $1/|\vec{r}_2| \sin \theta$.

The reciprocal basis vectors $\vec{w}_1$ and $\vec{w}_2$ define a reciprocal lattice in the uv-plane. The reciprocal lattice is a periodic arrangement of points in the uv-plane, where the points are defined by the position vectors $\vec{w}_{mn}$ given by

$$\vec{w}_{mn} = m\vec{w}_1 + n\vec{w}_2, \qquad m \text{ and } n = 0, \pm 1, \pm 2, \ldots \tag{12}$$

Figures 2c and 2d are reciprocal lattices for the sampling lattices in Figs. 2a and 2b, respectively.

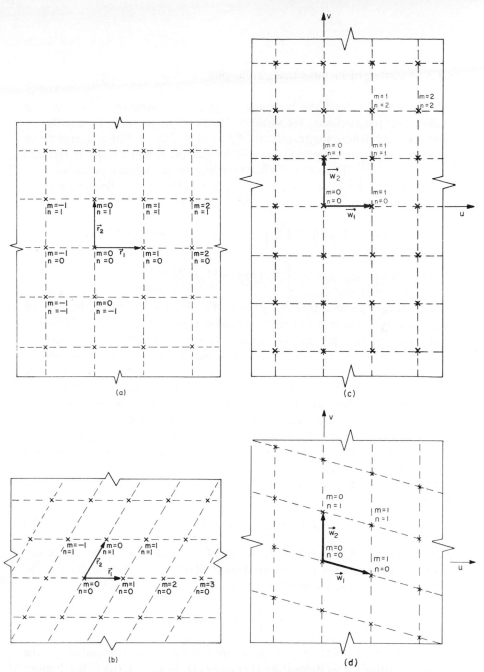

Fig. 2 (a) An example of a sampling lattice generated by the basis vectors $\vec{r}_1$ and $\vec{r}_2$ in the xy-plane. The sampling lattice is a two-dimensional periodic array of points marked x. (b) Another example of a sampling lattice in the xy-plane. (c) Reciprocal lattice in the spatial frequency plane for the sampling lattice in (a). Note that $\vec{w}_1$ and $\vec{w}_2$ are uniquely related to $\vec{r}_1$ and $\vec{r}_2$ in (a). (d) Reciprocal lattice for the sampling lattice in (b).

4.1.3   Sampling of Pictures Using a Sampling Lattice

We will now extend the preceding theory to two-dimensional functions defined over the xy-plane. We assume that these functions are Fourier transformable, and denote the transform of $f(x, y)$ by $F(u, v)$. For convenience in presentation in this subsection, we will denote $f(x, y)$ and $F(u, v)$ by $f(\vec{r})$ and $F(\vec{w})$, respectively. Here $\vec{r}$ is a position vector in the picture plane with coordinates (x, y) while $\vec{w}$ is a position vector in the spatial frequency plane with coordinates (u, v). Then by (12) of Chapter 2

$$F(\vec{w}) = F(u, v) = \int \int f(x, y) \exp[-j2\pi(xu+yv)] \, dx \, dy$$

$$= \int f(\vec{r}) \exp(-j2\pi\vec{w}\cdot\vec{r}) \, d\vec{r}$$

We will now assume that $f(x, y)$ is spatial frequency limited. By this we mean that $F(u, v)$ is zero outside a bounded region in the spatial frequency plane. We wish to represent the function $f(\vec{r})$ by its samples $f(\vec{r}_{mn})$ where $\vec{r}_{mn}$ are the points on a sampling lattice as defined in the preceding subsection. Evidently, the samples $f(\vec{r}_{mn})$ are a valid representation provided it is possible to reconstruct $f(\vec{r})$ from these samples. As in the one-dimensional case, we will use this reconstructability condition to derive constraints which should be satisfied by $f(\vec{r})$ so that it may be represented by its samples.

One way to reconstruct the original $f(\vec{r})$ from its samples $f(\vec{r}_{mn})$ would be to interpolate suitably between sampling points. Following the treatment for the one-dimensional case, this can be done by using a suitable interpolation function $g(\vec{r})$ such that $f(\vec{r})$ may be expanded as a linear function of its samples $f(\vec{r}_{mn})$:

$$f(\vec{r}) = \sum_m \sum_n f(\vec{r}_{mn}) \, g(\vec{r}-\vec{r}_{mn}) \tag{13}$$

Here $f(\vec{r}_{mn}) \, g(\vec{r}-\vec{r}_{mn})$ is the contribution of the sample $f(\vec{r}_{mn})$ to the reconstruction at the point $\vec{r}$. Employing steps similar to those used in deriving (8) from (1), we arrive at the result (for a detailed derivation see [13]):

$$F(\vec{w}) = \frac{G(\vec{w})}{Q} \sum_p \sum_q F(\vec{w}-\vec{w}_{pq}) \tag{14}$$

where the set of points $\vec{w}_{pq}$ is the reciprocal lattice corresponding to the sampling lattice $\vec{r}_{mn}$ as defined by (11) and (12). In (14), $G(\vec{w})$ is the Fourier transform of $g(\vec{r})$, and Q the area of the parallelogram formed by the vectors $\vec{r}_1$ and $\vec{r}_2$.

Assume that $f(\vec{r})$ is a spatial frequency-limited function so that $F(\vec{w})$ looks like Fig. 3a. For a given choice of $\vec{w}_1$ and $\vec{w}_2$, the functions $F(\vec{w}-\vec{w}_{pq})$ for

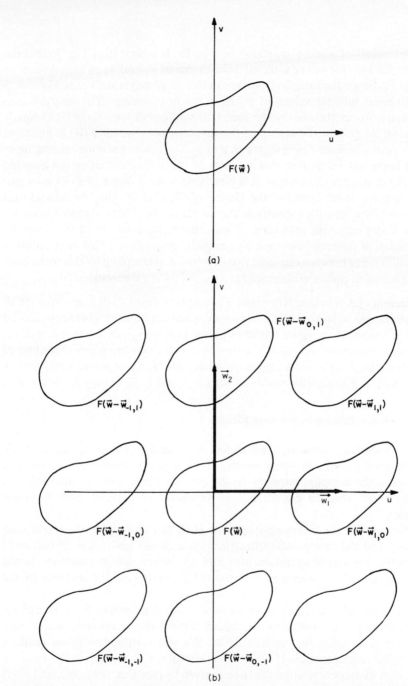

Fig. 3 (a) $F(\vec{w})$ for a two-dimensional bandlimited function $f(x, y)$. (b) Nonoverlapping copies of $F(u, v)$.

71

various values of p and q are shown in Fig. 3b. It is clear from Fig. 3b that the equality in Eq. (14) will be satisfied if the choice of $\vec{w}_1$ and $\vec{w}_2$ (which determine $\vec{r}_1$ and $\vec{r}_2$, hence the sampling strategy in the xy-plane) is such that $F(\vec{w} - \vec{w}_{pq})$ for different integral values of p and q do not overlap. The interpolation function $g(\vec{r})$ can then be chosen such that its Fourier transform $G(\vec{w})$ equals the constant Q over the region of the uv-plane over which $F(\vec{w})$ is nonzero, and is zero wherever $F(\vec{w} + \vec{w}_{pq})$, $p \neq 0$, $q \neq 0$, are nonvanishing; the value of $G(\vec{w})$ being arbitrary over that portion of the uv-plane that is not covered by $F(\vec{w})$ or its periodic images. It is clear that if the domain of $F(\vec{w})$ is irregularly shaped, then, even for the choice of $\vec{w}_1$ and $\vec{w}_2$ that permits closest packing of the repetitive spectra in the uv-plane, there may always be regions where $G(\vec{w})$ may take arbitrary values. Therefore, even for such a case the interpolation function may not be uniquely determined. This is certainly a basic difference between one- and two-dimensional sampling by this technique. The discussion presented here leads to the following theorem:

Theorem. A function $f(\vec{r})$ whose Fourier transform $F(\vec{w})$ vanishes over all but a bounded region of spatial frequency space can be everywhere reproduced from its values taken over a lattice of points $(m\vec{r}_1 + n\vec{r}_2)$, $m, n = 0, \pm 1, \pm 2, ...$, provided the vectors $\vec{r}_1$ and $\vec{r}_2$ are small enough to ensure nonoverlapping of the spectrum $F(\vec{w})$ with its images on a periodic lattice of points $(p\vec{w}_1 + q\vec{w}_2)$, $p, q = 0, \pm 1, \pm 2, ...$, with $\vec{r}_i \cdot \vec{w}_j = 0$ for $i \neq j$, and $\vec{r}_i \cdot \vec{w}_i = 1$ for $i, j = 1$ and 2.

4.1.4 Generalization to Random Fields

The preceding subsection presented a technique for sampling a single bandlimited picture. In other words, given a bandlimited picture, one can determine the sampling strategy in the picture plane and the interpolation function such that the picture can be reconstructed without error from the samples.

Now we pose the following question: How can one determine a sampling strategy and the interpolation function for a *family* or a *class* of pictures? Evidently, the sampling procedures and the interpolation function should now depend on the average or statistical properties of the pictures in the family.

In Section 2.4.1, a family of pictures was called a *random field*. Therefore, we now turn our attention to reformulating the previously introduced sampling techniques for random fields. We will assume the given random field to be homogeneous. If we use the notation $\mathbf{f}(\vec{r})$ to denote the random field, then its autocorrelation function is given by [see Eqs. (89), (99), and (101) of Chapter 2]:

$$R_{ff}(\vec{r}_1, \vec{r}_2) = E\{\mathbf{f}(\vec{r}_1)\mathbf{f}(\vec{r}_2)\} = R_{ff}(\vec{r}_1 - \vec{r}_2) = R_{ff}(\vec{r}_2 - \vec{r}_1) \qquad (15)$$

In keeping with the sampling technique introduced in the preceding subsections, we desire to represent each picture in the random field $\mathbf{f}(\vec{r})$ by samples that are values of the gray level at a set of periodic sampling points $\vec{r}_{mn}$ generated by basis vectors $\vec{r}_1$ and $\vec{r}_2$ as in (10). To be able to reconstruct each picture in the random field from its samples, we seek an expansion of form (13), that is,

$$\mathbf{f}(\vec{r}) = \sum_m \sum_n \mathbf{f}(\vec{r}_{mn}) \, g(\vec{r} - \vec{r}_{mn}) \tag{16}$$

This is a family of equations, one for each picture in the collection represented by the random field $\mathbf{f}(\vec{r})$.

Unless the interpolation function $g(\vec{r})$ is of an appropriate form and the random field $\mathbf{f}(\vec{r})$ satisfies certain constraints, the equality in (16) will not be satisfied. To determine such a $g(\vec{r})$ and such constraints, we will proceed as follows. In the absence of any *a priori* knowledge of the correct $g(\vec{r})$ in (16), we first seek an optimum function $g(\vec{r})$ such that the linear combination of the periodic samples of $\mathbf{f}(\vec{r})$ as given by

$$\mathbf{f}_e(\vec{r}) = \sum_m \sum_n \mathbf{f}(\vec{r}_{mn}) \, g(\vec{r} - \vec{r}_{mn}) \tag{17}$$

results in the minimization of the ensemble average error as given by

$$e = E\{(\mathbf{f}(\vec{r}) - \mathbf{f}_e(\vec{r}))^2\} \tag{18}$$

Substituting (17) in (18) and making use of (15), we obtain

$$e = R_{ff}(0) - 2 \sum_m \sum_n R_{ff}(\vec{r} - \vec{r}_{mn}) \, g(\vec{r} - \vec{r}_{mn})$$
$$+ \sum_m \sum_n \sum_p \sum_q R_{ff}(\vec{r}_{mn} - \vec{r}_{pq}) \, g(\vec{r} - \vec{r}_{mn}) \, g(\vec{r} - \vec{r}_{pq}) \tag{19}$$

To determine g which minimizes e at every $\vec{r}$, we assume an arbitrary variation $\varepsilon g'(\vec{r})$ in $g(\vec{r})$ where ε is a real number and $g'(\vec{r})$ an arbitrary function of $\vec{r}$, such that the function $g(\vec{r}) + \varepsilon g'(\vec{r})$ is "close" to the function $g(\vec{r})$. In Eq. (19), we replace $g(\vec{r})$ by $g(\vec{r}) + \varepsilon g'(\vec{r})$. Now if $g(\vec{r})$ is the optimum function we are after, then the derivative $\partial e / \partial \varepsilon$ must be equal to zero at $\varepsilon = 0$ for any $g'(\vec{r})$ (so long as the preceding conditions are satisfied).

If we substitute $g(\vec{r}) + \varepsilon g'(\vec{r})$ for $g(\vec{r})$ in (19), determine the derivative $\partial e / \partial \varepsilon$, substitute $\varepsilon = 0$ in the expression for $\partial e / \partial \varepsilon$, and equate the resulting expression to zero, we obtain

$$\sum_m \sum_n R_{ff}(\vec{r} - \vec{r}_{mn}) \, g'(\vec{r} - \vec{r}_{mn})$$
$$= \sum_m \sum_n \sum_p \sum_q R_{ff}(\vec{r}_{mn} - \vec{r}_{pq}) \, g(\vec{r} - \vec{r}_{mn}) \, g'(\vec{r} - \vec{r}_{pq}) \tag{20}$$

which is a condition for $g(\vec{r})$ to be the optimum interpolation function. This equation can be rewritten as

$$\sum_p \sum_q \left\{ R_{ff}(\vec{r}-\vec{r}_{pq}) - \sum_m \sum_n R_{ff}(\vec{r}_{mn}-\vec{r}_{pq})\, g(\vec{r}-\vec{r}_{mn}) \right\} g'(\vec{r}-\vec{r}_{pq}) = 0. \qquad (21)$$

If (21) is to be satisfied for any $g'(r)$, the expression within the curly brackets must vanish for all $\vec{r}_{pq}$. In particular for $\vec{r}_{pq} = 0$ we have

$$R_{ff}(\vec{r}) = \sum_m \sum_n R_{ff}(\vec{r}_{mn})\, g(\vec{r}-\vec{r}_{mn}). \qquad (22)$$

In light of Eq. (13), we can give the following interpretation to this condition: The optimum interpolation function $g(\vec{r})$ for the random field $\mathbf{f}(\vec{r})$ is one that exactly reproduces the nonrandom autocorrelation function $R_{ff}(\vec{r})$ from its samples. Since (22) is identical in form to (13), Fourier-transforming both sides leads to an equation similar to (14), that is,

$$S_{ff}(\vec{w}) = \frac{G(\vec{w})}{Q} \sum_p \sum_q S_{ff}(\vec{w}-\vec{w}_{pq}) \qquad (23)$$

where $S_{ff}(\vec{w})$ is the spectral density of the process (see Eq. (103) of Chapter 2) and where $G(\vec{w})$ is the Fourier transform of $g(\vec{r})$. The constant Q and the periodic points $\vec{w}_{pq}$ in the uv-plane are related to the vectors $\vec{r}_1$ and $\vec{r}_2$ as in (14). By arguments identical to those in Section 4.1.3, (23) is satisfied if we choose the reciprocal vectors $\vec{w}_1$ and $\vec{w}_2$ in the uv-plane such that the $S_{ff}(\vec{w}-\vec{w}_{pq})$, for different integral values of p and q, do not overlap. Also, the interpolation function $g(\vec{r})$ should be such that its Fourier transform $G(\vec{w})$ equals the constant Q over the region of the uv-plane over which $S_{ff}(\vec{w})$ is nonzero. Its value is arbitrary over that portion of the uv-plane that is not covered by $S_{ff}(\vec{w})$ or its periodic images, and it is zero elsewhere. Note that since $R_{ff}(\vec{r})$ is real and is symmetric with respect to 180° rotation, $S_{ff}(\vec{w})$, $G(\vec{w})$, and $g(\vec{r})$ are all real and possess the same symmetry.

We have shown that the sampling strategy and $g(\vec{r})$ as dictated by (22) minimize the mean square error e as given by (18). We will now show that for bandlimited random fields ($S_{ff}(\vec{w}) = 0$ outside a finite region $\mathscr{S}$ in the uv-plane) this error is indeed zero.

Substituting (22) in (19), we have for the minimum mean square error

$$e = R_{ff}(0) - \sum_m \sum_n R_{ff}(\vec{r}-\vec{r}_{mn})\, g(\vec{r}-\vec{r}_{mn}) \qquad (24)$$

which can be expressed as

$$e = R_{ff}(0) - \int R_{ff}(\vec{r}')\, g(\vec{r}') \left\{ \sum_m \sum_n \delta(\vec{r}'-\vec{r}+\vec{r}_{mn}) \right\} d\vec{r}' \qquad (25)$$

We will now examine the two terms on the right-hand side separately.

From Eq. (104) of Chapter 2, the first term is

$$R_{ff}(0) = \int S_{ff}(\vec{w}) \, d\vec{w} \tag{26}$$

As regards the second term, the Fourier transform of the product $R_{ff}(\vec{r}') \, g(\vec{r}')$ is given by

$$\int_{uv\text{-plane}} S_{ff}(\vec{w} - \vec{s}) \, G(\vec{s}) \, d\vec{s} \tag{27}$$

by the convolution property of the Fourier transform (Eq. (23a) of Chapter 2), while the Fourier transform of $\sum_m \sum_n \delta(\vec{r}' - \vec{r} + \vec{r}_{mn})$ is given by

$$\frac{1}{Q} \left(\sum_m \sum_n \delta(\vec{w} + \vec{w}_{mn}) \exp(-j2\pi\vec{w}\cdot\vec{r}) \right) \tag{28}$$

using Exercise 4 and Eq. (21a) of Chapter 2, where Q and $\vec{w}_{mn}$ are the same as in (25). By using Parseval's theorem as given by Eq. (24) of Chapter 2, Eqs. (27) and (28) lead to the following result:

$$\int R_{ff}(\vec{r}') \, g(\vec{r}') \left\{ \sum_m \sum_n \delta(\vec{r}' - \vec{r} + \vec{r}_{mn}) \right\} d\vec{r}'$$

$$= \frac{1}{Q} \int_{uv\text{-plane}} \left\{ \int_{uv\text{-plane}} S_{ff}(\vec{w} - \vec{s}) \, G(\vec{s}) \, d\vec{s} \right\}$$

$$\times \sum_m \sum_n \delta(\vec{w} + \vec{w}_{mn}) \exp(+j2\pi\vec{w}\cdot\vec{r}) \, d\vec{w}$$

$$= \int_{uv\text{-plane}} \frac{G(\vec{s})}{Q} \sum_m \sum_n \exp(+j2\pi\vec{r}\cdot\vec{w}_{mn}) S_{ff}(\vec{s} - \vec{w}_{mn}) \, d\vec{s} \tag{29}$$

Substituting (26) and (29) in (25), we obtain

$$e = \int_{uv\text{-plane}} \left[S_{ff}(\vec{w}) - \frac{G(\vec{w})}{Q} \sum_m \sum_n \exp(j2\pi\vec{r}\cdot\vec{w}_{mn}) S_{ff}(\vec{w} - \vec{w}_{mn}) \right] d\vec{w} \tag{30}$$

If $G(\vec{w})$ and $\vec{w}_{mn}$ are chosen to satisfy (23) by the method previously described, it is easy to see that then the expression within the brackets in (30) is zero, making the sampling error e equal to zero for the bandlimited process.

4.1.5 Aliasing Problems in Picture Sampling

To best illustrate the concept of aliasing, let us consider the following two experiments:

Experiment 1.   Consider a sampling lattice generated by basis vectors $\vec{r}_1$ and $\vec{r}_2$ in Fig. 2a. The reciprocal vectors $\vec{w}_1$ and $\vec{w}_2$ in the frequency plane for this sampling lattice are those given by Fig. 2c. Let us assume that this sampling lattice is used for sampling a picture $f(\vec{r})$ whose Fourier transform $F(\vec{w})$ is nonzero only within the dotted rectangle shown in Fig. 4. Also, to satisfy the conditions on the interpolation function $g(\vec{r})$ as given in Section 4.1.3, let $G(w)$ be equal to the constant Q within the dotted region and zero outside. (Note that the dotted rectangle is such that it does not overlap with its own periodic images, such as those in Fig. 3, on the reciprocal lattice.)

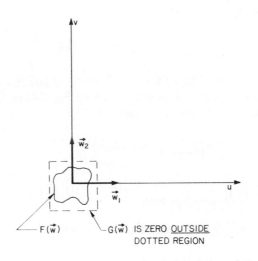

Fig. 4   The sampling lattice obtained from the choice of $\vec{w}_1$ and $\vec{w}_2$ shown here would lead to error free reproduction of all pictures whose Fourier transforms are zero outside the dotted rectangle.

The sampling lattice and the function $g(\vec{r})$ just described would result in error-free reproduction from samples of all pictures that have Fourier transform zero outside the dotted rectangle in Fig. 4.   ▮

Experiment 2.   Now let us apply the sampling lattice and the interpolation function in Experiment 1 to the picture in Fig. 5 which is described by

$$f(\vec{r}) = \cos 2\pi(\vec{w}_0 \cdot \vec{r})$$

and which has Fourier transform given by (Chapter 2, Exercise 3)

$$F(\vec{w}) = \tfrac{1}{2}[\delta(\vec{w} - \vec{w}_0) + \delta(\vec{w} + \vec{w}_0)]$$

The Fourier transform of the picture consists of two impulses in the spatial frequency plane, one at $\vec{w}_0$ and the other at $-\vec{w}_0$, as shown in Fig. 6a. Note

Fig. 5 $\text{Cos } 2\pi(\vec{w}_0 \cdot \vec{r})$.

that the periodicity of the sinusoidal pattern is such that the points $\vec{w}_0$ and $-\vec{w}_0$ fall outside the dotted rectangle.

Let us now attempt to reconstruct the picture in this experiment from its samples by using the interpolation formula in Eq. (13). The Fourier transform of the reconstructed picture is given by the right-hand side of (14), that is,

$$\frac{G(\vec{w})}{Q} \sum_m \sum_n F(\vec{w} - \vec{w}_{mn}) \qquad (31)$$

which, after substituting the expression for $F(\vec{w})$, becomes

$$\frac{G(\vec{w})}{2Q} \sum_m \sum_n [\delta(\vec{w} - \vec{w}_{mn} - \vec{w}_0) + \delta(\vec{w} - \vec{w}_{mn} + \vec{w}_0)]$$

Since $G(\vec{w})$ is nonzero (and equal to Q) only inside the dotted rectangle in Fig. 6a, it follows from the geometrical considerations in this particular case that the preceding expression has only two nonzero terms given by

$$\tfrac{1}{2}[\delta(\vec{w} - (\vec{w}_0 + w_{-1,-1})) + \delta(\vec{w} + (\vec{w}_0 - \vec{w}_{1,1}))]$$

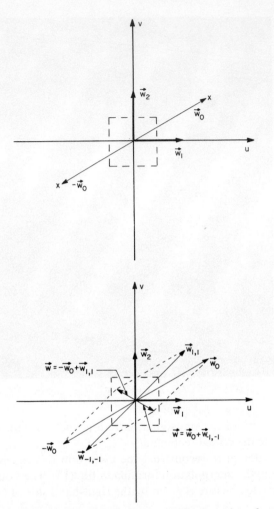

Fig. 6   (a) The Fourier transform of the picture in Fig. 5 consists of two impulses in the uv-plane, one at $\vec{w}_0$ and the other at $-\vec{w}_0$. (b) The Fourier transform of the picture reconstructed from the samples of the picture in Fig. 5 consists of two impulses, one at $\vec{w} = \vec{w}_0 + \vec{w}_{-1,-1}$ and the other at $\vec{w} = -\vec{w}_0 + \vec{w}_{1,1}$.

In other words, the Fourier transform of the reconstructed picture has two impulses, one at $\vec{w} = \vec{w}_0 + \vec{w}_{-1,-1}$, and the other at $\vec{w} = -\vec{w}_0 + \vec{w}_{1,1}$. This is shown in Fig. 6b. The reconstructed picture is the inverse Fourier transform of the previous expression and is given by

$$\cos 2\pi \left[(\vec{w}_0 - \vec{w}_{1,1}) \cdot \vec{r} \right]$$

which would look like the picture in Fig. 7.

Fig. 7   The reconstructed picture from the samples of the picture in Fig. 5. Note the change in frequency and orientation.

Thus we see that the picture in Fig. 5 after it is sampled and reconstructed can look like the picture in Fig. 7. Note that the original periodic pattern in Fig. 5 appears at a different frequency and orientation in Fig. 7, hence the name *aliasing* for this effect. The frequency $\vec{w}_0 - \vec{w}_{1,1}$ [$(\vec{w}_0 - \vec{w}_{mn})$ in a more general case] is called the *aliased spatial frequency*. ▌

In the preceding discussion on aliasing, we examined a picture with a pattern at a single spatial frequency $\vec{w}_0$ and it gave rise to a single aliased frequency $\vec{w}_0 - \vec{w}_{1,1}$. In practice, depending on the reciprocal vectors $\vec{w}_1$ and $\vec{w}_2$ and the region over which $G(\vec{w})$ is nonzero, a pattern at a frequency $\vec{w}_0$ may, in the reconstructed picture, give rise to a number of aliased frequencies $\vec{w}_0 + \vec{w}_{mn}$ for various m and n. The presence of these aliased frequencies in pictures reconstructed from their samples gives rise to what are called *Moiré patterns*.

Moiré patterns can best be illustrated by using Moiré pattern transparencies. Figure 8a is a rectangular grid transparency. When this transparency is placed

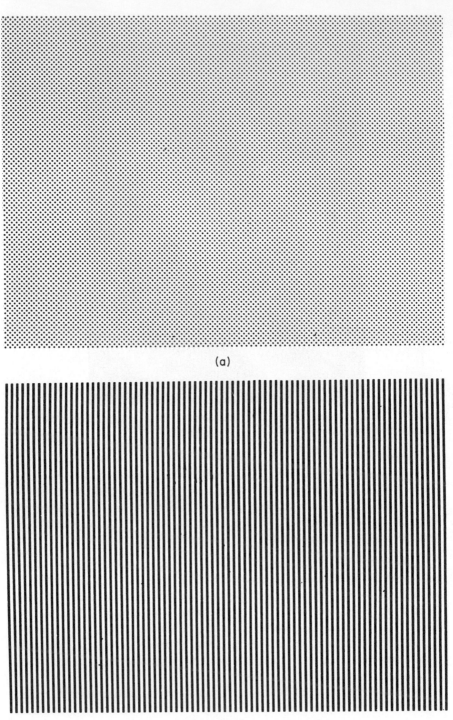

(a)

(b)

(c)

Fig. 8 (a) A rectangular grid transparency. When this transparency is placed on a picture, the light points serve as sampling points. (From [9].) (b) Bar pattern picture. (From [9].) (c) This figure illustrates the Moiré pattern caused by sampling the bar pattern in (b). (From [9].)

on a picture, the light points serve as sampling points. Figure 8b is the transparency of a bar pattern. The transparency in Fig. 8a is placed on top of the bar pattern transparency in Fig. 8b with the bar pattern slightly rotated in a clockwise direction with respect to the vertical direction of the sampling transparency. Figure 8c is a photograph of the transparencies on a light table. In this figure the low-frequency rotated Moire pattern is quite evident.

A picture must contain periodic structures and their relationship to the geometry of the sampling lattice must be just right for the Moiré patterns to occur. In practice, one does not very often run into pictures with strong periodic components. Cases that do arise include pictures of plowed fields, streets in high-altitude photographs of urban areas, ocean wave patterns, and wind patterns in sand.

For pictures with continuous spectra, aliasing can cause another kind of a problem whose effects are less dramatic and more difficult to interpret. We will illustrate this by the following example.

In the preceding discussion, it was pointed out that if a picture $f(\vec{r})$ is sampled by the sampling lattice in Experiment 1 and if its Fourier transform $F(\vec{w})$ is zero outside the dotted rectangle in the uv-plane in Fig. 4, then it can be reconstructed without error from its samples. Now consider a case where the same sampling lattice is used for a picture whose Fourier transform is nonzero over a much larger region, as in Fig. 9a. After the picture is sampled, we reconstruct it using the same interpolation function as in Experiment 1, i.e., $G(\vec{w})$ is equal to Q within the dotted rectangle in Fig. 9a and zero outside.

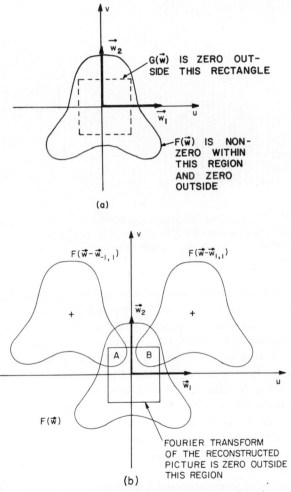

(a)

(b)

Fig. 9 (a) The sampling lattice in Experiment 1 is used to sample a picture whose Fourier transform is nonzero over a larger region than would lead to error-free reconstruction. (b) Three of the terms in Eq. (31) are pictorially illustrated here for $F(\vec{w})$ shown in Fig. 9a. These three terms correspond to (m, n) equal to $(0,0)$, $(1, -1)$, and $(1,1)$.

The Fourier transform of the reconstructed picture is again given by (31), which we rewrite here as

$$F(\vec{w}) + \sum_{m \neq 0} \sum_{n \neq 0} F(\vec{w} - \vec{w}_{mn})$$

for $\vec{w}$ within the dotted rectangle. Of course, the entire expression, hence the Fourier transform of the reconstructed picture, is zero outside the dotted rectangle (because $G(\vec{w})$ is zero there). Some of the terms in the previous expression are pictorially illustrated in Fig. 9b. In this particular example, because of the geometry of the various regions these three terms are the only ones making a nonzero contribution to the Fourier transform of the reconstructed picture.

Note that those frequencies in the original picture that are outside of the region over which $G(\vec{w})$ is nonzero irretrievably lost in reconstruction, leading to a loss of resolution. Within this region the transform is modified, leading to distortion. In the example under consideration, the frequencies in $F(\vec{w})$ outside the rectangle are lost to the reconstructed picture, and the frequencies within the rectangle are modified in regions A and B (Fig. 9b) by the periodic images of $F(\vec{w})$ on the reciprocal lattice. Figures 10a and 10b illustrate the aliasing effect just discussed. Notice the loss in resolution and other distortions.

4.2   SAMPLING USING ORTHONORMAL FUNCTIONS

The aim of sampling is to represent a picture by a finite string or array of numbers. As long as it is possible to reconstruct a picture from these numbers, the samples need not correspond to gray levels on a sampling lattice in the picture plane. In this section, we will show that if we expand a picture function in terms of a set of orthonormal functions, we may take the coefficients of the expansion as the picture samples.

4.2.1   Orthonormal Expansions

Let $f(x, y)$ be a real function defined over a region $\mathscr{S}$ of the xy-plane. The function $f(x, y)$ is assumed to be square integrable:

$$\iint_{\mathscr{S}} [f(x, y)]^2 \, dx \, dy < \infty \tag{32}$$

Suppose we are given a set of square integrable functions $\varphi_{mn}(x, y)$, $m = 0, 1, 2, \dots$; $n = 0, 1, 2, \dots$; defined over the same region $\mathscr{S}$.[†] The set of functions

---

[†] The need for using two indices for labeling the member functions in the set may not be quite evident now, and in fact, is not necessary for the development of the formal theory of orthonormal expansions in two dimensions. As will be shown presently, however, it makes the presentation of some of the practical two-dimensional sampling techniques using orthonormal expansions more convenient.

Fig. 10a Photographic print of a picture before sampling.

Fig. 10b Picture of Fig. 10a after sampling and reconstruction. (Courtesy Perkin-Elmer Corp. These pictures appear in the book cited in [9].)

called *orthogonal* if

$$\iint \varphi_{mn}(x, y)\, \varphi_{pq}^*(x, y)\, dx\, dy = 0 \tag{33}$$

for $m \ne p$ or $n \ne q$, $m, n, p, q = 0, 1, 2, \ldots$. If in addition to (33) the following property holds

$$\iint_{\mathscr{S}} |\varphi_{mn}(x, y)|^2\, dx\, dy = 1, \qquad m, n = 0, 1, 2, \ldots \tag{34}$$

then the set of functions is called *orthonormal*. The functions φ_{mn} may be either real- or complex-valued.

Exercise 1. Show that over the rectangle defined by $-A/2 \leqslant x \leqslant A/2$, $-B/2 \leqslant y \leqslant B/2$, the functions

$$\varphi_{mn}(x, y) = \frac{1}{\sqrt{AB}} \exp\left[j2\pi\left(\frac{mx}{A} + \frac{ny}{B} \right) \right], \qquad m, n = 0, 1, 2, \ldots$$

form an orthornormal set. ∎

We now wish to approximate the function $f(x, y)$ at all points within $\mathscr{S}$ by a sum of the form

$$\sum_{m=0}^{M-1} \sum_{n=0}^{N-1} a_{mn}\, \varphi_{mn}(x, y) \tag{35}$$

in such a way that the mean square error

$$e_{MN}^2 = \iint_{\mathscr{S}} \left| f(x, y) - \sum_{m=0}^{M-1} \sum_{n=0}^{N-1} a_{mn}\, \varphi_{mn}(x, y) \right|^2 dx\, dy \tag{36}$$

is minimized. The subscript MN in e_{MN}^2 is indicative of there being MN terms in the summation in (35). The following theorem addresses itself to this problem:

Theorem. The constants a_{mn} that minimize e_{MN}^2 are given by

$$a_{mn} = \iint_{\mathscr{S}} f(x, y)\, \varphi_{mn}^*(x, y)\, dx\, dy \tag{37}$$

Proof. We will prove the theorem by showing that if the constants a_{mn} are given by (37), then for any arbitrary selection b_{mn} of constants

$$\iint_{\mathscr{S}} \left| f(x, y) - \sum_{m=0}^{M-1} \sum_{n=0}^{N-1} a_{mn}\, \varphi_{mn}(x, y) \right|^2 dx\, dy$$

$$\leqslant \iint_{\mathscr{S}} \left| f(x, y) - \sum_{m=0}^{M-1} \sum_{n=0}^{N-1} b_{mn}\, \varphi_{mn}(x, y) \right|^2 dx\, dy \tag{38}$$

The right-hand side of (38) can be written as

$$
\iint_{\mathscr{S}} \left| f(x, y) - \sum_{m=0}^{M-1} \sum_{n=0}^{N-1} b_{mn}\, \varphi_{mn}(x, y) \right|^2 dx\, dy
$$

$$
= \iint_{\mathscr{S}} \left| f(x, y) - \sum_{m=0}^{M-1} \sum_{n=0}^{N-1} a_{mn}\, \varphi_{mn}(x, y) \right.
$$

$$
\left. + \sum_{m=0}^{M-1} \sum_{n=0}^{N-1} (a_{mn} - b_{mn})\, \varphi_{mn}(x, y) \right|^2 dx\, dy
$$

$$
= \iint_{\mathscr{S}} \left| f(x, y) - \sum_{m=0}^{M-1} \sum_{n=0}^{N-1} a_{mn}\, \varphi_{mn}(x, y) \right|^2 dx\, dy + \sum_{m=0}^{M-1} \sum_{n=0}^{N-1} |a_{mn} - b_{mn}|^2
$$

$$(39)$$

In arriving at the second equality in (39), we made use of (33), (34), and (37). The nonnegative nature of the terms in the second equality in (39) proves (38) and, hence, the theorem. ∎

We can now pose the following question: Does approximation (35) to $f(x, y)$ become increasingly accurate as the number of terms in (35) is increased? The dependence of the mean square error e_{MN}^2 on M and N depends on the nature of the orthonormal functions.

An orthonormal set of functions, $\varphi_{mn}(x, y)$, $m = 0, 1, 2, \ldots$; $n = 0, 1, 2, \ldots$; is called *complete* if for every square integrable function f we have

$$
\lim_{M \to \infty,\, N \to \infty} e_{MN}^2 = 0 \tag{40}
$$

— that is, if the mean square error in approximating $f(x, y)$ by (35) approaches zero as the number of terms in (35) approaches infinity. Henceforth, we will be concerned with complete orthonormal sets of functions only. A complete orthonormal set of functions is also called an *orthonormal basis*.

Summarizing the preceding discussion, we can say that given an orthonormal basis $\varphi_{mn}(x, y)$, $m = 0, 1, 2, \ldots$; $n = 0, 1, 2, \ldots$; defined over a region $\mathscr{S}$ of the xy-plane, then any function $f(x, y)$, square integrable over $\mathscr{S}$, can be expanded as

$$
f(x, y) = \sum_{m=0}^{\infty} \sum_{n=0}^{\infty} a_{mn}\, \varphi_{mn}(x, y) \tag{41a}
$$

where

$$
a_{mn} = \iint_{\mathscr{S}} f(x, y)\, \varphi_{mn}^{*}(x, y)\, dx\, dy \tag{41b}
$$

Of course, this expansion is valid only over the region $\mathscr{S}$ of the xy-plane.

Equations (41) have important practical implications. For example, suppose that in a communication link an orthonormal set of functions is available at both the transmitting and the receiving ends. Say that it is desired to transmit a picture $f(x, y)$. Then by (41) we need transmit the coefficients a_{mn} only, since by (41a) they can be used to reconstruct the picture at the receiving end. At the transmitting end the a_{mn} are obtained from $f(x, y)$ by using (41b). In practice, of course, one would transmit only a finite number of a_{mn}, even though, in general, an infinite number of them would be required for an error-free reconstruction of the picture at the receiving end. The orthonormal functions are usually ordered so that higher-order terms contribute to the fine detail in a picture and neglecting them may lead to a loss of resolution. Nothing is gained, however, by increasing M and N beyond the spatial resolution capabilities of the observer or the user at the receiving end.

Since a picture $f(x, y)$ can be represented by and reconstructed from the coefficients a_{mn}, we can call these coefficients "samples of the picture." Given a picture and an orthonormal set of functions defined over the same region of the xy-plane, these samples may be obtained by using (41b).

4.2.2 Sampling of a Random Field Using Orthonormal Expansions

If we sample a *single* picture $f(x, y)$ using an orthonormal basis, and retain only $M \times N$ samples, then the error in reconstructing the picture from these samples is given by (36).

We would now like to determine the following: Given a random field, i.e., an ensemble or a class of pictures (television pictures, for example), if we use an orthonormal basis to sample each picture in the class and retain only $M \times N$ samples for each picture, then what is the error e_{MN}^2 when averaged over all the pictures in the class? Can e_{MN}^2 when averaged over all the pictures be expressed as a function of the statistical properties of the class? We want to study the dependence of the average e_{MN}^2 on the statistical properties of the given random field. In what follows we will assume the random field to be real and homogeneous.

Let $\mathbf{f}(x, y)$ again denote a real homogeneous random field with auto-correlation function $R_{ff}(\alpha, \beta)$. Relationships (41) for the random field $\mathbf{f}(x, y)$ can be expressed as

$$\mathbf{f}(x, y) = \sum_{m=0}^{\infty} \sum_{n=0}^{\infty} \mathbf{a}_{mn}\, \varphi_{mn}(x, y) \tag{42a}$$

at all points (x, y) within $\mathscr{S}$. In (42a)

$$\mathbf{a}_{mn} = \int\!\!\int_{\mathscr{S}} \mathbf{f}(x, y)\, \varphi_{mn}^{*}(x, y)\, dx\, dy, \qquad m, n = 0, 1, 2, \ldots \tag{42b}$$

Note that each $\mathbf{a}_{mn}$ is now a random variable. This is because (42b) is actually a family of equations, one for each picture in the random field. The value of each $\mathbf{a}_{mn}$ will depend on which picture is selected. Similarly, (42a) represents a family of equations and can be used to reconstruct each picture in the random field from its samples.[†]

If only $M \times N$ samples are retained for every picture in the random field, then upon reconstruction of a picture from its samples, there will be a mean square error e_{MN}^2, as defined in the preceding subsection. This mean square error when averaged over all the pictures in the random field is called the sampling error for the random field $\mathbf{f}(\vec{r})$ and will be denoted by ε_{MN}^2. From (36) ε_{MN}^2 is given by

$$\varepsilon_{MN}^2 = E\left\{ \int \int_{\mathscr{S}} \left| \mathbf{f}(x, y) - \sum_{m=0}^{M-1} \sum_{n=0}^{N-1} \mathbf{a}_{mn}\,\varphi_{mn}(x, y) \right|^2 dx\, dy \right\} \qquad (43)$$

By making use of (42b), this equation can easily be shown to reduce to

$$\varepsilon_{MN}^2 = E\left\{ \int \int_{\mathscr{S}} [\mathbf{f}(x, y)]^2\, dx\, dy - \sum_{m=0}^{M-1} \sum_{n=0}^{N-1} |\mathbf{a}_{mn}|^2 \right\} \qquad (44)$$

Substituting (42b) in (44), we obtain for the sampling error

$$\varepsilon_{MN}^2 = E\left\{ \int \int_{\mathscr{S}} [\mathbf{f}(x, y)]^2\, dx\, dy - \sum_{m=0}^{M-1} \sum_{n=0}^{N-1} \left| \int \int_{\mathscr{S}} \mathbf{f}(x, y)\,\varphi_{mn}^*(x, y)\, dx\, dy \right|^2 \right\}$$

$$= E\left\{ \int \int_{\mathscr{S}} [\mathbf{f}(x, y)]^2\, dx\, dy - \sum_{m=0}^{M-1} \sum_{n=0}^{N-1} \int \int_{\mathscr{S}} \int \int_{\mathscr{S}} \mathbf{f}(x, y)\,\mathbf{f}(x', y') \right.$$

$$\left. \times \varphi_{mn}^*(x, y)\,\varphi_{mn}(x', y')\, dx\, dy\, dx'\, dy' \right\} \qquad (45)$$

The operation of expectation in (45) is ensemble averaging, while the integrations in (45) involve spatial averaging (see Section 2.4.6 for the difference). Since both operations are linear, the order in which they occur can be interchanged. Interchanging their order in (45) and making use of the definition of the autocorrelation $R_{ff}(\alpha, \beta)$ in Section 2.4.3, we can write

$$\varepsilon_{MN}^2 = S R_{ff}(0, 0) - \sum_{m=0}^{M-1} \sum_{n=0}^{N-1} \int \int_{\mathscr{S}} \int \int_{\mathscr{S}} R_{ff}(x - x', y - y')$$

$$\times \varphi_{mn}^*(x, y)\,\varphi_{mn}(x', y')\, dx\, dy\, dx'\, dy' \qquad (46)$$

[†] For mathematical precision, a more correct interpretation of the equality in (42a) is as follows. As the number of terms on the right-hand side in (42a) approaches infinity, the summation converges to $\mathbf{f}(x, y)$ in some sense. For example, in Section 4.2.3 for the case of Fourier sampling, the convergence is in the mean square sense, i.e.,

$$E\left\{ \left| \mathbf{f}(x, y) - \sum_{m=0}^{M-1} \sum_{n=0}^{N-1} \mathbf{a}_{mn}\,\varphi_{mn}(x, y) \right|^2 \right\}_{\substack{M \to \infty \\ N \to \infty}} = 0.$$

which expresses the dependence of the mean square error as averaged over all the pictures in the ensemble on the statistical properties of the ensemble with S as the area of the region $\mathcal{S}$ in the xy-plane.

At this point a few words about the autocorrelation function $R_{ff}(\alpha, \beta)$ are in order. Under the assumption that a class of pictures form a homogeneous random field, the autocorrelation function $R_{ff}(\alpha, \beta)$ is often assumed to be of the form

$$R_{ff}(\alpha, \beta) = [R_{ff}(0,0) - \eta^2] \exp[-c_1 |\alpha| - c_2 |\beta|] + \eta^2 \qquad (47)$$

where c_1 and c_2 are positive constants and where, by the definition of the autocorrelation in Section 2.4.3,

$$R_{ff}(0,0) = E\{[\mathbf{f}(x, y)]^2\} \qquad \text{and} \qquad \eta = E\{\mathbf{f}(x, y)\}$$

We will remind the reader that for a homogeneous random field, the mean value η of the gray level in the picture is a constant independent of position. Note that the autocorrelation function in (47) can be used to model pictures with different amounts of correlation in the horizontal and vertical directions by choosing different values of c_1 and c_2.

4.2.3   Examples of Sampling Using Orthonormal Expansions: Fourier Sampling

Suppose that the region in the xy-plane over which the pictures are to be sampled is a rectangle with sides A and B, and that over this region the orthonormal basis functions are of the form

$$\varphi_{mn}(x, y) = \frac{1}{\sqrt{AB}} \exp\left[+j2\pi\left(\frac{m}{A}x + \frac{n}{B}y\right) \right]. \qquad (48)$$

The sampling defined in this way is called *Fourier sampling.*

The Fourier samples of a picture $f(x, y)$ are obtained by substituting (48) in (41b):

$$a_{mn} = \frac{1}{\sqrt{AB}} \int_{-B/2}^{B/2} \int_{-A/2}^{A/2} f(x, y) \exp\left[-j2\pi\left(\frac{m}{A}x + \frac{ny}{B}\right) \right] dx \, dy \qquad (49)$$

where we have assumed the origin to be at the center of the picture. By making use of the Fourier transform relationship in Eq. (12) of Chapter 2, (49) can be written as

$$a_{mn} = \frac{1}{\sqrt{AB}} F\left(\frac{m}{A}, \frac{n}{B}\right) \qquad (50)$$

where $F(m/A, n/B)$ is the value of the Fourier transform $F(u, v)$ of the picture at $u = m/A$ and $v = n/B$.

We will now briefly describe an optical method of computing the Fourier samples a_{mn} of a picture. In Fig. 11 we have marked two planes, the xy-plane in front of a thin lens, and the $x'y'$-plane behind the lens. Each plane is one focal length d_0 away from the lens. A transparency $f(x, y)$ is placed in the xy-plane in front of the lens and a parallel beam of coherent light passed through it. Let $U(x', y')$ denote the light distribution in the focal plane behind the lens. $U(x', y')$ is given by (see Chapter 5 of Goodman [4])

$$U(x', y') = \frac{1}{j\lambda d_0} \int\limits_{-\infty}^{\infty}\int f(x, y) \exp\left[-j\frac{2\pi}{\lambda d_0}(xx' + yy') \right] dx\, dy \qquad (51)$$

where λ is the wavelength of the light used and d_0 the focal length of the lens. By Eq. (12) of Chapter 2, (51) can be written as

$$U(x', y') = \frac{1}{j\lambda d_0} F\left(\frac{x'}{\lambda d_0}, \frac{y'}{\lambda d_0} \right) \qquad (52)$$

In other words, except for a constant, the light distribution in the image at the point (x', y') is equal to the value of the Fourier transform $F(u, v)$ at $u = x'/\lambda d_0$ and $v = y'/\lambda d_0$. Thus the Fourier transform of a picture $f(x, y)$ may be obtained optically; it is represented by the light distribution, both in amplitude and phase, in the focal plane behind the lens. If a photographic plate is placed in the $x'y'$-plane, it will record only the intensity given by $|U(x', y')|^2$, which is proportional to the power spectrum of $f(x, y)$; the phase

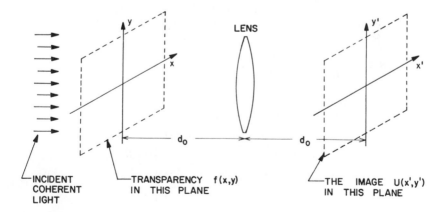

Fig. 11 If a two-dimensional transparency with transmittance $f(x, y)$ is placed in the xy-plane, then the light distribution behind the lens in the $x'y'$-plane is proportional to the Fourier transform of $f(x, y)$. Both planes are one focal length d_0 away from the lens.

information is lost. Ordinary photographic methods are unable to record phase, although techniques such as holographic methods are capable of doing so.

Combining (50) and (52), we obtain

$$a_{mn} = \frac{j\lambda d_0}{\sqrt{AB}}\, U\!\left(\frac{m\lambda d_0}{A}, \frac{n\lambda d_0}{B}\right) \tag{53}$$

This shows that the Fourier sample a_{mn} is given by the light level at the point $x' = m\lambda d_0/A$, $y' = n\lambda d_0/B$ in the $x'y'$-plane in Fig. 11.

In the rest of our discussion of Fourier sampling we will extend the general result for the sampling error as given by Eq. (46) to the particular case of Fourier sampling for the autocorrelation function given by (47). For simplicity we will assume the random field to be zero mean, i.e., $n = 0$ in (47). It is a simple matter to extend the results to the case of nonzero η. Substituting (47) with $\eta = 0$ and (48) in (46), we obtain

$$\varepsilon_{MN}^2 = ABR_{ff}(0,0) - \frac{R_{ff}(0,0)}{AB}$$

$$\times \sum_{m=0}^{M-1}\sum_{n=0}^{N-1} \iint_{\mathscr{S}}\iint_{\mathscr{S}} \exp(-c_1|x-x'| - c_2|y-y'|)$$

$$\times \exp\left[j2\pi\left(\frac{m}{A}x + \frac{n}{B}y\right)\right]\exp\left[-j2\pi\left(\frac{m}{A}x' + \frac{n}{B}y'\right)\right] dx\, dy\, dx'\, dy'$$

$$= ABR_{ff}(0,0) - \frac{R_{ff}(0,0)}{AB}\sum_{m=0}^{M-1}\sum_{n=0}^{N-1}\left\{\int_{-A/2}^{A/2}\int_{-A/2}^{A/2} \exp(-c_1|x-x'|)\right.$$

$$\times \exp\frac{j2\pi mx}{A}\exp\frac{-j2\pi mx'}{A}\, dx\, dx'\left.\right\}\left\{\int_{-B/2}^{B/2}\int_{-B/2}^{B/2} \exp(-c_2|y-y'|)\right.$$

$$\times \exp\frac{j2\pi ny}{B}\exp\frac{-j2\pi ny'}{B}\, dy\, dy'\left.\right\}$$

$$= ABR_{ff}(0,0)\left[1 - \sum_{m=0}^{M-1}\sum_{n=0}^{N-1}\left\{\frac{2c_1 A}{(c_1 A)^2 + (2\pi m)^2}\right.\right.$$

$$\left. + \frac{2[\exp(-c_1 A)-1][(c_1 A)^2 - (2\pi m)^2]}{[(c_1 A)^2 + (2\pi m)^2]^2}\right\}\left\{\frac{2c_2 B}{(c_2 B)^2 + (2\pi n)^2}\right.$$

$$\left.\left. + \frac{2[\exp(-c_2 B)-1][(c_2 B)^2 - (2\pi n)^2]}{[(c_2 B)^2 + (2\pi n)^2]^2}\right\}\right] \tag{54}$$

For given values of M and N, the normalized sampling error $\varepsilon_{MN}^2/ABR_{ff}(0,0)$ depends on $c_1 A$ and $c_2 B$ only. If we assume $M = N$ in (54), we obtain an expression for $\varepsilon_{N^2}^2/ABR_{ff}(0,0)$ which is the normalized sampling error using N^2

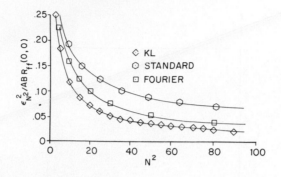

Fig. 12 Normalized sampling error for a random field using Fourier ⊡, standard ⊙, and optimum ◇ sampling techniques as a function of the number of samples N^2. (From [6].)

Fourier samples. In Fig. 12, the middle curve corresponds to $\varepsilon_{N^2}^2/ABR_{ff}(0,0)$ for Fourier sampling with $c_1 A = c_2 B = 1$. Under the conditions assumed, this curve indicates how rapidly the sampling error decreases as the number of samples is increased.

Standard Sampling

We now consider a sampling technique in which each sample is obtained by averaging a small portion of the picture. Let the region in the xy-plane over which the pictures are defined be again a rectangle with sides A and B. Over this region we define the orthonormal basis functions as

$$\varphi_{mn}(x, y) = \sqrt{\frac{MN}{AB}} \quad \begin{cases} \dfrac{mA}{M} \leqslant x < \dfrac{(m+1)A}{M} \\[2mm] \dfrac{nB}{N} \leqslant y < \dfrac{(n+1)B}{N} \end{cases}$$

$$= 0 \quad \text{elsewhere} \quad \begin{aligned} m &= 0, 1, 2, ..., M-1 \\ n &= 0, 1, 2, ..., N-1 \end{aligned} \tag{55}$$

where we have assumed the origin to coincide with the lower left corner of the rectangle. In other words, the picture area is divided into MN rectangular regions, and each $\varphi_{mn}(x, y)$ is constant over one of these regions and zero elsewhere. For example, if $M = 8$ and $N = 6$, the set of orthonormal functions generated by the previous definition would have 48 members. As an illustration the member $\varphi_{3,2}(x, y)$ of this set is shown in Fig. 13.

By substituting (55) in (41b), we obtain for the samples

$$a_{mn} = \sqrt{MN/AB} \int_{nB/N}^{(n+1)|B/N} \int_{mA/M}^{(m+1)A/N} f(x, y) \, dx \, dy \tag{56}$$

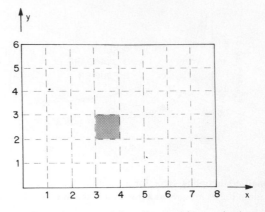

Fig. 13 An example of an orthonormal function used in standard sampling. The function $\varphi_{3,2}(x, y)$ shown here belongs to the orthonormal set generated by Eq. (55) with $M = 8$ and $N = 6$.

It is seen that except for a constant factor, a_{mn} is the average gray level of the picture over the area where φ_{mn} is nonzero.

We will now extend the general result for the sampling error for a random field as given by (46) to the case of standard sampling. We will again assume a zero mean real homogeneous random field with the autocorrelation function given by (47) with $\eta = 0$. Substituting (55) and (47) with $\eta = 0$ in (46), we obtain for the sampling error

$$
\begin{aligned}
\varepsilon_{MN}^2 &= ABR_{ff}(0,0) - \frac{MN}{AB} \\
&\quad \times R_{ff}(0,0) \sum_{m=0}^{M-1} \sum_{n=0}^{N-1} \int_{nB/N}^{(n+1)B/N} \int_{mA/M}^{(m+1)A/M} \int_{nB/N}^{(n+1)B/N} \\
&\quad \times \int_{mA/M}^{(m+1)A/M} \exp(-c_1|x-x'|-c_2|y-y'|)\, dx\, dy\, dx'\, dy' \\
&= ABR_{ff}(0,0) - \frac{MN}{AB} R_{ff}(0,0) \sum_{m=0}^{M-1} \sum_{n=0}^{N-1} \left\{ \int_{mA/M}^{(m+1)A/M} \int_{mA/M}^{(m+1)A/M} \right. \\
&\quad \times \left. \exp(-c_1|x-x'|\, dx\, dx' \right\} \\
&\quad \times \left\{ \int_{nB/N}^{(n+1)B/N} \int_{nB/N}^{(n+1)B/N} \exp(-c_2|y-y'|)\, dy\, dy' \right\} \\
&= ABR_{ff}(0,0) \left[1 - \frac{4MN}{(c_1 A)(c_2 B)} \left\{ 1 + \frac{M}{c_1 A} \left[\exp\left(\frac{-c_1 A}{M}\right) - 1 \right] \right\} \right. \\
&\quad \times \left. \left\{ 1 + \frac{N}{c_2 B} \left[\exp\left(\frac{-c_2 B}{N}\right) - 1 \right] \right\} \right]
\end{aligned} \tag{57}
$$

For a given number of samples MN, the normalized sampling error $\varepsilon_{MN}^2/ABR_{ff}(0,0)$ depends on $c_1 A$ and $c_2 B$ only. If we assume $M = N$ in (57), we obtain the normalized sampling error using N^2 samples. In Fig. 13 the top curve corresponds to the normalized sampling error for this case. It is clear from the figure that for a given number of samples, Fourier sampling is more efficient than standard sampling.

4.2.4 Optimal Sampling

Optimal sampling of a random field defined over a region $\mathscr{S}$ of the xy-plane is achieved by that set of orthonormal functions $\varphi_{mn}(x, y)$ which for every M and N yields the minimum value for the sampling error ε_{MN}^2. Since the first term in (46) is independent of the choice of the orthonormal basis, ε_{MN}^2 is minimized when the second term, i.e., the nonnegative quantity

$$\sum_{m=0}^{M-1} \sum_{n=0}^{N-1} \int_{-B/2}^{B/2} \int_{-A/2}^{A/2} \int_{-B/2}^{B/2} \int_{-A/2}^{A/2} R_{ff}(x-x', y-y') \varphi_{mn}^*(x, y)$$

$$\times \varphi_{mn}(x', y')\, dx\, dy\, dx'\, dy' \tag{58}$$

is maximized. In (58) we have assumed the region $\mathscr{S}$ to consist of a rectangle with sides A and B, the origin of the xy-plane coinciding with the center of the rectangle.

The assumption of homogeneity of the random field that is incorporated in this equation is not necessary for the present section. For the general case, which includes inhomogeneous random fields, $R_{ff}(x-x', y-y')$ in (46) can be replaced by the more general $R_{ff}(x, y, x', y')$ and (58) can then be written as

$$\sum_{m=0}^{M-1} \sum_{n=0}^{N-1} \int_{-B/2}^{B/2} \int_{-A/2}^{A/2} \int_{-B/2}^{B/2} \int_{-A/2}^{A/2} R_{ff}(x, y, x', y') \varphi_{mn}^*(x, y)$$

$$\times \varphi_{mn}(x', y')\, dx\, dy\, dx'\, dy' \tag{59}$$

The problem then is to find functions $\varphi_{mn}(x, y)$ that for a given $R_{ff}(x, y, x', y')$ maximize (59) subject to the orthonormality condition

$$\int_{-B/2}^{B/2} \int_{-A/2}^{A/2} \varphi_{mn}(x, y)\, \varphi_{pq}^*(x, y)\, dx\, dy = \begin{cases} 0 & m \neq n \quad \text{or} \quad p \neq q \\ 1 & m = n \quad \text{and} \quad p = q \end{cases} \tag{60}$$

A straightforward extension of the theory given by Brown [2] to the case of two dimensions shows that the functions $\varphi_{mn}(x, y)$ that maximize (59) subject to the orthonormality conditions (60) are the solutions of the following integral equation:

$$\int_{-B/2}^{B/2} \int_{-A/2}^{A/2} R_{ff}(x, y, x', y')\, \varphi(x', y')\, dx'\, dy' = \gamma \varphi(x, y) \tag{61}$$

Questions pertaining to the existence of the solutions of (61) and their dependence on the form of $R_{ff}(x, y, x', y')$ belong to the study of integral equations [18] and will not be pursued here. Suffice it to say that for continuous $R_{ff}(x, y, x', y')$ the integral equation in (61) has nonzero solutions $\varphi_{mn}(x, y)$ (eigenfunctions) for certain values γ_{mn} of γ (eigenvalues).

Since the orthonormal functions $\varphi_{mn}(x, y)$ that we seek are a solution of (61) for certain values γ_{mn} of γ, the functions $\varphi_{mn}(x, y)$ and the corresponding eigenvalues must satisfy

$$\int_{-B/2}^{B/2} \int_{-A/2}^{A/2} R_{ff}(x, y, x', y')\, \varphi_{mn}(x', y')\, dx'\, dy' = \gamma_{mn}\, \varphi_{mn}(x, y) \qquad (62)$$

Substituting (62) in (59) and making use of the orthonormality relationship in (60), (59) reduces to

$$\sum_{m=0}^{M-1} \sum_{n=0}^{N-1} \gamma_{mn} \qquad (63)$$

The sampling error for the case of optimal sampling is then obtained by substituting (63) for the double summation in (46), giving[†]

$$\varepsilon_{MN}^2 = ABR_{ff}(0, 0, 0, 0) - \sum_{m=0}^{M-1} \sum_{n=0}^{N-1} \gamma_{mn} \qquad (64)$$

Often the autocorrelation function can be considered to be separable in its x and y dependence, i.e.,

$$R_{ff}(x, y, x', y') = R'(x, x')\, R''(y, y') \qquad (65)$$

One example of such a separable autocorrelation function is given by Eq. (47) with $\alpha = x - x'$, $\beta = y - y'$ for the case where $\eta = 0$. For separable autocorrelation functions as in (65), one may look for solutions $\varphi(x, y)$ of (61) that are also separable in their x and y dependence. The function $\varphi(x, y)$ in (61) can then be expressed as

$$\varphi(x, y) = \varphi'(x)\, \varphi''(y) \qquad (66)$$

Substituting (65) and (66) in (61), the integral equation can be decomposed into two one-dimensional integral equations:

$$\int_{-A/2}^{A/2} R'(x, x')\, \varphi'(x')\, dx' = \gamma' \varphi'(x) \qquad (67a)$$

$$\int_{-B/2}^{B/2} R''(y, y')\, \varphi''(y')\, dy' = \gamma'' \varphi''(y) \qquad (67b)$$

[†] While for the case of homogeneous random fields, the value of the autocorrelation function at the origin can be denoted by $R_{ff}(0, 0)$, for the more general case the autocorrelation function $R_{ff}(x, y, x', y')$ is a function of four variables, and its value at the origin must be denoted by $R_{ff}(0, 0, 0, 0)$.

where we have expressed γ in (61) as a product $\gamma'\gamma''$. The integral equation (67a) has solutions $\varphi_m'(x)$ for certain values γ_m' of γ'. Similarly, (67b) has solutions $\varphi_n''(y)$ for certain values γ_n'' of γ''. The orthonormal functions over the region $\mathcal{S}$ of the xy-plane are then given by the products

$$\varphi_{mn}(x, y) = \varphi_m'(x)\,\varphi_n''(y) \tag{68}$$

and to each function $\varphi_{mn}(x, y)$ there corresponds an eigenvalue

$$\gamma_{mn} = \gamma_m'\gamma_n'' \tag{69}$$

An important question presents itself at this point. How should the orthonormal functions $\varphi_{mn}(x, y)$ and the corresponding eigenvalues γ_{mn} be ranked? In other words, we would like to know, of all $\varphi_{mn}(x, y)$ and the corresponding γ_{mn}, which one should be assigned the indices $m = 0$, $n = 0$; which one $m = 0$, $n = 1$; and so on. Let us first consider the case of separable autocorrelation functions, that is, when the eigenfunctions and eigenvalues are obtained by solving (67a) and (67b). In order that the sampling error ε_{MN}^2 in (64) be a minimum for $M = 1$, $N = 1$ (only one sample), it is clear that the indices $m = 0$, $n = 0$ should be assigned to that solution that results in the largest product $\gamma_m'\gamma_n''$. Once γ_0' and γ_0'' have been picked, the indices $m = 0$, $n = 1$ should be assigned to that solution to which corresponds the largest number $\gamma_0'\gamma_n''$ except for $\gamma_0'\gamma_0''$. The process of ranking can be continued in this way.

For random fields for which the autocorrelation function is not separable, the use of double indices for the orthonormal functions is really meaningless (see footnote on p. 83). The solutions of (61) and their corresponding eigenvalues can now be labeled by a single index. In this case there will be a single summation on the right-hand side in (64). By making use of the equation corresponding to (64) in this case and by the arguments regarding minimization of ε_{MN}^2 presented previously, it is easy to see that the solution with the largest eigenvalue will get the first rank, the solution with the next largest value will be ranked second, and so on.

So far we have indicated that given a random field one can, in general, find an optimum set of orthonormal functions that minimize the sampling error by solving (61) [or (67a) and (67b) for the special case of a separable autocorrelation function]. There exist useful situations in which closed form solutions to these integral equations can be obtained. One such case is when the autocorrelation function is of the form given by (47) with $\eta = 0$. In this case, the solutions to both (67a) and (67b) can be obtained by solving the integral equations

$$\int_{-T}^{T} k(s, t)\,\varphi(t)\,dt = \gamma\varphi(s) \tag{70}$$

where T equals $A/2$ for (67a) and $B/2$ for (67b), and where

$$k(s,t) = k_0 e^{-c|s-t|} \tag{71}$$

c being equal to c_1 for (67a) and c_2 for (67b). We can arbitrarily set $k_0 = \sqrt{R(0,0)}$ for both the cases.

The theoretical techniques for solving (70) with $k(s,t)$ given by (71) can be found in many places in the literature (see, for example, pp. 186–190 of [19]). In Fig. 12, the bottom curve corresponds to the case in which the sampling error was calculated by first solving (70) with $k(s,t)$ given by (71), writing down the corresponding solutions for (67a) and (67b), substituting these solutions and their corresponding eigenvalues in (68) and (69) to generate the orthonormal functions $\varphi_{mn}(x,y)$ and their corresponding eigenvalues γ_{mn}, ranking the solutions and their eigenvalues by the method just discussed, and finally substituting the eigenvalues in (64). The values of c_1 and c_2 were chosen such that $c_1 A = c_2 B = 1$. As expected, this procedure results in minimum error sampling of the random field. This optimality, however, is gained at the expense of considerable complexity in implementation of this technique.

The orthonormal functions obtained by the method indicated in this section are called *Karhunen–Loève* (K–L) *functions*, and the sampling procedure that uses these functions is referred to as *Karhunen–Loève sampling*.

4.3   QUANTIZATION OF PICTURE SAMPLES

In digital processing the picture samples must be quantized. This means that the range of values of the samples must be divided into intervals and all the values within an interval must be represented by a single level. In order that a picture reconstructed from the quantized samples be acceptable, it may sometimes be necessary to use 100 or more quantizing levels. When the samples are obtained by using an array of points as in Section 4.1 or standard sampling as in Section 4.2.3, a fine degree of quantization is particularly important for samples taken in regions of a picture across which the gray level changes slowly. In this situation large parts of these regions will be quantized to constant grey levels, while between these parts, there will be curves along which there is an abrupt gray level jump. These curves will tend to appear as conspicuous "false contours" cutting across the regions and may make the quantized approximation to the original picture unacceptable since they define spurious "objects" which may compete with or conceal the real objects shown in the picture. In Fig. 14a we have shown a picture reconstructed from samples that were quantized to 16 discrete levels. The original picture was sampled on a 128×128 square array of points. In the reconstructed pictures in Figs. 14b–14d, the samples were quantized to 8, 4, and 2 levels, respectively.

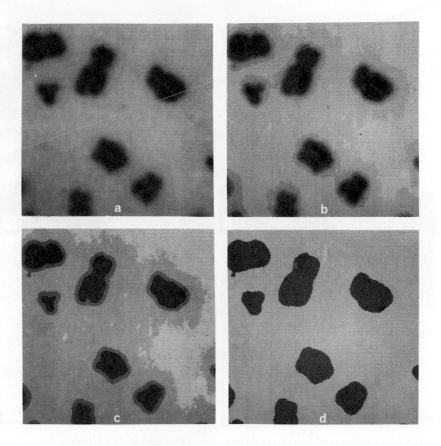

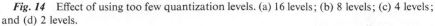

Fig. 14 Effect of using too few quantization levels. (a) 16 levels; (b) 8 levels; (c) 4 levels; and (d) 2 levels.

As in the case of sampling, it is usually simplest to choose quantization levels that are evenly spaced, but unequally spaced levels may sometimes be preferable. For example, suppose that the sample values in a certain range occur frequently, while the others occur rarely. In such a case one might wish to use quantization levels that are finely spaced inside this range and coarsely spaced outside it; this increases the average accuracy of the quantization without increasing the number of levels. This method, known as "tapered quantization," is illustrated in Fig. 15.

In the next section we will present the theory of optimum quantization. It will be seen that this theory legitimizes the reasons given here for tapering the quantization scale.

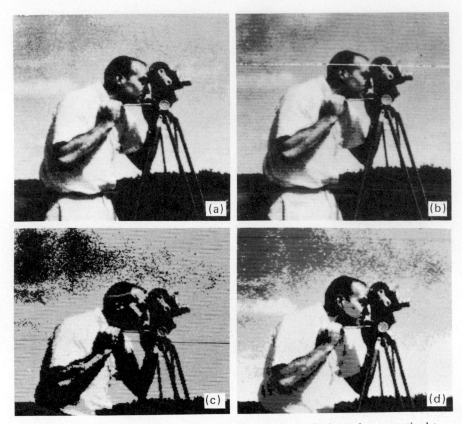

Fig. 15 Tapered quantization: A picture of a cameraman is shown here quantized to: (a) 16 equally spaced levels; (b) 16 tapered levels; (c) 4 equally spaced levels; and (d) 4 tapered levels. (From [8].)

4.3.1 Optimal Quantization

Let the range of values of the samples be represented by the line $H_1 H_2$ in Fig. 16. We wish to quantize the samples to K discrete levels, represented in Fig. 16 by $q_1, q_2, ..., q_K$. These are the output levels of the quantizer. Now let the input intervals be represented by the decision levels $z_1, z_2, ..., z_{K+1}$. This means that if a sample at the input to the quantizer has a value anywhere between z_k and z_{k+1} it is assigned the value q_k at the output.

Let δ_q^2 denote the mean square quantization error between the input and the output of the quantizer for a given choice of the output levels $q_1, q_2, ..., q_K$, and the input intervals as represented by the decision levels $z_1, z_2, ..., z_{K+1}$.

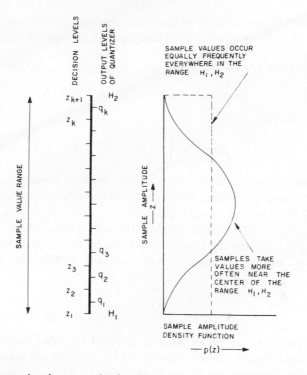

Fig. 16   If a sample takes any value between z_k and z_{k+1}, it is assigned a fixed value q_k at the output of the quantizer. $p(z)$ is the sample value probability density function.

Thus $\delta_q^{\,2}$ is given by

$$\delta_q^{\,2} = \sum_{k=1}^{K} \int_{z_k}^{z_{k+1}} (z - q_k)^2 p(z)\, dz \tag{72}$$

where $p(z)$ is the probability density function for the input sample value which is represented by the continuous variable z. Equation (72) is obtained by applying Eq. (60) of Chapter 2 to each of the input intervals.

For a given number K of output levels, we would now like to determine the output levels q_k's and the decision levels z_k's such that the mean square quantization error $\delta_q^{\,2}$ is a minimum. If we wish to minimize $\delta_q^{\,2}$ for a given K, we can derive the necessary conditions by partially differentiating $\delta_q^{\,2}$ with respect to the z_k's and q_k's and setting the derivatives equal to zero:

$$\frac{\partial}{\partial z_k}\delta_q^{\,2} = (z_k - q_{k-1})^2 p(z_k) - (z_k - q_k)^2 p(z_k) = 0, \qquad k = 2, \ldots, K \tag{73}$$

$$\frac{\partial}{\partial q_k}\delta_q^{\,2} = -2\int_{z_k}^{z_{k+1}} (z - q_k)\, p(z)\, dz = 0, \qquad k = 1, 2, \ldots, K \tag{74}$$

Note that in (73) we have, for obvious reasons, not differentiated with respect to z_1 and z_{k+1}. These are the fixed endpoints of the sample amplitude range at the input.

From (73), for $p(z_k)$ not equal to zero, we obtain

$$z_k = (q_{k-1} + q_k)/2, \qquad k = 2, 3, ..., K \qquad (75)$$

while (74) implies

$$q_k = \int_{z_k}^{z_{k+1}} z p(z) \, dz \left/ \int_{z_k}^{z_{k+1}} p(z) \, dz \right., \qquad k = 1, 2, ..., K \qquad (76)$$

We see that for an optimum quantizer the decision levels (z_k's) are located halfway between the output levels (q_k's), while each q_k is the centroid of the portion of $p(z)$ between z_k and z_{k+1}.

If the sample values occur equally frequently everywhere in the range $H_1 H_2$ in Fig. 16, then the sample values are uniformly distributed over the range $H_1 H_2$ and $p(z)$ is equal to some constant. In this case, (75) and (76) reduce to

$$z_k = (q_{k-1} + q_k)/2, \qquad k = 2, 3, ..., K$$

$$q_k = (z_k + z_{k+1})/2, \qquad k = 1, 2, ..., K$$

The two equations are simultaneously satisfied provided the decision levels as well as the output levels are equally spaced, with the output levels midway between the decision levels and vice versa. This is called "uniform quantization."

If the sample values occur more frequently in some part of the range $H_1 H_2$ in Fig. 16, then the density function $p(z)$ is not equal to a constant and the solutions of (75) and (76) become more complex.

Given the sample value density function $p(z)$ and the end values z_1 and z_{K+1}, a method of solving (75) and (76) is as follows. Pick q_1, then calculate the succeeding z_k's and q_k's by using (75) and (76). If the calculated value of the last output level q_K is the centroid of the area between z_K and z_{K+1}, then the calculated z_k's and q_k's represent the correct solution. If q_K is not the appropriate centroid, the calculation must be repeated with a different choice of q_1. The search for the correct value of q_1 can be systematized so as to yield the correct solutions in a short time. Max [11] employed this procedure to calculate the decision levels (z_k's) and the output levels (q_k's) for the optimum quantizer for the case when the sample value has Gaussian density [$p(z) = (1/\sqrt{2\pi}) \exp(-z^2/2)$] with zero mean and unit variance, and when the end values z_1 and z_{K+1} of the input range are given by $z_1 = -\infty$ and $z_{K+1} = +\infty$, with the restriction that $z_{(K/2)+1} = 0$ for K even, and $q_{(K+1)/2} = 0$ for K odd. This procedure yields symmetric results. In Table 1 we have shown the results of such a computation for $K = 16$. Note that the levels tend to be closer near

TABLE 1

Optimum Quantization for $K = 16^a$ [11]

Decision levels	Output levels	Decision levels	Output levels
$z_1 = -\infty$	$q_1 = -2.733$	$z_{10} = 0.2582$	$q_{10} = 0.3881$
$z_2 = -2.401$	$q_2 = -2.069$	$z_{11} = 0.5224$	$q_{11} = 0.6568$
$z_3 = -1.844$	$q_3 = -1.618$	$z_{12} = 0.7996$	$q_{12} = 0.9424$
$z_4 = -1.437$	$q_4 = -1.256$	$z_{13} = 1.099$	$q_{13} = 1.256$
$z_5 = -1.099$	$q_5 = -0.9424$	$z_{14} = 1.437$	$q_{14} = 1.618$
$z_6 = -0.7996$	$q_6 = -0.6568$	$z_{15} = 1.844$	$q_{15} = 2.069$
$z_7 = -0.5224$	$q_7 = -0.3881$	$z_{16} = 2.401$	$q_{16} = 2.733$
$z_8 = -0.2582$	$q_8 = -0.1284$	$z_{17} = \infty$	
$z_9 = 0.0$	$q_9 = 0.1284$		

$^a p(z) = (1/\sqrt{2\pi}) \exp(-z^2/2)$, $z_1 = -\infty$, $z_{K+1} = \infty$.

the center of the scale, where $p(z)$ takes its maximum value, so that the sample values occur more frequently there.

Nonuniform quantization is usually accomplished by *companding* [16]. This involves passing each sample through a nonlinear device called a "compressor" and then uniformly quantizing the output of the compressor. Of course, now the quantized samples must pass through the inverse of the compressor, called the "expander," before reconstruction of the picture.

4.3.2   Sampling, Quantization, and Picture Detail

Let a picture be sampled on a rectangular array of $M \times N$ points, so that it is represented by $M \times N$ samples. Let the samples be quantized to K levels. If $K = 2^b$ and if the natural binary code is used to transform the quantized samples into binary words, then each sample is represented by a b-bit binary word. Therefore, the total number of bits necessary to represent the picture is $M \times N \times b$.

The following question now arises: For a given total number of bits, how should one choose the values for N, M, and b in order that the reconstruction error is a minimum? In this section it will be shown that the proper choice of N, M, and b strongly depends on picture detail. We will not endeavor to present any systematic techniques for determining the optimum values of N, M, and b, for a given total number of bits, since such techniques do not yet exist.

Figure 17 shows two pictures of a face reconstructed with different values of N, M, and b but with the same total number of bits. It is clear that the picture

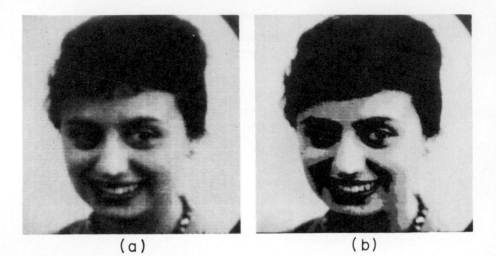

(a)   (b)

Fig. 17 Tradeoff between sampling and quantization. (a) Picture of a face, 128×128 samples, 64 quantization levels. (b) Same picture as in (a), 256×256 samples, 16 quantization levels. (From [7].)

(a)   (b)

Fig. 18 Tradeoff between sampling and quantization. (a) Picture of a crowd, 128×128 samples, 64 quantization levels. (b) Same picture as in (a), 256×256 samples, 16 quantization levels. (From [7].)

in Fig. 17a is more pleasing (esthetically) than the picture in Fig. 17b, even though the former has lower spatial resolution. Similar results for a picture of a crowd are presented in Fig. 18. Again, both the pictures in the figure were reconstructed from the same total number of bits. This example yields a result opposite to that for the picture of the face; here the picture with higher spatial resolution and lower gray scale resolution is preferable to the picture with lower spatial resolution and higher gray scale resolution.

In a recent study the pictures of the face and the crowd, as shown in Figs. 17 and 18, were represented by $N \times N$ samples and the samples quantized into 2^b levels for different values of N and b. The reconstructed pictures were ranked by observers according to their subjective quality. The results were plotted in the form of isopreference curves (solid curves) in the Nb-plane (Fig. 19). [Each point in the Nb-plane represents a picture with $N \times N$ samples and 2^b quantization levels]. The points on an isopreference curve represent

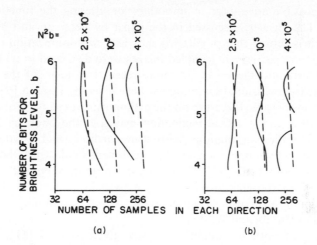

Fig. 19 Isopreference curves (a) for the pictures in Fig. 17, (b) for the pictures in Fig. 18. The dotted lines are the lines of constant total number of bits required to represent a picture. (From [7].)

pictures of the same quality as judged by the observers. Results for the face are shown in Fig. 19a while those for the crowd are in Fig. 19b. It is seen that the isopreference curves depend markedly on the picture types and differ greatly from the curves for constant total number of bits (the dashed curves in the figure).

The discussion and the examples presented point to the following important observation: In a slowly changing scene it is important to have fine quantization, but the sampling can be coarse; while in a scene with a large amount of detail, it is necessary to sample finely, but quantization can be coarse.

4.4 BIBLIOGRAPHICAL NOTES

The theory of sampling using a periodic lattice of points in two or more dimensions was given in a classic paper by Peterson and Middleton [13]. The case of a rectangular sampling lattice has also been treated by Bracewell [1].

In Section 4.1 it was shown that a two-dimensional function can be reproduced exactly from its samples taken over a suitable sampling lattice, provided the function is bandlimited, i.e., its Fourier transform (power spectrum for the case of random fields) vanishes outside some region in the frequency plane. The question now arises: Suppose a function $f(x, y)$ is not bandlimited, is there any possible advantage in smoothing or filtering the function before sampling? The answer, discussed in the paper by Peterson and Middlteon, is yes, and it is shown that prefiltering may lead to minimization of aliasing problems. Moiré patterns in sampled images were discussed in a now classic paper by Mertz and Grey [12]. A more recent treatment of the effects of aliasing in two-dimensional sampling has been given by Legault [9].

There is considerable literature on the orthonormal expansion of functions. Readers with a mathematical inclination may find the book by Davis [3] enjoyable. Sampling of random fields using orthonormal expansions is a generalization to two dimensions of the one-dimensional theory that is given in the book by Van Trees [19]. The theory of optimum sampling is based on the treatment for the one-dimensional case given by Brown [2]. The results for the sampling error using Fourier, standard, and optimum sampling are to be found in Habibi and Wintz [6].

The theory of optimum quantization was given by Max [11]. Optimum quantization for constant output entropy, rather than constant number of output levels, has been discussed by Wood [20] based on an approximation to Max's results given by Roe [15]. Nonuniform quantization through the use of a nonlinear gain characteristic followed by uniform quantization has been described by Smith [16]. The false contours that result from using too few quantization levels can be "broken up" by adding an irregular "dither" to the grey level prior to quantization; see Roberts [14] and Lippel [10]. The effects of using too few quantization levels can also be minimized by using optimally designed pre- and postfilters around the quantizer; see Graham [5].

A picture that is reconstructed from samples contains, in general, both sampling and quantization errors. A consequence of the analysis given by Totty and Clark [17] is that if a picture is sampled using orthogonal basis functions, the total mean square error between the original picture and the reconstructed picture is equal to the sum of the mean square sampling error and the mean square quantization error for any quantization process.

The subjective tradeoff between sampling and quantization has been discussed by Huang [7].

REFERENCES

1. R. N. Bracewell, Two-dimensional aerial smoothing in radio astronomy, *Aust. J. Phys.* **9**, 1956, 297–314.
2. J. L. Brown, Mean square truncation error in series expansion of random functions, *J. SIAM* **8**, 1960, 28–32.
3. P. Davis, "Interpolation and Approximation," Ginn (Blaisdell), Waltham, Masssachusetts, 1966.
4. J. W. Goodman, "Introduction to Fourier Optics," McGraw-Hill, New York, 1968.
5. D. N. Graham, Optimal Filtering to Reduce Quantization Noise, Master of Science Thesis, Dep. Elec. Eng., Massachusetts Inst. of Technol., Cambridge, Massachusetts, June 1966.
6. A. Habibi and P. A. Wintz, Image coding by linear transformation and block quantization, *IEEE Trans. Comm. Technol.* **COM-19**, 1971, 50–62, 1971.
7. T. S. Huang, O. J. Tretiak, B. Prasada, and Y. Yamaguchi, Design considerations in PCM transmission of low resolution monochrome still pictures, *Proc. IEEE* **55**, 1967, 331–335.
8. T. S. Huang, PCM picture transmission, *IEEE Spectrum* **2**, 1965, 57–63.
9. R. Legault, The aliasing problems in two-dimensional sampled imagery, "Perception of Displayed Information," *in* (L. M. Biberman, ed.). Plenum Press, New York, 1973.
10. B. Lippel, Effect of dither on luminance quantization of pictures, *IEEE Trans. Comm. Technol.* COM-19, 1971, 879–888.
11. J. Max, Quantizing for minimum distortion, *IRE Trans. Informat. Theory* **IT-6**, 1960, 7–12.
12. P. Mertz and F. Grey, A theory of scanning and its relation to the characteristics of the transmitted signal in telephotography and television, *BSTJ* **13**, 1934, 464–515.
13. D. P. Peterson and D. Middleton, Sampling and reconstruction of wave-number-limited functions in n-dimensional euclidean spaces, *Informat. Contr.* **5**, 1962, 279–323.
14. L. G. Roberts, Picture coding using pseudo-random noise, *IRE Trans Informat. Theory* **IT-8**, 1962, 145–154.
15. G. M. Roe, Quantizing for minimum distortion, *IEEE Trans. Informat. Theory* **IT-10**, 1964, 384–385.
16. B. Smith, Instantaneous companding of quantized signals, *BSTJ* **36**, 1957, 653–709.
17. R. E. Totty and G. C. Clark, Reconstruction error in waveform transmission, *IRE Trans. Informat. Theory* **IT-13**, 1967, 336–338.
18. F. G. Tricomi, "Integral Equations," Wiley (Interscience), New York, 1957.
19. H. L. Van Trees, "Detection, Estimation, and Modulation Theory," Part 1. Wiley, New York, 1968.
20. R. C. Wood, On optimum quantization, *IEEE Trans. Informat. Theory* **IT-15**, 1969, 248–252.

Chapter 5

Compression

Suppose we intend to transmit a picture over a channel or store it in an electronic device. For given gray scale and spatial resolution capabilities of a receiver or a user, the aim of picture compression is to represent the picture by as few bits as possible for the purpose of transmission or storage. Consider, for example, a 5 cm × 5 cm photograph with a resolution of 50 μ in all directions. Let us assume that the signal-to-noise ratio in the photograph is such that one can extract 256 ($=2^8$) gray levels from the photograph. (By this we mean that the standard deviation of the noise fluctuations is approximately 1/256 of the maximum range of gray scale in the photograph.) If fixed-length binary code words are used, then the gray level at each resolution element in the photograph can be represented by a binary word 8 bits long. It would, therefore, take 8×10^6 bits to store or transmit this picture in a "distortion free" form.

Suppose the given picture is to be received by a system the spatial and gray scale resolution of which exceeds or at least matches those of the picture. It must then be possible to reproduce the picture faithfully from the digital data. For the picture just described this would require 8×10^6 bits. A question that now arises is: Can we represent the picture in this example by less than 8×10^6 bits and still be able to reproduce it faithfully? If some gray levels occur more often than others, then the answer is yes, if we use shorter binary words for gray levels that occur more often. The techniques by which this can be done form an important subject in information theory [6, 16, 33]. The degree of data compression that can be achieved by these techniques for pictures is rather limited and we will not discuss them any further here. Sampling a picture on a grid of points and representing the samples by binary code words (which may be of unequal length, to achieve compression) is called *pulse code modulation* (PCM).

If the spatial and gray scale resolution of the receiver (or the user) are inferior to those of the original picture, then one can take liberties with the sampling and quantization of the picture and thereby represent it by a reduced number of bits. This means that now a certain amount of distortion is permitted between the original picture and the version that can be reproduced from the binary data. This leads us to the following questions: For a given amount and type of distortion, what techniques would allow us to represent the picture by a reduced number of bits? For a given amount of distortion, is there a minimum number of bits that are needed to represent the picture? The first question is answered in the following sections where we discuss various two-dimensional picture-compression techniques that (at the cost of a small amount of distortion, often unnoticeable) significantly economize on the number of bits required for a picture. The second question is answered by the rate-distortion function, which will be discussed in conceptual terms at the end of the chapter.

5.1 TRANSFORM COMPRESSION

In this section we will first justify qualitatively that in picture compression one needs to represent a picture by uncorrelated data, and we will present some reversible linear transformations that achieve this. The data must then be ranked according to degree of significance of their contribution to both the information content and the subjective quality of the picture. Once such a ranking is achieved then those elements of the data that are unimportant from the point of view of the gray scale and spatial resolution capability of the receiver can be neglected. This makes possible a major degree of picture compression.

5.1.1 Karhunen–Loève Compression: The Continuous Case

As a rather trivial but conceptually illustrative example of picture compression, consider the case where we know *a priori* that all the pictures in a collection are sinusoidal patterns. If we sampled the pictures on a fine grid and quantized the resulting samples, we would need a large number of bits to represent each picture. On the other hand, however, the only information we need to extract from these pictures in order to reproduce them distortion free is the spatial frequency, the orientation, the amplitude, and the phase (with respect to a fixed origin, which could be the center of the pictures) of the sinusoidal pattern in each picture. The total number of bits required for this is relatively small, resulting in data compression.

The simple example just presented has several important features. The reason why data compression was achieved in this example is that the gray levels at all points in each picture are highly correlated. By this we mean that the gray levels at all points in a picture can be predicted from the level at any one point in the same picture given the spatial frequency, the phase, the orientation, and the amplitude. Note, also, that the data (spatial frequency, orientation, amplitude, and phase) that are used to represent a picture and from which the picture can be reproduced are uncorrelated. For example, a picture with a given spatial frequency may have any phase, any amplitude, and any orientation.

This discussion leads us to a basic idea in picture compression. One attempts to represent a picture by uncorrelated data because then each element of the data is a unique property of the picture—unique in the sense that it cannot be predicted (at least from the point of view of linear mean square estimation theories) from the rest of the data. The example just discussed was of a rather restricted nature because of our *a priori* knowledge that the pictures were sinusoidal patterns. For a more general class of pictures, when an attempt is made to represent each picture by uncorrelated data, the number of bits required for the uncorrelated data may be quite large. In such a case the elements of the uncorrelated data should be ranked in order of their importance. When this is done, the higher-order terms may, for example, represent high spatial resolution effects in the picture. If the receiver or user of the picture has limited spatial resolution capability, these high-resolution terms may as well be deleted from the data. In fact, as we will see presently, such deletions from the uncorrelated data contribute significantly to picture compression.

For a restricted class of pictures such as in the previous example, it is easy to find a set of uncorrelated parameters that represent a picture and from which it can be reproduced. For a more general class of pictures, the problem becomes more difficult. In Chapter 4, we discussed various techniques for representing pictures by samples. One technique involved expanding a picture in terms of a family of orthonormal functions and taking the coefficients of expansion as picture samples. In what follows we will seek a set of orthonormal functions that result in uncorrelated samples for a given class of pictures (a random field $\mathbf{f}(x, y)$) defined in terms of its autocorrelation function $R(\vec{r}_1, \vec{r}_2)$ (see

Eq. (89) of Chapter 2).[†] We recall that if (x, y) and (x', y') denote the co-ordinates of the position vectors $\vec{r}_1$ and $\vec{r}_2$, respectively, then $R(\vec{r}_1, \vec{r}_2)$ can also be expressed as $R(x, y, x', y')$. If the random field $\mathbf{f}(\vec{r})$ is homogeneous, then the autocorrelation function becomes a function of only two variables and can be expressed as $R(x-x', y-y')$ as discussed in Section 2.4.3.

The problem of obtaining uncorrelated samples from a picture is answered by the following theorem:

Theorem (Karhunen–Loève transformation for continouus pictures): Let $-A/2 \leqslant x \leqslant A/2$, $-B/2 \leqslant y \leqslant B/2$ define a region $\mathcal{S}$ of the xy-plane. Let $\varphi_{mn}(x, y)$ be a complete family of orthonormal functions defined over the region $\mathcal{S}$. A random field $\mathbf{f}(x, y)$ may then be expanded in region $\mathcal{S}$ as follows [see Eqs. (42a) and (42b) of Chapter 4]:

$$\mathbf{f}(x, y) = \sum_{m=0}^{\infty} \sum_{n=0}^{\infty} \mathbf{a}_{mn}\, \varphi_{mn}(x, y) \tag{1}$$

where the summation on the right-hand side converges to $f(x, y)$ in some sense and where

$$\mathbf{a}_{mn} = \int_{-B/2}^{B/2} \int_{-A/2}^{A/2} \mathbf{f}(x, y)\, \varphi_{mn}^{*}(x, y)\, dx\, dy \tag{2}$$

For zero-mean random fields the functions $\varphi_{mn}(x, y)$ that result in uncorrelated samples $\mathbf{a}_{mn}$ must satisfy the following integral equation:

$$\int_{-B/2}^{B/2} \int_{-A/2}^{A/2} R(x, y, x', y')\, \varphi_{mn}(x', y')\, dx'\, dy' = \gamma_{mn}\, \varphi_{mn}(x, y) \tag{3}$$

for $-A/2 \leqslant x \leqslant A/2$, $-B/2 \leqslant y \leqslant B/2$, where

$$\gamma_{mn} = E\{|\mathbf{a}_{mn}|^2\} \tag{4}$$

Proof: We do not want to eliminate the possibility that the desired orthonormal functions $\varphi_{mn}(x, y)$ may be complex. An examination of (2) reveals that even for real random fields $\mathbf{a}_{mn}$ can be complex. Uncorrelatedness for complex random variables $\mathbf{a}_{mn}$ means

$$E\{\mathbf{a}_{mn}\,\mathbf{a}_{ij}^{*}\} = E\{\mathbf{a}_{mn}\}\, E\{\mathbf{a}_{ij}^{*}\}, \qquad m \neq i \ \ \text{or} \ \ n \neq j \tag{5}$$

If we take the expectation of both sides of (2) and use the assumption that the random field $\mathbf{f}(x, y)$ has zero mean, we obtain

$$E\{\mathbf{a}_{mn}\} = \int_{-B/2}^{B/2} \int_{-A/2}^{A/2} E\{\mathbf{f}(x, y)\}\, \varphi_{mn}^{*}(x, y)\, dx\, dy$$

$$= 0 \qquad \text{for} \quad m, n = 0, 1, 2, 3, \ldots \tag{6}$$

[†] In this chapter, since we will be concerned with only one random field at a time, the autocorrelation function of a homogeneous random field $\mathbf{f}(\vec{r})$ will be denoted by $R(\alpha, \beta)$ instead of $R_{ff}(\alpha, \beta)$.

Thus for a zero-mean random field the samples also have zero mean. Substituting (6) in (5), the uncorrelatedness of the $\mathbf{a}_{mn}$ implies that

$$E\{\mathbf{a}_{mn}\mathbf{a}_{ij}^*\} = 0, \qquad m \neq i \quad \text{or} \quad n \neq j \tag{7}$$

To prove the theorem, we must now show that the functions $\varphi_{mn}(x, y)$ that yield samples satisfying (7) must be solutions of the integral equation (3). Multiplying both sides of (1) by $\mathbf{a}_{ij}^*$, we obtain

$$\mathbf{f}(x, y)\,\mathbf{a}_{ij}^* = \sum_{m=1}^{\infty} \sum_{n=1}^{\infty} \mathbf{a}_{mn}\mathbf{a}_{ij}^* \, \varphi_{mn}(x, y) \tag{8}$$

Taking the expectation of both sides and making use of the linearity of expectation and Eq. (7), we obtain

$$E\{\mathbf{f}(x, y)\,\mathbf{a}_{ij}^*\} = E\{|\mathbf{a}_{ij}|^2\}\,\varphi_{ij}(x, y) \tag{9}$$

The complex conjugate of (2) can be rewritten as

$$\mathbf{a}_{ij}^* = \int_{-B/2}^{B/2} \int_{-A/2}^{A/2} \mathbf{f}(x', y')\,\varphi_{ij}(x', y')\,dx'\,dy' \tag{10}$$

where we have made use of the fact that the random field is real.[†] Multiplying both sides of (10) by $\mathbf{f}(x, y)$ and taking the expectation, we obtain

$$E\{\mathbf{f}(x, y)\,\mathbf{a}_{ij}^*\} = \int_{-B/2}^{B/2} \int_{-A/2}^{A/2} R(x, y, x', y')\,\varphi_{ij}(x', y')\,dx'\,dy' \tag{11}$$

where we have made use of the definition of the autocorrelation function. Comparing (9) and (11), we see that $\varphi_{ij}(x, y)$ must satisfy

$$\int_{-B/2}^{B/2} \int_{-A/2}^{A/2} R(x, y, x', y')\,\varphi_{ij}(x', y')\,dx'\,dy' = E\{|\mathbf{a}_{ij}|^2\}\,\varphi_{ij}(x, y) \tag{12}$$

which proves the theorem. ∎

Exercise 1. Prove the converse to the previous theorem, that is, given a set of orthonormal functions that are the solutions of the integral equation (3), then if a zero-mean random field is expanded in terms of these orthonormal functions, the coefficients of expansion are uncorrelated. ∎

By comparing Eq. (61) of Chapter 4 with (3) we see that the orthonormal functions that give uncorrelated samples also minimize the sampling error.

Thus far we have seen how we may represent a continuous picture by uncorrelated data. This method could perhaps be implemented in hardware, but this has never been done. A more realistic approach, however, is first to

---

[†] The assumption of real random fields is not necessary to the theorem provided one makes use of the definition of autocorrelation for complex fields, which is $R(x, y, x', y') = E\{\mathbf{f}(x, y)\,\mathbf{f}^*(x', y')\}$.

digitize the picture by sampling it on a fine sampling lattice (see Section 4.1.3) and then use a digital computer to transform the samples into un-correlated data. We will now address ourselves to this problem.

5.1.2 Karhunen-Loève Compression: The Discrete Case

Let a picture belonging to a random field $\mathbf{f}(x, y)$ be sampled on an $N \times N$ square sampling lattice. We assume that N is large enough not to create any aliasing problems (see Section 4.1.5). The samples thus obtained may be denoted by $\mathbf{f}(m, n)$, where m and n both take integer values from 0 through $N-1$. The matrix having elements $\mathbf{f}(m, n)$ will be denoted by $[\mathbf{f}]$. Our aim at this point is to find a reversible transformation such that the elements of the transform are uncorrelated. The transformation must be reversible because we want to be able to reconstruct $[\mathbf{f}]$ from the uncorrelated data.

By the discussion in Section 2.2, given two $N \times N$ deterministic nonsingular matrices $[P]$ and $[Q]$, we can transform the matrix $[\mathbf{f}]$ into another matrix $[\mathbf{F}]$ by

$$[\mathbf{F}] = [P][\mathbf{f}][Q] \tag{13}$$

The inverse transform is given by (Eq. (33) of Chapter 2)

$$[\mathbf{f}] = [P]^{-1}[\mathbf{F}][Q]^{-1} \tag{14}$$

Let us denote the matrix $[P]^{-1}$ by $[P']$ and $[Q]^{-1}$ by $[Q']$; then (14) may be written in an expanded form as

$$\mathbf{f}(m, n) = \sum_{u=0}^{N-1} \sum_{v=0}^{N-1} \mathbf{F}(u, v) P'(m, u) Q'(v, n) \tag{15}$$

for $m = 0, \ldots, N-1$, $n = 0, \ldots, N-1$. Some thought indicates that (15) may be rewritten as

$$[\mathbf{f}] = \sum_{u=0}^{N-1} \sum_{v=0}^{N-1} [\varphi^{(u, v)}] \mathbf{F}(u, v) \tag{16}$$

where $[\varphi^{(u, v)}]$ is an $N \times N$ matrix whose (m, n)th element is $P'(m, u) Q'(v, n)$.

As an example of the representation in (16), note that a comparison of Eq. (40) of Chapter 2 with (15) reveals that for the discrete Fourier transform the (m, n)th element of this matrix $[\varphi^{(u, v)}]$ is

$$\exp\left[j2\pi \left(\frac{mu}{M} + \frac{nv}{N} \right) \right]$$

For the rest of this section we consider an $N \times N$ matrix to be a vector in an N^2-dimensional space. The first component of this vector for the matrix $[\varphi]$ is $\varphi(0, 0)$; the second component is $\varphi(0, 1), \ldots$, the Nth component

$\varphi(0, N-1)$; the $(N+1)$st component $\varphi(1,0)$; the $(N+2)$nd component $\varphi(1,1), \ldots$; and finally the N^2th component $\varphi(N-1, N-1)$. One now has a framework for defining the dot product of two matrices. By the definition of dot product in a complex vector space, the dot product of two $N \times N$ matrices $[\varphi]$ and $[\Gamma]$ is equal to

$$[\varphi] \cdot [\Gamma] = \sum_{m=0}^{N-1} \sum_{n=0}^{N-1} \varphi(m,n) \, \Gamma^*(m,n) \tag{17}$$

where Γ^*_{mn} is the complex conjugate of the element $\Gamma(m,n)$ of the matrix $[\Gamma]$. Note that the dot product of two matrices is a single number, real or complex.

Going back to (16), we have a set of N^2 matrices $[\varphi^{u,v}]$. These matrices form an orthonormal set provided[†]

$$[\varphi^{(u,v)}] \cdot [\varphi^{(r,s)}] = \begin{cases} 0 & u \neq r \quad \text{or} \quad v \neq s \\ 1 & u = r \quad \text{and} \quad v = s \end{cases} \tag{18}$$

Assume for a moment that the matrices $[\varphi^{(u,v)}]$ in (16) are indeed orthonormal; then multiplying (in the sense of the dot product) both sides by $[\varphi^{(r,s)}]$ and making use of (18), we obtain

$$[\mathbf{f}] \cdot [\varphi^{(r,s)}] = \mathbf{F}(r,s) \tag{19}$$

Summarizing the previous discussion: given a set of N^2 orthonormal matrices $[\varphi^{(u,v)}]$, we may expand an aribtrary $N \times N$ matrix $[\mathbf{f}]$ as

$$[\mathbf{f}] = \sum_{u=0}^{N-1} \sum_{v=0}^{N-1} \mathbf{F}(u,v) \, [\varphi^{(u,v)}] \tag{20a}$$

where the coefficients of expansion $\mathbf{F}(u,v)$ are given by

$$\mathbf{F}(u,v) = [\mathbf{f}] \cdot [\varphi^{(u,v)}], \qquad u = 0,1,2,\ldots,N-1, \quad v = 0,1,2,\ldots,N-1 \tag{20b}$$

Equations (20a) and (20b) provide a conceptually convenient framework for transforming a matrix $[\mathbf{f}]$ into another matrix $[\mathbf{F}]$. Of course, given $[\mathbf{F}]$, we may recover $[\mathbf{f}]$ by using (20a), so that the transformation is reversible.

The orthonormal expansion in (1) for the continuous case may be compared to that in (20) for the discrete case. Note that the expansion in (20) involves a finite number of terms only.

Theorem (Karhunen–Loève transformation for discrete pictures): Let $R(m,n,p,q)$ be the autocorrelation function of $[\mathbf{f}]$, that is,

$$R(m,n,p,q) = E\{\mathbf{f}(m,n)\mathbf{f}(p,q)\} \tag{21}$$

[†] Note that in an N^2-dimensional space, we cannot have more than N^2 matrices (defined as vectors in this space) that are mutually orthogonal.

For zero-mean random fields, the orthonormal matrices $[\varphi^{(u,v)}]$ that result in uncorrelated $\mathbf{F}(u,v)$ in (20) satisfy the equation

$$\sum_{p=0}^{N-1}\sum_{q=0}^{N-1} R(m,n,p,q)\,\varphi^{(u,v)}(p,q) = \gamma_{uv}\,\varphi^{(u,v)}(m,n) \tag{22}$$

where $\varphi^{(u,v)}(p,q)$ and $\varphi^{(u,v)}(m,n)$ are the (p,q)th and the (m,n)th elements, respectively, of the matrix $[\varphi^{(u,v)}]$, and where

$$\gamma_{uv} = E\{|\mathbf{F}(u,v)|^2\} \tag{23}$$

The matrices $[\varphi^{(u,v)}]$ are called the *eigenmatrices* or the *basis* matrices of $R(m,n,p,q)$.

Proof: By making use of the definition of the dot product in (17), (20b) can be written as

$$\mathbf{F}(u,v) = \sum_{m=0}^{N-1}\sum_{n=0}^{N-1} \mathbf{f}(m,n)\,\varphi^{(u,v)*}(m,n) \tag{24}$$

Taking the expectation of both sides, we obtain

$$E\{\mathbf{F}(u,v)\} = \sum_{m=0}^{N-1}\sum_{n-0}^{N-1} E\{\mathbf{f}(m,n)\}\,\varphi^{(u,v)*}(m,n)$$

For a zero-mean random field $E\{\mathbf{f}(m,n)\} = 0$. Substituting this in the preceding equation, we have

$$E\{\mathbf{F}(u,v)\} = 0, \qquad u = 0,1,...,N-1, \quad v = 0,1,...,N-1 \tag{25}$$

Now suppose that the $\mathbf{F}(u,v)$ are uncorrelated. We want to retain the possibility that $\mathbf{F}(u,v)$ may be complex even for real random fields. Uncorrelatedness of $\mathbf{F}(u,v)$, therefore, means (see (5))

$$E\{\mathbf{F}(u,v)\,\mathbf{F}^*(u',v')\} = E\{\mathbf{F}(u,v)\}\,E\{\mathbf{F}^*(u',v')\}, \qquad u \neq u' \quad \text{or} \quad v \neq v'$$

which by (25) reduces to

$$E\{\mathbf{F}(u,v)\,\mathbf{F}^*(u',v')\} = 0, \qquad u \neq u' \quad \text{or} \quad v \neq v' \tag{26}$$

Now (20a) can be written as

$$\mathbf{f}(m,n) = \sum_{u=0}^{N-1}\sum_{v=0}^{N-1} \mathbf{F}(u,v)\,\varphi^{(u,v)}(m,n) \tag{27}$$

for $m = 0,...,N-1$, $n = 0,...,N-1$. Multiplying both sides by $\mathbf{F}^*(u',v')$, taking expectation, and making use of (26), we obtain

$$E\{\mathbf{f}(m,n)\,\mathbf{F}^*(u,v)\} = E\{|\mathbf{F}(u',v')|^2\}\,\varphi^{(u',v')}(m,n) \tag{28}$$

On the other hand, from (24), we have

$$\mathbf{F}^*(u',v') = \sum_{p=0}^{N-1}\sum_{q=0}^{N-1} \mathbf{f}(p,q)\,\varphi^{(u',v')}(p,q) \tag{29}$$

where we have assumed the random field [**f**] to be real (the footnote on p. 112 also applies here). Multiplying both sides of (29) by $\mathbf{f}(m, n)$, taking the expectation, and using (21), we obtain

$$E\{\mathbf{f}(m, n)\,\mathbf{F}^*(u', v')\} = \sum_{p=0}^{N-1} \sum_{q=0}^{N-1} R(m, n, p, q)\,\varphi^{(u', v')}(p, q) \qquad (30)$$

Comparing (28) with (30), we see that the basis matrices $[\varphi^{(u,\,v)}]$ must satisfy (22) and (23), which proves the theorem. ∎

We thus see that the basis matrices $[\varphi^{(u,\,v)}]$ that yield an uncorrelated representation of a picture matrix [**f**] are a solution of Eq. (22). By a slight change in notation, (22) can be put in a more convenient form for the purpose of solution on a digital computer. The trick is to represent a two-dimensional array of numbers by a one-dimensional string of numbers by replacing indices of matrix elements by a single index as shown in Table 1. With this change in notation, the picture matrix [**f**] will become a one-dimensional string denoted by a vector $\vec{\mathbf{f}}$ as

$$
\begin{bmatrix}
\mathbf{f}(0, 0) & \mathbf{f}(0, 1) & \cdots & \mathbf{f}(0, N-1) \\
\vdots & \vdots & \vdots & \vdots \\
\mathbf{f}(N-1, 0) & \mathbf{f}(N-1, 1) & \cdots & \mathbf{f}(N-1, N-1)
\end{bmatrix}
\Rightarrow
\begin{bmatrix}
\mathbf{f}_0 \\
\mathbf{f}_1 \\
\vdots \\
\mathbf{f}_{N-1} \\
\mathbf{f}_N \\
\mathbf{f}_{N+1} \\
\vdots \\
\mathbf{f}_{N^2-1}
\end{bmatrix}
$$

and the orthonormal matrices can also be represented by equivalent one-dimensional strings as

$$
\begin{bmatrix}
\varphi^{(u, v)}(0, 1) & \cdots & \varphi^{(u, v)}(0, N-1) \\
\vdots & \vdots & \vdots \\
\varphi^{(u, v)}(N-1, 0) & \cdots & \varphi^{(u, v)}(N-1, N-1)
\end{bmatrix}
\Rightarrow
\begin{bmatrix}
\varphi_0^s \\
\vdots \\
\varphi_{N-1}^s \\
\varphi_N^s \\
\vdots \\
\varphi_{N^2-1}^s
\end{bmatrix}
$$

where s is the equivalent index in the new notation of the double index (u, v). In the new notation, $R(m, n, p, q)$ will become an $N^2 \times N^2$ matrix $R(i, j)$, where i is the equivalent index of (m, n) and j the equivalent of (p, q). In order to avoid confusion between this $R(i, j)$ and the two-dimensional autocorrelation

TABLE 1

Double indices of matrix elements	Equivalent single index
$0,0$	0
$0,1$	1
$0,2$	2
$\vdots$	$\vdots$
$0, N-1$	$N-1$
$1,0$	N
$1,1$	$N+1$
$\vdots$	$\vdots$
$1, N-1$	$2N-1$
$2,0$	$2N$
$2,1$	$2N+1$
$\vdots$	$\vdots$
$N-1, N-2$	N^2-2
$N-1, N-1$	N^2-1

function $R(\alpha, \beta)$ of a homogeneous random field, we will denote $R(i, j)$ as obtained from $R(m, n, p, q)$ (by change in index notation) by $K(i, j)$. Evidently,

$$K(i, j) = E\{\mathbf{f}_i \mathbf{f}_j\}, \qquad i = 0, 1, ..., N^2 - 1, \quad j = 0, 1, ..., N^2 - 1 \quad (31)$$

where $\mathbf{f}_i$ and $\mathbf{f}_j$ are the ith and the jth components, respectively, of the picture matrix $[\mathbf{f}]$ in its string representation.

It is easy to see that in this notation (22) can be written as

$$\sum_{j=0}^{N^2-1} K(i, j)\, \varphi_j^s = \gamma_s\, \varphi_i^s; \qquad i = 0, 1, ..., N^2 - 1 \quad (32)$$

or equivalently,

$$
\begin{bmatrix}
K(0,0) & K(0,1) & \cdots & K(0, N-1) \\
K(1,0) & K(1,1) & \cdots & K(1, N-1) \\
\vdots & \vdots & \cdots & \vdots \\
K(N-1,0) & K(N-1,1) & \cdots & K(N-1, N-1)
\end{bmatrix}
\begin{bmatrix}
\varphi_0^s \\
\varphi_1^s \\
\vdots \\
\varphi_{N^2-1}^s
\end{bmatrix}
= \gamma_s
\begin{bmatrix}
\varphi_0^s \\
\varphi_1^s \\
\vdots \\
\varphi_{N^2-1}^s
\end{bmatrix}
$$

$$(33)$$

Therefore, each vector $\vec{\varphi}^s = [\varphi_0^s, \varphi_1^s, ..., \varphi_{N^2-1}^s]$ is transformed into a scalar multiple of itself when it is multiplied by the matrix $[K]$. The vectors $\vec{\varphi}^s$, $s = 0, 1, ..., N^2 - 1$ are called the *eigenvectors* of the matrix $[K]$. To each eigenvector φ^s there corresponds an eigenvalue γ_s. Techniques for obtaining eigenvectors and eigenvalues of a matrix are well known in the theory of linear transformations; see, for example, the book by Carnahan *et al.* [8]. These methods require inverting the matrix $[K]$ which is $N^2 \times N^2$. Even for small N, this requires a large number of computations.

5.1.3   Karhunen–Loève Compression: Application to Pictures

We will now present an example of picture compression using the concepts previously discussed. For this example we will use the picture in Fig. 1. This picture is a 256×256 matrix each element of which is an 8-bit number. Therefore, the total number of bits required to represent the picture is $2^8 \times 2^8 \times 2^3 = 2^{19}$. Let us partition this picture into 256 16×16 subpictures for convenience in numerical implementation (Fig. 2). Each subpicture may now be transformed into uncorrelated coefficients by first finding the eigenvectors and the corresponding eigenvalues of the $[K]$ matrix, which is 256×256. Note that for a homogeneous random field $[K]$ would be the same for each subpicture. Even for subpictures of this small size, finding the eigenvectors and the eigenvalues of the $[K]$ matrix is computationally expensive since it involves inverting a 256×256 matrix. For the example under discussion, a slightly different approach as now described was used.

The autocorrelation of Fig. 1 (in zero-mean form) can be modeled by (see Habibi and Wintz [22])

Fig. 1   The picture of the cameraman is a 256×256 matrix each element of which is an 8-bit number. (From [65].)

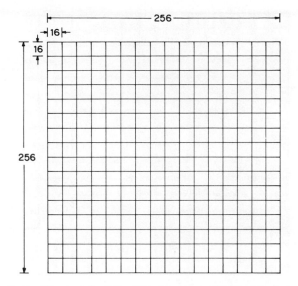

Fig. 2   Partitioning of a 256 × 256 picture into 256 16 × 16 subpictures.

$$R(x, y, x', y') \simeq \exp[-0.125(x-x') - 0.249(y-y')] \qquad (34)$$

In Section 4.2.4 we indicated that the analytical solutions of (3) for this auto-correlation function are known. Let these solutions for an aribrtary sub-picture (which we take to be centered at the origin) be $\varphi_{uv}(x, y)$, where u and v take positive integer values. These solutions may be ranked either on the basis of eigenvalues as discussed in Section 4.2.4 or on the basis of sequency. The sequency of a zero-mean picture in any given direction is defined as half the average number of sign changes per unit distance in that direction.

If each of the orthonormal functions $\varphi_{uv}(x, y)$ is sampled on a 16 × 16 grid and the resulting matrix denoted by $[\varphi^{(u, v)}]$, it is clear that we will have a set of matrices that are "approximately" orthonormal, i.e., (18) will not hold exactly. The finer the sampling of the orthonormal functions $\varphi_{uv}(x, y)$, the more accurately orthonormal the corresponding matrices $[\varphi^{(u, v)}]$ will be. In Fig. 3 we have shown 256 sequency ordered 16 × 16 matrices $[\varphi^{(u, v)}]$ for the autocorrelation function of (34). Note that since the matrices are sequency ordered, the matrix $[\varphi^{(0, 0)}]$, which is at the top left corner in Fig. 3, has no variations; the matrix $[\varphi^{(0, 1)}]$, which is second from the left in the top row, has no sign changes in one direction but has one sign change in the other; and so on.

Once the matrices $[\varphi^{(u, v)}]$ are determined, the uncorrelated coefficients

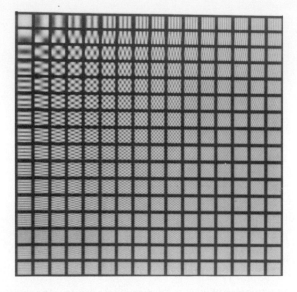

Fig. 3 The 256 16×16 sequency ordered basis pictures for the Karhunen–Loève
transformation. (From [65].)

$F(u,v)$ for each subpicture can be found by using (20b). Of course, the coefficients $F(u,v)$ will not be completely uncorrelated because the matrices
$[\varphi^{(u,v)}]$ are not exactly orthonormal; however, the correlations between them
should be much less than those between the 256 picture elements in each
16×16 subpicture. It follows from (20b) that for each subpicture there will
be 256 coefficients $F(u,v)$. The subpicture may be reconstructed from these
coefficients by using (20a).

Each $F(u,v)$ is a coefficient in the orthonormal expansion in (20a). Since,
as we see in Fig. 3, the matrices $[\varphi^{(u,v)}]$ with higher values of u and v have
more rapid variations in their elements, the coefficients for larger u and v
represent the higher spatial resolution effects in each 16×16 subpicture. It is
clear that if the spatial resolution capabilities of the receiver or the user of
these pictures are limited, we need retain only those $F(u,v)$ that are compatible
with this resolution—contributing, thereby, to data compression.

Suppose that we retain only the first 128 out of the 256 $F(u,v)$ for each
subpicture. These 128 coefficients are chosen on the basis of their ranking by
the eigenvalues γ_{uv}. (As mentioned earlier, the ranking achieved in this fashion
is roughly the same as that based on sequency considerations.) If we use (20a)
to reconstruct each subpicture from the 128 coefficients for that subpicture,
the overall reconstructed picture is as shown in Fig. 4a. We see that half the

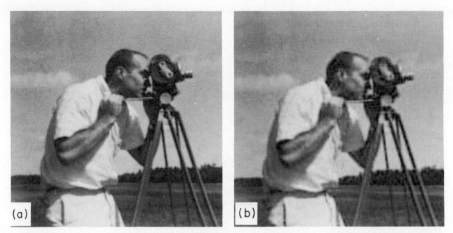

Fig. 4 (a) Reconstructed picture obtained by retaining the first 128 of the 256 Karhunen–Loève coefficients for each of the subpictures of Fig. 1. (b) Reconstructed picture obtained by retaining the first 64 of the 256 Karhunen–Loève coefficients for each of the subpictures. (From [65].)

coefficients for each subpicture can be disregarded with no visible degradation in picture quality. In the reconstruction in Fig. 4b where only 64 out of the 256 coefficients $F(u, v)$ for each subpicture were retained, the spatial resolution is lower. This is revealed by a careful examination of the camera and other edges in the picture.

After each subpicture has been transformed into uncorrelated coefficients, the coefficients must be quantized and binary code words assigned to the quantized coefficients. Before we can discuss quantization, however, the following property of the coefficients becomes important. Note that most of each subpicture is covered with slowly varying gray levels. Therefore, the largest contribution comes from those coefficients that represent slow variations. This means that when the coefficients $F(u, v)$ are sequency ranked, those with small u and v will have large magnitudes. This is also true when the coefficients $F(u, v)$ are ranked on the basis of the eigenvalues γ_{uv} of the corresponding orthonormal matrices $[\varphi^{(u, v)}]$. This is because by Eqs. (23) and (25), for a zero-mean picture the expected value of each $F(u, v)$ is zero, but the variance of $F(u, v)$ is equal to the eigenvalue γ_{uv}. This implies that if γ_{uv} is large for a given u and v, then the corresponding $F(u, v)$ generally has large magnitude. In Fig. 5 we have plotted the variance of the $F(u, v)$ as obtained by averaging each $|F(u, v)|^2$ over all the subpictures. In this figure, the coefficients have been ranked by the magnitudes of their associated eigenvalues. Note that the vertical scale is logarithmic.

Since the variances of the coefficients vary widely, as illustrated in Fig. 5,

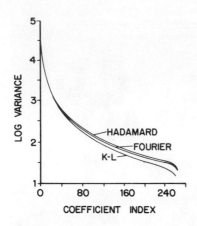

Fig. 5 The variance of the coefficients $\mathbf{F}(u,v)$ as obtained by averaging $|\mathbf{F}(u,v)^2|$ over all the subpictures of Fig. 1. Note that $E\{\mathbf{F}(u,v)\} = 0$ for zero-mean pictures. (From [65].)

it would be inefficient to use the same quantizer for all the coefficients. That is, if the quantizer output levels are adjusted to span the expected range of the coefficient with the largest variance, then the coefficients with smaller variances will fall in smaller ranges with the result that most of the quantizer levels will not be used. One can get around this difficulty by first normalizing each coefficient by dividing it by its standard deviation and then quantizing the normalized coefficients. Note that the normalized coefficients have unit variance.

Returning to the example under discussion, let us assume that we retain 128 out of 256 coefficients for each subpicture. If we normalize each coefficient by its variance and quantize the resulting normalized coefficients to 16 levels (4 bits), the total number of bits required to represent the picture is $256 \times 128 \times 4$ ($= 2^{17}$). Since we have 256×256 ($= 2^{16}$) picture elements in Fig. 1, the number of bits averages out to be 2 bits per picture element. Note that the direct representation of the picture, in which each picture element was coded separately, required 8 bits per picture element. Therefore, we have a compression ratio of 4. To examine the quality of the picture for this compression ratio, the 128 coefficients for each subpicture must be denormalized by multiplying them by their standard deviations. Each subpicture can then be reconstructed by using (20a) and inserted into its proper place in the reconstructed picture. The result is shown in Fig. 6a.

We can try to increase the compression ratio by quantizing each of the retained coefficients to only 4 levels (2 bits), yielding a compression ratio of 8 (1 bit per picture element). The resulting reconstruction is shown in Fig. 6b. Note the degradation in picture quality.

One can achieve a greater degree of compression for the same quality of reconstructed pictures by nonuniform quantization as discussed in Section 4.3.1. The optimal quantization scheme described there would minimize the

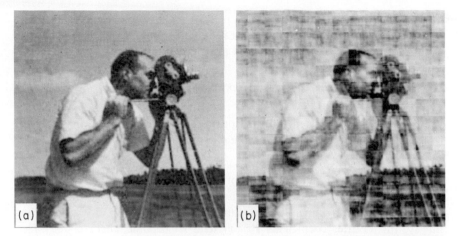

Fig. 6 (a) Reconstructed picture obtained by quantizing each of the 128 Karhunen–Loève coefficients of the picture in Fig. 4a to 16 levels (4 bits). Note that with this representation the picture shown here needs 2 bits per picture element on the average as compared to 8 bits per picture element in Fig. 1. (b) Same as (a), but quantizing to only 4 levels. (From [65].)

mean square quantization error for each coefficient taken singly; it would have to be applied to each coefficient individually. Quantization strategies for minimizing the total mean square quantization error for all the retained coefficients have been discussed by Huang and Schultheiss [32]. In the context of picture compression these methods have been briefly discussed by Wintz [65]. Even though these methods result in higher compression efficiencies, their disadvantages lie in the problems inherent in handling binary words of unequal length.

5.1.4   Fourier and Hadamard Compression

In the preceding section we indicated that a first step in picture compression is to represent a picture by uncorrelated data. To do this we presented the Karhunen–Loève transformations for both continuous and discrete pictures. The shortcomings of this method are primarily in the need for precise estimation of the covariance matrix for the random field and for determination of the eigenfunctions of the matrix, and the large number of operations required in (20a) and (20b). One can get around some of these difficulties by using transformations that may not necessarily be optimal. Fourier and Hadamard transforms, which fall in this category and which give performances not significantly different from Karhunen–Loève transforms, will now be discussed.

The Fourier transform of a discrete picture is defined in Section 2.2.1. If it is rewritten in forms (20a) and (20b), it is easy to show that the basis matrices $[\varphi^{(u,\,v)}]$ for the Fourier transformation are given by

$$\varphi^{(u,\,v)}(m,n) = \frac{1}{N}\exp\left[\frac{j2\pi}{N}(mu+nv)\right] \tag{35}$$

where the factor $1/N$ is needed for the purpose of the normalization condition in (18) and where m and n designate the mth column and nth row, respectively, of the basis matrix $[\varphi^{(u,\,v)}]$.

The Fourier transformation is suboptimum in the sense that the coefficients $F(u,v)$ are not uncorrelated. It can be shown, however, that it is asymptotically equivalent to the Karhunen–Loève transform [21]. As N in (35) approaches infinity, the Fourier coefficients $F(u,v)$ tend to become uncorrelated. This indicates that if the picture size is greater than the distances over which significant correlations between picture elements exist, the performance of the Fourier transform in a picture compression scheme should not be very different from that of the Karhunen–Loève transform.

Among the advantages of Fourier compression are that: estimation of the covariance matrix is not necessary (one does not have to calculate the eigenmatrices); and the transformation can be implemented very rapidly by the use of fast Fourier transform algorithms.

For an example of Fourier compression, we will again consider the discrete picture in Fig. 1. As in the preceding case this picture was divided into 256 16×16 subpictures for convenience in numerical implementation. Upon Fourier transformation, each subpicture was transformed into 256 Fourier coefficients. The variance of each coefficient $F(u,v)$ was estimated by averaging $|F(u,v)|^2$ over all the subpictures. The 256 complex Fourier coefficients in each subpicture were ranked on the basis of their variances (Fig. 5) and only the highest ranking 128 were retained in each subpicture. Since the picture is real, of the original 256 complex coefficients, only 128 are distinct. It follows that of the 128 retained complex coefficients only 64 complex coefficients should be distinct. Of course, these 64 complex coefficients amount to representing a subpicture by 128 real coefficients.

When each subpicture is reconstructed from its 128 Fourier coefficients by using (20a) and (35), the result is as shown in Fig. 7a. If only 64 coefficients are retained for each subpicture, the reconstructed picture looks like Fig. 7b. Note that the reconstructions in Figs. 7a and 7b are similar to those in Figs. 4a and 4b, respectively, for the case of Karhunen–Loève compression. The next step is to normalize each of the retained coefficients by its variance (for the same reasons as before) and quantize the normalized coefficients. When a uniform quantizer is used and the output levels of the quantizer are rep-

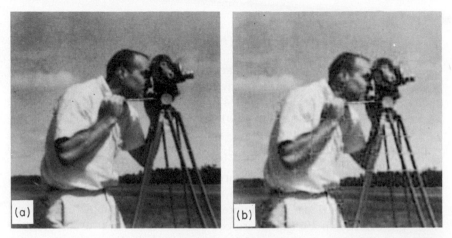

Fig. 7 (a) Reconstructed picture obtained by retaining the first 128 of the 256 Fourier coefficients for each of the subpictures of Fig. 1. (b) Same as (a), but using only 64 coefficients. (From [65].)

resented by equal length binary code words, the results are very similar to those shown in Figs. 6a and 6b.

For the Karhunen–Loève transformation, once the eigenmatrices are determined, the number of computer operations required to determine all the coefficients is proportional to N^4. For the Fourier transformation, on the other hand, if it is implemented by a fast Fourier transform algorithm, the number of computer operations is proportional to only $N^2 \log_2 N$. By a computer operation in both cases is meant a multiplication or a division and an addition.

From a computational standpoint, an even more efficient transform is the Hadamard transform described in Section 2.2.2. The Hadamard matrix is composed of $+1$'s and -1's only. Therefore, computation of a Hadamard transform does not require multiplications. For an $N \times N$ picture, the Hadamard transform coefficients can be calculated with a number of additions or subtractions proportional to $N^2 \log_2 N$. Of course, like the Fourier coefficients, the Hadamard transform coefficients are not completely uncorrelated, but they are more uncorrelated than the picture elements.

When the Hadamard transformation given by Eqs. (52) of Chapter 2 is expressed in the form described by Eqs. (20a) and (20b), the basis matrices are given by (see Pratt *et al.* [48])

$$\varphi^{(u,\,v)}(m,n) = (1/N)(-1)^{b(u,\,v,\,m,\,n)} \qquad (36)$$

for N a power of 2, where $1/N$ is needed for the normalization condition in

(18). Here

$$b(u, v, m, n) = \sum_{h=0}^{\log_2 N - 1} [b_h(u) b_h(v) + b_h(m) b_h(n)] \qquad (37)$$

where $b_h(\cdot)$ is the hth bit in the binary representation of $(\cdot)$. This result is valid only when the Hadamard matrices in Eqs. (52) of Chapter 2 are symmetric, as they are if Eqs. (46) and (47) there are employed in constructing them. When (36) is substituted in (20b), and if we use (17), we obtain

$$\mathbf{F}(u, v) = (1/N) \sum_{m=0}^{N-1} \sum_{n=0}^{N-1} \mathbf{f}(m, n) \cdot (-1)^{b(u, v, m, n)} \qquad (38)$$

In Fig. 8 we have shown 256 Hadamard basis matrices, $[\varphi^{(u, v)}]$, each 16×16, for $u = 0, 1, \ldots, 15$, and $v = 0, 1, \ldots, 15$.

As an example of Hadamard compression, each 16×16 subpicture of Fig. 1 was transformed into 256 Hadamard coefficients $\mathbf{F}(u, v)$. The variance of each of these coefficients was estimated by averaging each $|\mathbf{F}(u, v)|^2$ over all the subpictures. The coefficients in each subpicture were then ranked according to their variance (Fig. 5) and only the first 128 retained. When we reconstruct each subpicture from the retained 128 coefficients by using (20a) and (36), the result is as shown in Fig. 9a. As in the Karhunen–Loève and Fourier cases, deleting half the coefficients in each subpicture does not result in any visible degradation of the picture. If, however, we retain only the first 64

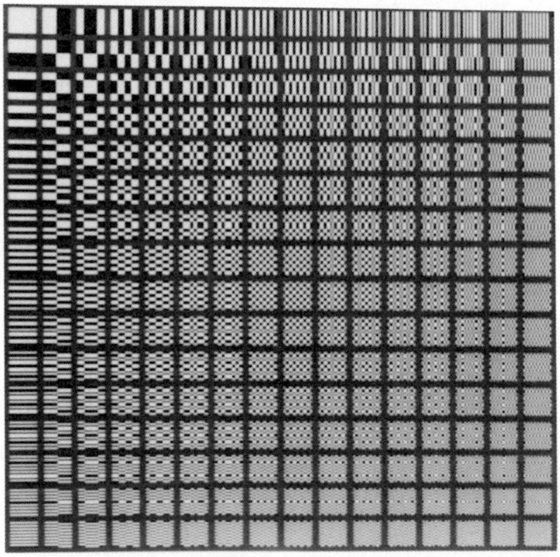

Fig. 8 The 256 16×16 sequency ordered basis pictures for the Hadamard transformation. (From [65].)

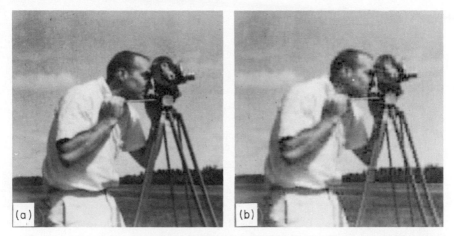

Fig. 9   (a) Reconstructed picture obtained by retaining the first 128 of the 256 Hadamard coefficients for each of the subpictures of Fig. 1. (b) Same as (a), but using only 64 coefficients. (From [65].)

coefficients for each subpicture, the reconstruction looks like Fig. 9b and a certain loss in spatial resolution is apparent. If in each of the two cases the retained Hadamard coefficients are normalized, quantized, and equal length binary code words are assigned to the output of the quantizer, then for the same quantization strategy the results are very similar to those in the cases of Karhunen–Loève and Fourier compression.

Other transformations that possess the same computational simplicity as the Fourier and the Hadamard transforms have been proposed recently. These include the slant transformation [9, 50] and the cosine transformation [2]. For the same number of coefficients, these transforms reportedly have a better mean square error performance than either the Fourier or the Hadamard transform.

5.2   PREDICTIVE COMPRESSION

As pointed out earlier, the first step in picture compression is to represent a picture by uncorrelated data. We proved in Section 5.1 that of all the linear orthogonal transformations, the Karhunen–Loève transformation achieves this result. If we do not limit ourselves to linear orthogonal transformations, there are other techniques that achieve the same result. One such technique that has the advantage of easy implementation is predictive compression, which we will now discuss.

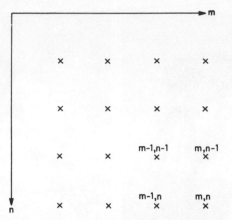

Fig. 10 The points $(m-1, n)$, $(m-1, n-1)$, and $(m, n-1)$ are used to form the estimate at the point (m, n). Note the direction of increasing m and n.

Let a digitized picture be represented by the matrix $[\mathbf{f}]$. Let $\mathbf{f}(m, n)$ be the element of this matrix that is in the mth column and the nth row (Fig. 10). Let $R(m, n, p, q)$ be the autocorrelation function of the real random field to which the picture belongs, i.e.,

$$R(m, n, p, q) = E\{\mathbf{f}(m, n)\mathbf{f}(p, q)\} \tag{39}$$

Because the picture elements $\mathbf{f}(m, n)$ are correlated, it is possible to derive an estimate or prediction $\hat{\mathbf{f}}(m, n)$ for a given element $\mathbf{f}(m, n)$ in terms of the rest of the picture elements. The difference $\mathbf{f}(m, n) - \hat{\mathbf{f}}(m, n)$ is the estimation error $\mathbf{e}(m, n)$ for that picture element and will be called the *differential signal*. Since the prediction $\hat{\mathbf{f}}(m, n)$ is directly dependent on the correlations between $\mathbf{f}(m, n)$ and the rest of the picture, it is reasonable that the sequence of random variables formed by the differential signal $\mathbf{e}(m, n)$ should be less correlated than the elements in the original picture.

It is known that the best estimate $\hat{\mathbf{f}}(m, n)$ (best in the sense that it minimizes the mean square estimation error) is in general a nonlinear function of the picture elements that are used to form the estimate [46]. Often, for reasons of mathematical tractability, the additional constraint of linearity is imposed on the form of the estimate. In such cases the estimate obtained is the best *linear* estimate.

We will further assume that as the picture is scanned row by row from top to bottom, only the three nearest neighboring elements that have already been scanned are used to form the linear estimate for a picture element $\mathbf{f}(m, n)$ (Fig. 10). In other words, $\hat{\mathbf{f}}(m, n)$ is of the following form:

$$\hat{\mathbf{f}}(m, n) = a_1 \mathbf{f}(m-1, n) + a_2 \mathbf{f}(m-1, n-1) + a_3 \mathbf{f}(m, n-1) \tag{40}$$

where the unknowns a_1, a_2, and a_3 are such that the mean square estimation error

$$E\{\mathbf{f}(m,n) - \hat{\mathbf{f}}(m,n)^2\} \qquad (41)$$

is minimized. Substituting (40) in (41), differentiating the resulting expression with respect to a_1, a_2, and a_3 separately, equating each derivative to zero, and using (39), we obtain the following three equations:

$$a_1 R(m-1,n,m-1,n) + a_2 R(m-1,n-1,m-1,n)$$
$$+ a_3 R(m,n-1,m-1,n) = R(m,n,m-1,n)$$

$$a_1 R(m-1,n,m-1,n-1) + a_2 R(m-1,n-1,m-1,n-1)$$
$$+ a_3 R(m,n-1,m-1,n-1) = R(m,n,m-1,n-1) \qquad (42)$$

$$a_1 R(m-1,n,m,n-1) + a_2 R(m-1,n-1,m,n-1)$$
$$+ a_3 R(m,n-1,m,n-1) = R(m,n,m,n-1)$$

which can be solved for a_1, a_2, and a_3. If we further assume that the random field $\mathbf{f}$ is homogeneous, has zero mean, and has autocorrelation function given by Eq. (47) of Chapter 4 (with $\eta = 0$), these equations reduce to

$$a_1 R(0,0) + a_2 R(0,1) + a_3 R(1,1) = R(1,0)$$
$$a_1 R(0,1) + a_2 R(0,0) + a_3 R(1,0) = R(1,1) \qquad (43)$$
$$a_1 R(1,1) + a_2 R(1,0) + a_3 R(0,0) = R(0,1)$$

where

$$R(\alpha,\beta) = R(0,0) \exp(-c_1|\alpha| - c_2|\beta|) \qquad (44)$$

Note that for this autocorrelation function $R(1,1) = R(1,0)R(0,1)/R(0,0)$. Using this relationship and solving (43) for a_1, a_2, and a_3, we obtain

$$a_1 = R(1,0)/R(0,0), \qquad a_2 = -R(1,1)/R(0,0), \qquad a_3 = R(0,1)/R(0,0) \qquad (45)$$

The differential signal $\mathbf{e}(m,n)$ at each picture element is given by

$$\mathbf{e}(m,n) = \mathbf{f}(m,n) - \hat{\mathbf{f}}(m,n)$$
$$= \mathbf{f}(m,n) - [a_1 \mathbf{f}(m-1,n) + a_2 \mathbf{f}(m-1,n-1) + a_3 \mathbf{f}(m,n-1)] \qquad (46)$$

It is clear that if the random field $\mathbf{f}$ has zero mean, then

$$E\{\mathbf{e}(m,n)\} = 0 \qquad (47)$$

Therefore, the variance of the differential signal at every point is given by

$$E\{\mathbf{e}^2(m,n)\} = E\{[\mathbf{f}(m,n) - (a_1 \mathbf{f}(m-1,n) + a_2 \mathbf{f}(m-1,n-1)$$
$$+ a_3 \mathbf{f}(m,n-1))]^2\} \qquad (48)$$

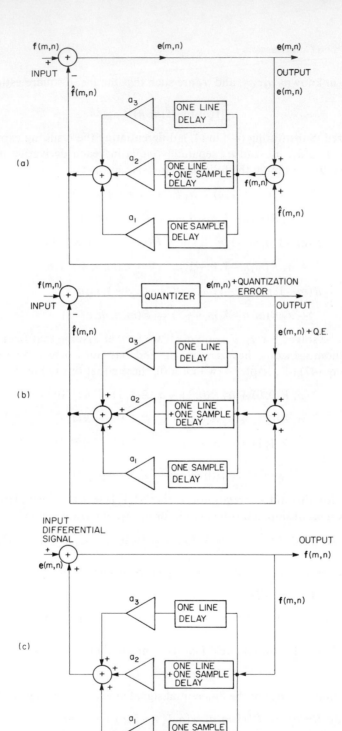

130

By expanding the right-hand side of (48) and making use of (43), it can be shown that

$$E\{e^2(m,n)\} = R(0,0) - [a_1 R(1,0) + a_2 R(1,1) + a_3 R(0,1)] \qquad (49)$$

Since f is homogeneous, the variance of each element is the same, namely $R(0,0)$. Moreover (44) implies that $R(1,0)$, $R(0,1)$, and $R(1,1)$ cannot be greater than $R(0,0)$. By substituting the values of a_1, a_2, and a_3 from (45) in (49), it is thus seen that the variance of the differential data $e(m,n)$ is less than the variance of the picture elements $f(m,n)$.

By substituting the values of a_1, a_2, and a_3 from (45) in (46) it can be shown that for zero-mean random fields with autocorrelation function given by (44), the following result holds:

$$E\{e(m,n)e(p,q)\} = 0, \qquad m \neq p \quad \text{or} \quad n \neq q \qquad (50)$$

i.e., the differential data $e(m,n)$ are uncorrelated.

It is clear that if the gray levels in a picture are known on the topmost row and the leftmost column (Fig. 10), the entire picture can be reconstructed from the differential data $e(m,n)$.

Figure 11a shows a system that generates the differential data $e(m,n)$ from the digitized picture. The picture is scanned row by row and the picture elements introduced serially into the adder at top left in the figure. Since in practice one would like to have the differential data quantized, in Fig. 11b we have shown the system with a quantizer in the feedback loop. The output of the system is now equal to e_{mn} + quantization error. In Fig. 11c we have shown the decoder which reconverts the differential data back into the picture. Of course, the reconstructed picture is now corrupted by the effects of the noise introduced by the quantizer in Fig. 11b. At the end of this section we briefly discuss some aspects of the distortions caused by the quantizer.

It should be evident in a qualitative sense that the differential data for a picture represent a "smaller amount of information" than the gray levels in the original digitized picture. This is because pictures generally contain areas of almost constant or slowly varying gray levels, and in these regions $e(m,n)$ is very small. When these differences are small enough to be undetectable by the user of the pictures, they may be approximated by zero. In fact that is one of the results achieved by the quantizer which we will discuss later in the section.

If the autocorrelation function cannot be modeled by (44), then one may need a larger number of picture elements for estimation in order for the

Fig. 11 (a) Block diagram of a system that generates the differential signal $e(m,n)$. (b) System in Fig. 11a with a quantizer in the feedback loop. (c) Block diagram of a system that reconstructs the picture from the differential data.

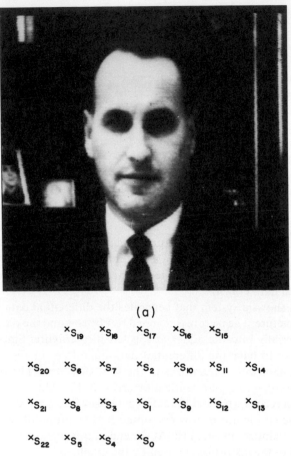

(a)

$^{x}S_{19}$ $^{x}S_{18}$ $^{x}S_{17}$ $^{x}S_{16}$ $^{x}S_{15}$

$^{x}S_{20}$ $^{x}S_{6}$ $^{x}S_{7}$ $^{x}S_{2}$ $^{x}S_{10}$ $^{x}S_{11}$ $^{x}S_{14}$

$^{x}S_{21}$ $^{x}S_{8}$ $^{x}S_{3}$ $^{x}S_{1}$ $^{x}S_{9}$ $^{x}S_{12}$ $^{x}S_{13}$

$^{x}S_{22}$ $^{x}S_{5}$ $^{x}S_{4}$ $^{x}S_{0}$

(b)

Fig. 12 (a) Picture used for obtaining the results shown in Figs. 13–15. (From Habibi [23].) (b) As the picture is scanned row by row, S_0 represents the current sample, S_1 the sample one line above, and so on as shown in the figure.

differential data to be uncorrelated completely. In practice, however, the number of picture elements required for considerable decorrelation is usually very small. As an example consider the picture in Fig. 12a. The picture was sampled on a 208×250 matrix. Suppose the picture is scanned line by line starting at the top. At a given instant in time, let S_0 represent the current sample, S_1 the sample one line above, and so on as shown in Fig. 12b. Let $R_s(i, j)$ denote the expectation $E\{S_i S_j\}$. We have used the subscript s on $R_s(i, j)$ to avoid confusion with the autocorrelation $R(\alpha, \beta)$ of a homogeneous random field.

A linear estimate $\hat{S}_0$ of the current sample S_0 in terms of the samples $S_1, S_2, ..., S_n$ has the form

$$\hat{S}_0 = a_1 S_1 + a_2 S_2 + \cdots + a_n S_n \tag{51}$$

where $a_1, a_2, ..., a_n$ are such that the mean square estimation error $E\{(S_0 - \hat{S}_0)^2\}$ is minimized. By using steps identical to those employed in arriving at (42) it can be shown that the mean square estimation error is minimized provided $a_1, a_2, ..., a_n$ are a solution of the following n simultaneous equations:

$$R_s(0, i) = \sum_{j=1}^{n} a_j R_s(i, j), \qquad i = 1, 2, ..., n \tag{52}$$

Table 2 shows the values of $R_s(0, i)$ as obtained by spatially averaging the products $S_0 S_i$ over the entire picture, after the digitized picture in Fig. 12a is normalized by first subtracting the mean from each picture element and then dividing by the standard deviation, which makes $R_s(0, 0) = 1$. If ergodicity conditions were satisfied, the $R_s(0, i)$, as calculated by this spatial averaging, would equal the ensemble averages.

TABLE 2

Correlations between S_0 and S_i for the picture in Fig. 12a [23]

Correlations	Interlaced	Noninterlaced
$R_s(0, 1)$	0.9771	0.9866
$R_s(0, 2)$	0.9534	0.9771
$R_s(0, 3)$	0.9479	0.9570
$R_s(0, 4)$	0.9648	0.9648
$R_s(0, 5)$	0.9026	0.9026
$R_s(0, 6)$	0.8679	0.8901
$R_s(0, 7)$	0.9197	0.9479
$R_s(0, 8)$	0.8901	0.8981
$R_s(0, 9)$	0.9483	0.9573
$R_s(0, 10)$	0.9193	0.9483
$R_s(0, 11)$	0.7569	0.8927
$R_s(0, 12)$	0.7120	0.8997
$R_s(0, 13)$	0.8416	0.8463
$R_s(0, 14)$	0.8254	0.8416
$R_s(0, 15)$	0.8421	0.8828
$R_s(0, 16)$	0.8866	0.9338
$R_s(0, 17)$	0.9079	0.9591
$R_s(0, 18)$	0.8878	0.9337
$R_s(0, 19)$	0.8406	0.8800
$R_s(0, 20)$	0.8186	0.8371
$R_s(0, 21)$	0.8371	0.8439
$R_s(0, 22)$	0.8478	0.8478

TABLE 3

A List of Numerical Values of the Coefficients a_i Used for Each Predictor [23]

Predictor:	1st Order horizontal	1st Order vertical	2nd Order	3rd Order	4th Order	6th Order	12th Order	18th Order	22nd Order
a_1		0.977	0.617	0.826	0.897	0.855	0.664	0.648	0.630
a_2					−0.059	−0.057	−0.073	−0.063	−0.081
a_3	0.965			−0.594	−0.574	−0.534	−0.460	−0.446	−0.429
a_4			0.379	0.746	0.729	0.827	0.834	0.831	0.865
a_5						−0.108	−0.135	−0.131	−0.261
a_6						0.006	0.002	0.008	0.006
a_7							0.021	0.037	0.035
a_8							−0.001	−0.011	0.139
a_9							0.117	0.142	0.002
a_{10}							0.051	−0.005	−0.124
a_{11}							−0.125	−0.107	0.069
a_{12}							0.096	0.048	0.029
a_{13}								0.041	0.006
a_{14}								0.001	−0.049
a_{15}								−0.057	0.123
a_{16}								0.112	−0.076
a_{17}								−0.040	0.071
a_{18}								−0.014	−0.086
a_{19}									0.061
a_{20}									−0.086
a_{21}									0.129
a_{22}									0.016

If the random field is homogeneous, the other $R_s(i,j)$'s can be simply determined from $R_s(0,i)$'s. For example, from the definition of homogeneous random fields (Section 2.4.3) and Fig. 12b, it is clear that $R_s(1,7)$ must be equal to $R_s(0,3)$. Thus from the information in Table 2, one can compute the constants a_i in (52).

The predictive compression system will be said to employ a first-order horizontal predictor if only the picture element S_4 is used for estimation in (51). This means that the summation on the right-hand side in (51) has only one term $a_4 S_4$. The predictor will be called first-order vertical if only S_1 is

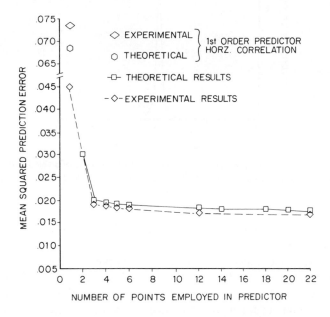

Fig. 13 Mean square estimation error versus the order of the predictor. (From [23].) For the first-order predictor using horizontal correlation the experimental ($\Diamond$) and theoretical ($\bigcirc$) points are also shown.

used to estimate S_0. In a second-order predictor S_1 and S_4 are used for estimation, while in a third-order predictor S_1, S_3, and S_4 are used for estimation. For $n > 3$, the nth order predictor uses $S_1, S_2, ..., S_n$ for estimation.

Table 3 lists the numerical values of the constants a_i that each predictor employs. These values were obtained by solving (52) for each predictor. Figure 13 shows the mean square estimation error obtained for each predictor by averaging $e^2(m,n)$ over all the picture elements in Fig. 12a. These curves show a significant reduction in estimation error when the number of points used in prediction is increased to 3, and no significant reduction past this

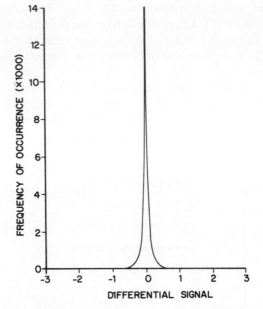

Fig. 14 Histogram of the differential signal $e(m, n)$ for the case of a third-order predictor for the picture in Fig. 12a after it is normalized (zero mean and unit variance). (From [23].)

point. Of course, the results in Table 3 and Fig. 13 are valid only for the picture in Fig. 12a.

Figure 14 shows the histogram of the differential signal $e(m, n)$ in the case of a third-order predictor for the picture in Fig. 12a after it is normalized (zero mean and unit variance). Expect for a normalization factor, this histogram is an estimate of the density function of the differential data. With such a highly peaked density function, it is clear by the arguments presented in Section 4.3.1 that the differential data would be best quantized by a nonuniform quantizer. Given the histogram in Fig. 14, the output levels of this nonuniform quantizer can be calculated using the formulas in Section 4.3.1.

Nonuniform quantization of the differential data can also be justified on the basis of properties of the human visual system. In Section 4.3.2 we indicated that in a slowly changing region it is important to have fine quantization. The differential signals will be small in such regions, so it is necessary that the quantization levels be more densely packed. This is achieved by the nonuniform quantizer.

The result of 3-bit compression of the picture in Fig. 1, using the concepts previously presented, is shown in Fig. 15a. (In this and the other results in Fig. 15, only the first-order predictor using the nearest horizontal neighbor

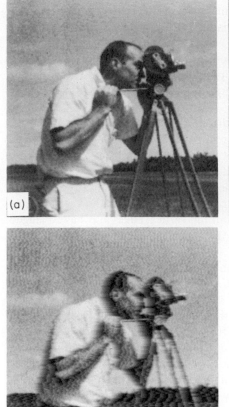

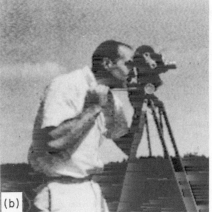

Fig. 15 (a) Result of 3-bit predictive compression of Fig. 1. The first-order predictor was used and the differential data quantized to 8 levels (3 bits). Note that this picture needs 3 bits per picture element, as compared to 8 bits per picture element in Fig. 1. (b) Reconstruction for 2-bit predictive compression. (c) Reconstruction for 1-bit predictive compression.

on the left was used.) By 3-bit predictive compression we mean that the number of bits required for the reconstruction in Fig. 15a is on the average 3 bits per picture element. Equal length binary code words were used for the output levels of the nonuniform quantizer so that for a 3-bit predictive compression system the quantizer has 8 output levels. Figures 15b and 15c, respectively, show the reconstructions for 2-bit and 1-bit predictive compression systems, again using equal length binary code words for the output levels of the quantizer.

The degree of compression achieved by a predictive compression system can be increased by using variable-length binary code words for the output levels of the quantizer. This is because the probabilities of occurrence of the various output levels of the quantizer are not the same, and the levels which occur more often can be assigned shorter code words. These variable length codes can be derived on the basis of information-theoretic concepts and will not be discussed here. In picture transmission, however, their use requires buffer storage.

We conclude this section by briefly describing several types of visual impairments in a picture that are unique to predictive compression. These include slope overload noise, edge busyness, granularity noise, and contouring noise [5, 42].

At sharp transitions in a picture, the error $e(m, n)$ is usually very large. If it is larger than the greatest quantizer step, the system is said to be in "slope overload." The result is that an abrupt transition may appear smeared, since when the system gets into slope overload, it may take several samples to recover. A nonuniform quantizer may be designed with large outer steps to force quick recovery. If, however, the quantizer structure is too coarse and the sampling is not synchronized with horizontal scanning, vertical edges may have a jagged appearance and may flicker and look "busy". This is known as "edge busyness."

Granularity noise and contouring noise are associated with flat regions or regions with gently sloping gray levels. (As mentioned in Section 4.3.2, the visual detectability of noise is greatest in such a region.) Granularity noise is caused in flat regions by the output of the qunatizer oscillating back and forth between the smallest output levels. This oscillation depends on the picture noise in the regions and the past history of the predictor. In gently sloping regions these random oscillations do not occur. Instead, the quantizer output continues to increase in uniform steps in the direction of the slope. If, however, the sampling interval multiplied by any one of the quantizer steps is not equal to the rate of change of brightness level, then periodic corrections become necessary. If the step used to make these corrections is too large, the resulting overshoots give rise to visible contouring patterns called "contouring noise."

5.3   MORE ON COMPRESSION TECHNIQUES

Both the transform and the predictive compression techniques have advantages and disadvantages. If the aim of picture compression is to reduce the data transfer rate in a transmission system, then the susceptibility of a compression technique to channel noise becomes an important consideration. The transform compression techniques are less vulnerable to channel noise than the predictive compression techniques. This is because if a transform coefficient is incorrectly received at the receiving end, its effect on the reconstructed picture is distributed all over the picture, making it less objectionable from a human visual standpoint. On the other hand, in a predictive compression technique, erroneous reception of an element of the differential data will result in incorrect reproduction of a picture element, and because of the predictor at the receiver, this error will propagate to neighboring picture elements [5]. This creates quite an unpleasant effect in the reconstructed picture. Other advantages of transform compression techniques include their lesser sensitivity to variations from one picture to another and their superior coding performance at lower bit rates. One of the major advantages of predictive compression techniques is the ease and the economy with which they can be implemented in hardware. Recently there have been attempts at combining the attractive features of both transform and predictive compression techniques [25, 26].

We will now briefly mention some other methods that have been suggested for picture compression. For details see the references that are cited.

For pictures in which the number of possible gray levels is small and which are composed of only a few regions each having a constant gray level, an efficient method of compression is contour coding. This involves tracing the contours or boundaries between the constant gray level regions and sending only that information to the receiver which would enable it to reproduce these contours. In the context of pciture compression, Wilkins and Wintz [63] give an algorithm for tracing these contours.

Another method of compression for pictures composed of few regions each having constant or slowly varying gray level is run-length coding [10, 18, 54]. Here raster scanning of the picture followed by quantization will give rise to a relatively small number of "runs" of constant gray level, and the picture can be encoded by specifying the lengths (or, equivalently, the positions) of these runs.

Another method of picture compression separates a picture into "highs" and "lows" [19, 57]. The "lows" picture is obtained by low-pass spatial filtering of the picture. This results in essentially an out-of-focus picture with no sharp edges. By the two-dimensional sampling theorem in Section 4.1.3, this "lows" picture can be represented by many fewer samples than would have been needed for the original picture. The "highs" signal is obtained by taking either the gradient or the Laplacian of the picture (Sections 6.4.1 and 6.4.2) and consists of essentially the edges in the picture. A two-dimensional high-frequency picture (also called the "synthetic highs") can be synthesized from this edge information. This high-frequency information, when combined with the "lows" picture, gives back essentially the original picture. The "highs" picture may be efficiently transmitted by contour coding the edge information [19].

Roberts [53] has suggested a pseudorandom noise modulation technique for picture compression. If a continuous picture is sampled at an array of points, the samples usually need to be quantized to between 16 and 256 gray levels depending on the requirements of the user of the picture. For example, the digitized picture in Fig. 1 was sampled at a 256×256 array of points and each sample represented by 8 bits (256 levels of gray). If an attempt is made to reduce the number of bits by making the quantization too coarse, the result is the appearance of artificial discontinuities (false contours) in the picture. These discontinuities are a result of the quantization noise, which is correlated with the picture samples. If uncorrelated random noise of the same rms value as the quantization noise is added to the original continuous picture, these discontinuities do not appear. Roberts used this observation in his pseudo-random noise modulation technique in which noise of uniform amplitude distribution, and peak-to-peak value equal to one quantum step, is added to the picture samples before quantization, and identical noise is subtracted at the receiver. The result looks like an unquantized output, in which random noise of the same rms value has replaced the quantization noise. Usable pictures are produced at 2 bits per picture element, fair ones at 3 bits, and good ones at 4 bits. Roberts' system also has excellent performance in the presence of channel noise.

5.4 THE RATE DISTORTION FUNCTION

An important question was raised at the beginning of this chapter: If the user or receiver of pictures belonging to a certain class is willing to accept an average amount of distortion D, then what is the least average number of bits required to represent pictures in that class? The answer to this question is, in principle, provided by the rate distortion function $R(D)$.

We must first decide just what we mean by distortion. The mean square error measure of distortion has been extensively used in the past primarily because of its mathematical tractability. The fact that this type of measure is a poor criterion is clear from an observation we made in the last section. We indicated there that if a picture is quantized too coarsely, artificial contours appear in the picture, but if uncorrelated noise of the same rms value as the quantization noise is added to the original picture, the objectionable artificial contours do not appear. This means that for a given amount of mean square error between a picture and its reconstruction, the quality of the reconstructed picture depends on other factors, such as the nature of the correlations between the errors and the gray levels, and the context of the errors.

Specification of a distortion measure that is both physically meaningful and analytically tractable constitutes one of the major difficulties in our ability to use the rate distortion function. Another source of difficulty is the fact that the computation of the rate-distortion function requires a detailed knowledge of the statistics of the random field to which the pictures belong. Even if these statistics could be determined in practice, their form may render the computation of any results a formidable if not an impossible task. Up to this time the performance achieved by various data compression systems has been compared with absolute bounds derived from rate distortion theory only in relatively simple cases [7, 13, 24–26, 34, 54, 58, 59]. Attempts are currently being made to extend the theory to the highly redundant, non-stationary sources and subtle distortion criteria often encountered in practice.

In spite of the previously mentioned limitations, rate-distortion theory does provide, from a conceptual standpoint, a theoretical framework that deals directly, in quantitative terms, with the concept of redundancy and its reduction to achieve data compression. For these reasons, we have included here a very brief introduction to rate-distortion theory, and have discussed its applications to pictorial data in conceptual terms. The theory is extensively discussed in the books by Berger [7], Gallager [16], and Jelinek [33]. We will first introduce some relevant information-theoretic concepts.

5.4.1   Some Information-Theoretic Concepts[†]

Assume that a source X can produce any of the m possible symbols belonging to a set $A_m = \{a_1, a_2, ..., a_m\}$. It is customary to call this set the "alphabet." Let the probability of the event that the source will produce a_j be $\mathscr{P}(a_j)$; then the $\mathscr{P}(a_j), j = 1, ..., M$, satisfy the condition

$$\sum_{j=1}^{m} \mathscr{P}(a_j) = 1 \qquad (53)$$

Let φ be a function defined over the alphabet. Suppose that the source operates over a long period of time, and for every a_i produced by the source, the corresponding $\varphi(a_i)$ is recorded. Then the mean or expected value of φ is given by (see Section 2.3.2)

$$E\{\varphi\} = \sum_{j=1}^{m} \varphi(a_j)\mathscr{P}(a_j) \qquad (54)$$

Suppose now that we are told that the source has produced the symbol a_i; we can then ask how much information $I(a_i)$ we have received.[†] Let us suppose that the amount of information associated with the symbol a_i is a continuous function of the probability $\mathscr{P}(a_i)$. Equivalently, we can assume that this information is a continuous function of the logarithm of $\mathscr{P}(a_i)$:

$$I(a_i) = h(\log \mathscr{P}(a_i)) \qquad (55)$$

Let us further assume that if two independent symbols a_i and a_j are both

---

[†] The symbol I was used in Section 2.4.1 to denote an interval. Here it is used as a measure of information. Also, the symbols a_i were used in Section 5.2 to denote prediction coefficient; here they represent the output states of an information source. We hope this will not cause any confusion.

produced, then the resulting information is the sum $I(a_i)+I(a_j)$. Now the probability of receiving both a_i and a_j, when they are independent, is $\mathscr{P}(a_i)\cdot\mathscr{P}(a_j)$; thus the log of this probability is

$$\log(\mathscr{P}(a_i)\cdot\mathscr{P}(a_j)) = \log(\mathscr{P}(a_i)) + \log(\mathscr{P}(a_j)) \tag{56}$$

We thus have

$$h\big(\log(\mathscr{P}(a_i))+\log(\mathscr{P}(a_j))\big) = I(a_i) + I(a_j)$$
$$= h(\log\mathscr{P}(a_i)) + h(\log\mathscr{P}(a_j)) \tag{57}$$

In other words, h is an additive function of $\log\mathscr{P}(a_i)$. Now it can be shown that the only continuous additive functions are the linear homogeneous functions, i.e., we have

$$h\big(\log(\mathscr{P}(a_i))\big) = c\cdot\log\mathscr{P}(a_i) \tag{58}$$

Here c should be a negative constant, since $\log(\mathscr{P}(a_i))$ is negative, and we naturally want I to be positive. Since changing the base to which the log is taken is equivalent to multiplication by a positive constant, we can set $c = -1$ without loss of generality, obtaining

$$I(a_i) = -\log\mathscr{P}(a_i) \tag{59}$$

This result can be further justified along the following lines. Suppose $\mathscr{P}(a_j) = 1$ for some particular value of j, which means the source will always produce only that a_j. Now if we are told that the source has produced that particular a_j, then we really have not received any information because we knew *a priori* what the source was going to produce. Equation (59) thus very correctly tells us that if $\mathscr{P}(a_j) = 1$, then $I(a_j) = 0$.

The base of the logarithm in (59) determines the information unit. The unit of information for base e logarithms is called a *nat*, while that for base 2 logarithms is called a *bit*.

The function $I(a_i)$ defined in (59) is called the *self-information*. The expected value of self-information is called the *entropy* of the source X and by (54) is given by

$$H(X) = -\sum_{j=1}^{m} \mathscr{P}(a_j)\log\mathscr{P}(a_j) \tag{60}$$

Clearly, $H(X)$ is a measure of the average amount of information one receives upon being told what symbol the source X has produced. Alternatively, $H(X)$ may be interpreted as a measure of the average *a priori* uncertainty regarding the source output.

Exercise 2.   What are the entropies in bits in the following cases?

(i) $\mathscr{P}(a_1) = \mathscr{P}(a_2) = \mathscr{P}(a_3) = \mathscr{P}(a_4) = \frac{1}{4}$;

(ii) $\mathscr{P}(a_1) = \frac{1}{2}$, $\mathscr{P}(a_2) = \frac{1}{4}$, $\mathscr{P}(a_3) = \frac{1}{8}$, $\mathscr{P}(a_4) = \frac{1}{8}$. ∎

Exercise 3. Prove that the entropy of the ensemble of four messages is greatest when $\mathscr{P}(a_1) = \mathscr{P}(a_2) = \mathscr{P}(a_3) = \mathscr{P}(a_4)$. ∎

Now suppose that the source outputs a_j are transmitted over a noisy channel. Let the output of the receiver Y be represented by the alphabet $B_n = \{b_1, b_2, ..., b_n\}$. One may define probabilities $\mathscr{P}(a_j, b_k)$ on the product space $A_m \times B_n$ consisting of all ordered pairs (a_j, b_k). Here $\mathscr{P}(a_j, b_k)$ is the probability that the source output is a_j while the receiver output is b_k. Associated with $\mathscr{P}(a_j, b_k)$ are the probabilities

$$\mathscr{P}(a_j) = \sum_{k=1}^{n} \mathscr{P}(a_j, b_k) \tag{61}$$

and

$$\mathscr{Q}(b_k) = \sum_{j=1}^{m} \mathscr{P}(a_j, b_k) \tag{62}$$

and the conditional probabilities

$$\mathscr{P}(a_j | b_k) = \mathscr{P}(a_j, b_k)/\mathscr{Q}(b_k) \tag{63}$$

$$\mathscr{Q}(b_k | a_j) = \mathscr{P}(a_j, b_k)/\mathscr{P}(a_j) \tag{64}$$

where $\mathscr{Q}(b_k)$ is the probability that the receiver output is b_k; it is analogous to $\mathscr{P}(a_j)$ at the source output. $\mathscr{P}(a_j | b_k)$ is the probability of the source output being a_j given that the receiver output is b_k, and $\mathscr{Q}(b_k | a_j)$ the probability of the receiver output being b_k given that the source output was a_j.

One can now define *conditional self-information* $I(\cdot | \cdot)$ as

$$I(a_j | b_k) = -\log \mathscr{P}(a_j | b_k) \tag{65}$$

The number $I(a_j | b_k)$ measures the information one receives upon being told that the source has produced a_j if one already knows that the receiver has produced b_k. The conditional self-information

$$I(b_k | a_j) = -\log \mathscr{Q}(b_k | a_j) \tag{66}$$

has a similar interpretation.

Another important, perhaps the most important, measure of information is the *mutual information*, which is specified by the relation

$$I(a_j; b_k) = I(a_j) - I(a_j | b_k) \tag{67}$$

Clearly, $I(a_j; b_k)$ is the difference between the amount of information that the occurrence of a_j at the source output conveys to someone who is ignorant of what is happening at the receiver, and that which it conveys to someone who already knows that b_k has occurred at the receiver output. By using (59)–(67), the mutual information may be expressed as

$$I(a_j; b_k) = \log \mathscr{P}(a_j \,|\, b_k)/\mathscr{P}(a_j) \tag{68}$$

$$= \log \mathscr{Q}(b_k \,|\, a_j)/\mathscr{Q}(b_k) \tag{69}$$

$$= \log \mathscr{P}(a_j, b_k)/\mathscr{P}(a_j)\,\mathscr{Q}(b_k) \tag{70}$$

Note that it follows from these equations that

$$I(a_j; b_k) = I(b_k; a_j) \tag{71}$$

Just as entropy was defined to be the average value of self-information, one can define the *conditional entropy* as the average value of conditional self-information. If $H(X\,|\,Y)$ and $H(Y\,|\,X)$ denote the conditional entropies for the source X and the receiver Y, then

$$H(X\,|\,Y) = -\sum_{j,k} \mathscr{P}(a_j, b_k) \log \mathscr{P}(a_j \,|\, b_k) \tag{72}$$

$$H(Y\,|\,X) = -\sum_{j,k} \mathscr{P}(a_j, b_k) \log \mathscr{Q}(b_k \,|\, a_j) \tag{73}$$

The conditional entropy $H(X\,|\,Y)$ is the average information conveyed about the source by a specified receiver output. $H(X\,|\,Y)$ can also be considered to be a measure of average uncertainty regarding which symbol the source has produced after the symbol produced by the receiver has been specified. $H(Y\,|\,X)$ has a similar interpretation.

The expected value of (67) (or, equivalently, (70)) is called the *average mutual information* between the source X and the receiver Y and is denoted by $I_A(X;Y)$. Thus

$$H(X;Y) = \sum_{j,k} \mathscr{P}(a_j, b_k) \log \frac{\mathscr{P}(a_j, b_k)}{\mathscr{P}(a_j)\,\mathscr{Q}(b_k)} \tag{74}$$

This expression for $H(X;Y)$ may also be written, using (60), (61), (63), and (72), as

$$H(X;Y) = H(X) - H(X\,|\,Y) \tag{75}$$

which indicates that the average mutual information may be considered to be the average *a priori* uncertainty about the source output minus the uncertainty that remains after the receiver output is specified.

It should be clear that the conditional probabilities $\mathscr{P}(a_j \,|\, b_k)$ completely specify the channel with respect to the behavior of the receiver and the source. It is also clear that $H(X\,|\,Y)$ is the average information (in bits or nats per letter) that is lost in going from the source to the receiver and, therefore, is related to the noise in the channel. In an ideal system, this would be zero and the average mutual information would then be equal to the average source information $H(X)$. In this case the information flowing from the source to the receiver is $H(X)$ bits or nats per source symbol. If $H(X|Y)$ is not zero, then the average information per symbol that is flowing from the source to the receiver must be equal to the average information per letter $H(X)$ that is being produced at the source minus the average information per symbol that is being lost in the channel as given by $H(X|Y)$. Hence, by (75), $H(X;Y)$ is really the rate of information transfer (in bits or nats per symbol) through the channel. From this discussion the following definition for the *channel capacity* C seems justified:

$$C = \max H(X;Y) \tag{76}$$

where the maximum is taken with respect to all possible choices of input distribution $\mathscr{P}(a_j)$.

5.4.2   The Rate Distortion Function

Let us assume that the information is produced by a discrete memoryless source. By this we mean that the successive symbols generated by the source are independent and identically distributed, i.e., $\mathscr{P}(a_j)$ does not depend on time. Next, we assume the channel to be a discrete memoryless channel, which means that the channel processes successive letters of an input word (i.e., an input sequence of symbols) independently of one another. Note that a channel is characterized by the set of conditional probabilities $\mathscr{Q}(b_k|a_j)$. We will denote this set by $\mathscr{Q}_C$. Then

$$\mathscr{Q}_C = \left\{ \mathscr{Q}(b_k|a_j) \,\middle|\, \begin{array}{l} k = 1, 2, ..., n \\ j = 1, 2, ..., m \end{array} \right\} \tag{77}$$

Now let us consider the problem of reconstructing the output of the source to within a certain accuracy at the receiving end of the channel. To determine whether or not the required accuracy has been achieved, we need a quantitative measure of the distortion that may exist between the source output and the channel output at the receiver. Let us assume that there is given for this purpose a nonnegative matrix $\rho(a_j, b_k)$ that specifies the penalty charged for reproducing the source symbol a_j by the symbol b_k. We will use ρ as our distortion measure. The expected value of the distortion depends on the $\mathscr{Q}(b_k|a_j)$, so we denote it by

$$\begin{aligned} d(\mathscr{Q}_C) &= \sum_{j,k} \rho(a_j, b_k) \mathscr{P}(a_j, b_k) \\ &= \sum_{j,k} \rho(a_j, b_k) \mathscr{P}(a_j) \mathscr{Q}(b_k|a_j) \end{aligned} \tag{78}$$

Here $d(\mathscr{Q}_C)$ is the average distortion associated with the channel. A channel is said to be *D-admissible* if $d(\mathscr{Q}_C) \leqslant D$ for all possible choices of the $\mathscr{P}(a_j)$. The set of all D-admissible conditional probability assignments is denoted by

$$\mathcal{Q}_D = \{\mathcal{Q}_C \mid d(\mathcal{Q}_C) \leqslant D\} \tag{79}$$

Each conditional probability assignment gives rise not only to an average distortion $d(\mathcal{Q}_C)$ but also to an average mutual information

$$H(X;Y) = \sum_{j,k} \mathscr{P}(a_j)\,\mathcal{Q}(b_k \mid a_j) \log \frac{\mathcal{Q}(b_k \mid a_j)}{\mathcal{Q}(b_k)} \tag{80}$$

We define the rate-distortion function $R(D)$ as

$$R(D) = \min_{\mathcal{Q}_C \in \mathcal{Q}_D} H(X;Y) \tag{81}$$

for the distortion measure $\rho(a_j, b_k)$. $R(D)$ is measured in nats (or bits) per source symbol.

Note that in order to determine the channel capacity, the mutual information is maximized with respect to the source distribution, while on the other hand the rate-distortion function requires minimization with respect to the channel conditional probabilities $\mathcal{Q}(b_k \mid a_j)$.

As was mentioned before, $H(X;Y)$ is the rate of information transfer (in bits or nats per source symbol) from the source to the receiver. Thus by (81), $R(D)$ is the minimum rate at which information about the source must be supplied to the user in order that the user may reproduce it with a prescribed fidelity. It is this interpretation of the rate-distortion function that makes it of considerable practical importance.

The rate-distortion function $R(D)$ can be shown to have the following properties:

1. $R(D)$ is not defined for $D < 0$.
2. There exists a $D_{\max} \geqslant 0$ given by

$$D_{\max} = \min_k \sum_j \mathscr{P}(a_j)\,\rho(a_j, b_k)$$

such that $R(D) = 0$ for $D \geqslant D_{\max}$.
3. $R(D)$ is positive, strictly decreasing, and continuous for $0 < D < D_{\max}$.
4. $R(D)$ is convex, that is, for all D' and D'' and all λ with $0 \leqslant \lambda \leqslant 1$ we have

$$R(\lambda D' + (1-\lambda) D'') \leqslant \lambda R(D') + (1-\lambda) R(D'').$$

Proofs of these properties can be found in Berger [7]. These properties indicate that an $R(D)$ curve in general must look like Fig. 16.

It should be emphasized that the concepts of the source, the channel, and the receiver in the preceding sections are purely formal and the existence of a communication link is not necessarily implied. For example, when we deal with a picture digitizer, the source can be identified with the original picture,

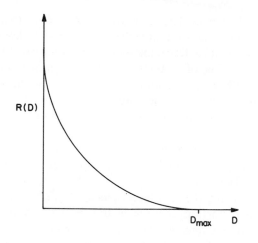

Fig. 16 A typical $R(D)$ curve.

the receiver with the user of the quantized samples, and the channel with the sampler and the quantizer. (Of course, in this case since the source is continuous, one would have to extend the rate distortion function to the continuous case.) In this case the rate distortion function gives the minimum rate at which information must be extracted from the original picture by the sampler–quantizer so that the user may reconstruct the picture within a prespecified distortion.

The rate-distortion function developed here does lend itself, at least conceptually, to the case of digitized pictures. Suppose that a picture has been "adequately" sampled and quantized. Let the picture be composed of $M \times N$ elements, the brightness level at each element taking on one of 2^B values. Given this picture, our aim is to represent it by the least number of bits such that one could reproduce the original within a certain prespecified distortion.

Following the excellent philosophical discussion by Landau and Slepian [36], one way to apply the rate-distortion function to the previously mentioned problem is to regard each symbol a_i of the set A_m as an entire picture. The set A_m is finite because there can only be $2^{B \times M \times N}$ pictures. The distortion measure $\rho(a_j, b_k)$ now expresses the penalty charged for reproducing the picture a_j as a_k. Conceptually such a measure exists, but we know little of it. We would have to prescribe such a measure for $2^{2B \times N \times M}$ pairs of pictures. To appreciate the magnitude of this number, consider the typical values $N = M = 256$, $B = 8$; $2^{B \times N \times M} = 2^{524288}$. Also, to compute the rate-distortion function, one would have to know the probability distribution over the $2^{B \times N \times M}$ pictures, since not all pictures in such an ensemble will occur equally often. For example, if we are dealing with television, then pictures in which the brightness level changes chaotically from one extreme to another extreme, from one picture element to another picture element should never occur. It is clear, however, that to determine such a probability distirbution function for such a large number of pictures would be an impossibly difficult task. If these practical difficulties did not exist, one could use this interpretation of the alphabet and the distortion measure and use the rate-distortion function to determine the minimum number of bits per picture for a given distortion in reproduction.

A more tractable application of rate-distortion theory to our problem might come from regarding the symbols a_j of the alphabet A_m as the gray levels of picture elements. The distortion matrix $\rho(a_j, b_k)$ now measures the penalty charged for reproducing the gray level a_j as a_k. In light of Section 4.3.2, it is clear that such a distortion measure would be excessively local in nature, because the human eye does not view a picture element independently of its context. As was pointed out there, the eye is more tolerant to distortion of a picture element in a region of high detail, but rather sensitive in a region where the gray level changes slowly.

It is clear that any meaningful results from rate-distortion theory for picture compression must await proper formulation of the distortion measure and adequate statistical description of pictures. Both these topics are of current research interest. Some recent results, under simplifying assumptions, appear in [28, 55, 61].

5.5 BIBLIOGRAPHICAL NOTES

Pictures as they are usually produced by a source have more detail and resolution than the observer can utilize. The phrase "redundant information" is often used to denote the information in a picture that may not be intelligible to the observer. The picture compression schemes that we have discussed are, therefore, called *redundancy reduction techniques*. Another commonly used name for these methods is *source encoding*, although this name is slightly more general in the sense that it includes techniques derived from information theory that reconstruct a digitized picture perfectly but still reduce the number of bits required by making use of the statistics of the gray levels [31, 52]. Some investigations have been directed at determining the various statistics of pictures [35, 43, 56].

In a picture transmission system, a source encoder would, in general, be followed by a channel encoder at the transmitting end. While the aim of the source encoder may be to remove the source redundancy, the job of the channel encoder is to insert controlled redundancy to combat channel noise. An excellent discussion of this point is found in [7].

The Karhunen–Loève transformation [41] is also called the "method of principal components" [30]. A number of investigators have studied the application of linear orthogonal transformations in picture compression [2, 3, 9, 14, 22, 23, 32, 36, 48, 50, 60, 65]. While most of these contributions involve the application of the Karhunen–Loève, Fourier, Hadamard, and Haar transforms, some of the more recent work has considered the application of slant transforms [9, 50] and cosine transforms [2] to pictures, with reportedly better mean square error performance than the Fourier and the Hadamard transforms. Andrews [4] and Pearl [47] have investigated "distances" between the Karhunen–Loève, Fourier, Hadamard, and Haar transformations.

Predictive compression systems, more commonly known as differential pulse code modulation (DPCM), are primarily based on an invention by Cutler [12]. In his original patent in 1952, Cutler used one or more integrators to perform the prediction function. From a theoretical standpoint, predictive compression has its roots in Wiener's theory of optimum linear prediction [62]. Graham [20] applied this theory to the system described by Cutler. O'Neal [44, 45] has analyzed DPCM and delta modulation (DPCM with only two quantizing levels) using one-dimensional prediction. Millard and Maunsell [42] and Abbott [1] describe DPCM for the Picturephone® system. Persuasive arguments in favor of using two-dimensional prediction have been presented by Connor et al. [11] and Habibi [23]. Connor et al. showed that two-dimensional prediction resulted in dramatic improvements in the rendition of vertical edges. Application of DPCM to encoding of television signals has also been discussed by Limb and Mounts [38] and Estournet [15]. The noise characteristics of DPCM systems in slope overload have been investigated by Protonotarios [51]. The effect of channel errors has been studied by Arguello et al. [5]

Both transform and predictive compression techniques have their advantages and limitations. Recently there have been attempts at combining the attractive features of both [25, 26]. In the Karhunen–Loève transformation, one uses an orthogonal operator to transform a picture into uncorrelated data. Habibi [24] has shown that if one instead uses a lower triangular operator, one can obtain a generalized DPCM system that reduces to simple DPCM for Markov data.

In general, pictures are not homogeneous. Areas of soft texture can lie adjacent to flat areas or contrasting areas. In Section 4.4 we have indicated that the visual response to noise changes markedly between these areas. In view of this, greater compression efficiencies may be achieved if the parameters of a compression algorithm are changed according to the local characteristics of the picture. Such adaptive compression techniques have been investigated by Anderson and Huang [3], Tasto and Wintz [60], Hayes [29], and Limb [37].

Other topics that were not covered in this chapter or were only briefly mentioned include compression of color images [9, 17, 39, 49], compression of two-valued images, run length coding [10, 18, 54], the methods of "highs" and "lows" [19, 57], and interframe coding [27, 40].

We have by no means been exhaustive in listing the references. The picture compression bibliography of Wilkins and Wintz [64] lists some 600 items. The reader's attention is also directed to the several special issues of the IEEE journals [66–68] dealing with picture compression.

REFERENCES

1. R. P. Abbott, A differential pulse code modulation coded for video telephony using four bits per sample, *IEEE Trans. Comm. Technol.* **COM-19**, 1971, 907–912.
2. N. Ahmed, T. Natarajan, and K. R. Rao, Discrete cosine transform, *IEEE Trans. Comput.* **C-23**, 1974, 90–93.
3. G. B. Anderson and T. S. Huang, Picture bandwidth compression by piecewise Fourier transformation, *IEEE Trans. Comm. Technol.* **COM-19**, 1971, 133–140.
4. H. C. Andrews, Some unitary transformations in pattern recognition and image processing, *in* "Information Processing 71," pp. 155–160, North Holland. Publ., Amsterdam, 1972.
5. R. J. Arguello, H. R. Sellner, and J. A. Stuller, The effect of channel errors in the differential pulse-code modulation transmission of sampled imagery, *IEEE Trans. Comm. Technol.* **COM-19**, No. 6, 1971, 926–933.
6. R. Ash, "Information Theory," Wiley, New York, 1965.
7. T. Berger, "Rate Distortion Theory: A Mathematical Basis for Data Compression," Prentice-Hall, Englewood Cliffs, New Jersey, 1971.
8. B. Carnahan, H. A. Luther, and J. O. Wilkes, "Applied Numerical Methods," Wiley, New York, 1969.
9. W. H. Chen and W. K. Pratt, Color Image Coding with the Slant Transform, *Proc. 1973 Symp. Appl. Walsh Functions* pp. 155–161, April 1973.
10. C. Cherry, M. H. Kubba, D. E. Pearson, and M. P. Barton, An experimental study of the possible bandwidth compression of visual image signals, *Proc. IEEE.* **51**, 1963, 1507–1517.
11. D. J. Connor, R. F. W. Pease, and W. G. Scholes, Television coding using two-dimensional spatial prediction, *BSTJ* **50**, 1971, 1049–1061.
12. C. C. Cutler, Differential Quantization of Communication Signals, Patent No. 2,605,361, July 29, 1952.
13. L. D. Davisson, Rate distortion theory and application, *Proc. IEEE* **60**, 1972, 800–808,
14. H. Enomoto and K. Shibata, Orthogonal transform coding system for television signals, *Television J. Inst. TV. Eng. Japan* **24**, 1970, 99–108; also in *Proc. 1971 Symp. Appl. Walsh Functions* pp. 11–17, April 1971.
15. D. Estournet, Compression d'information de signaux d'images par les systemes differential codes, *Onde Elec.* **49**, 1969, 858–867.
16. R. G. Gallager, "Information Theory and Reliable Communication." Wiley, New York, 1968.
17. L. S. Golding and R. Garlow, Frequency interleaved sampling of a color television signal, *IEEE Trans. Comm. Technol.* **COM-19**, 1971, 972–979.
18. G. G. Gouriet, Bandwidth compression of television signals, *IEE Proc. (London)* **104B**, Pt. 8, 1957, 256–272.

19. D. N. Graham, Image transmission by two-dimensional contour coding, *Proc. IEEE* **55**, 1967, 336–346.
20. R. E. Graham, Predictive quantizing of television signals, *IRE Wescon Convent. Record* Pt. 4, 1958, 147–156.
21. V. Grenander and G. Szego, "Toeplitz Forms and their Applications." Springer-Verlag, New York, 1969.
22. A. Habibi and P. A. Wintz, Image coding by linear transformation and block quantization, *IEEE Trans. Comm. Technol.* **COM-19**,1971, 50–62.
23. A. Habibi, Comparison of nth-order DPCM encoder with linear transformation and block quantization techniques, *IEEE Trans. Comm. Technol.* **COM-19**, 1971, 948–956.
24. A. Habibi and R. S. Hershel, A unified representation of differential pulse code modulation (DPCM) and transform coding systems, *IEEE Trans. Comm.* **COM-22**, 1974, 692–296.
25. A. Habibi and G. S. Robinson, A survey of digital picture coding, *Computer* **7**, 1974, 22–34.
26. A. Habibi, Hybrid coding of pictorial data, *IEEE Trans. Comm.* **COM-22**, 1974, 614–621.
27. B. G. Haskell, F. W. Mounts, and T. C. Candy, Interframe coding of videotelephone pictures, *Proc. IEEE* **60**, 1972, 792–800.
28. J. F. Hayes, A. Habibi, and P. A. Wintz, Rate distortion function for a Gaussian source model of images, *IEEE Trans. Informat. Theory* **IT-16**, 1970, 507–508.
29. J. F. Hayes, Experimental results on picture bandwidth compression, *Proc. UMR-Mervin J. Kelly Comm. Conf.* Univ. of Missouri, Rolla, Rolla, Missouri, Oct. 1970.
30. H. Hotelling, Analysis of a complex of statistical variables into principal components, *Educ. Psychol.* **24**, 1933, 417–441, 498–520.
31. T. S. Huang, Digital picture coding, *Proc. Nat. Electron. Conf.* 1966, 793–797.
32. T. T. Y. Huang and P. M. Schultheiss, Block quantization of correlated Gaussian random variables, *IRE Trans. Comm. Syst.* **CS-11**, 1963, 289–296.
33. F. Jelinek, "Probabilistic Information Theory," Chapter 11. McGraw-Hill, New York, 1968.
34. A. N. Kolmogorov, On the Shannon theory of information transmission in the case of continuous signals, *IRE Trans. Informat. Theory* **IT-2**, 1956, 102–108.
35. E. R. Kretzmer, Statistics of television signals, *BSTJ* **31**, 1952, 751–763.
36. H. J. Landau and D. Slepian, Some computer experiments in picture processing for bandwidth reduction, *BSTJ* **50**, 1971, 1525–1540.
37. J. O. Limb, Adaptive encoding of picture signals, *Symp. Picture Bandwidth Compression*, Massachusetts Inst. Technol., Cambridge, Massachusetts, April 1969.
38. J. O. Limb and F. W. Mounts, Digital differential quantizer for television, *BSTJ* **48**, 1969, 2583–2599.
39. J. O. Limb, C. B. Rubinstein, and K. A. Walsh, Digital coding of color picturephone signals by element-differential quantization, *IEEE Trans. Comm. Technol.* **COM-19**, 1971, 992–1006.
40. J. O. Limb, Buffering of data generated by the coding of moving images, *BSTJ* **51**, 1972, 239–261.
41. M. Loève, Fonctions aleatoires de seconde ordre, *in* "Processus stochastiques et Mouvement Brownien" (P. Levy, ed.), Hermann, Paris, 1948.
42. J. B. Millard and H. I. Maunsell, Digital encoding of the video signal, *BSTJ* **50**, 1971, 459–497.
43. A. Nishikawa, R. J. Massa, and J. C. Mott-Smith, Area properties of television pictures, *IEEE Trans. Informat. Theory* **IT-11**, 1965, 348–352.

44. J. B. O'Neal, Jr., Delta modulation quantizing noise analytical and computer simulation results for Gaussian and TV input signals, *BSTJ* **45**, 1966, 117–142.

45. J. B. O'Neal, Jr., Predictive quantizing (differential pulse code modulation) for the transmission of television signals, *BSTJ* **45**, 1966, 689–722.

46. A. Papoulis, "Probability, Random Variables and Stochastic Processes," McGraw-Hill, New York, 1971.

47. J. Pearl, Basis-restricted transformations and performance measures for spectral representations, *IEEE Trans. Informat. Theory* **IT-17**, 1971, 751–752.

48. W. K. Pratt, J. Kane, and H. C. Andrews, Hadamard transform image coding, *Proc. IEEE* **57**, 1969, 58–68.

49. W. K. Pratt, Spatial transform coding of color images, *IEEE Trans. Comm. Technol.* **COM-19**, 1971, 980–992.

50. W. K. Pratt, L. R. Welch, and W. Chen, Slant transform for image coding, *IEEE Trans. Comm.* **COM-22**, 1974, 1075–1093.

51. E. N. Protonotarios, Slope overload noise in differential pulse code modulation systems, *BSTJ* **46**, 1966, 689–721.

52. R. F. Rice and J. R. Plaunt, Adaptive variable-length coding for efficient compression of spacecraft television data, *IEEE Trans. Comm. Technol.* **COM-19**, 1971, 889–897.

53. L. G. Roberts, Picture coding using pseudo-random noise, *IRE Trans. Informat. Theory* **IT-8**, 1962, 145–154.

54. A. H. Robinson and C. Cherry, Results of prototype television bandwidth compression scheme, *Proc. IEEE* **55**, 1967, 356–564.

55. D. J. Sakrison and V. R. Algazi, Comparison of line-by-line and two-dimensional encoding of random images, *IEEE Trans. Informat. Theory* **IT-17**, 1971, 386–398.

56. W. F. Schreiber, The measurement of third order probability distributions of television signals, *IRE Trans. Informat. Theory* **IT-2**, 1956, 94–105.

57. W. F. Schreiber, C. F. Knapp, and N. D. Kay, Synthetic highs, an experimental TV bandwidth reduction system, *J. Soc. Motion Picture TV Eng.* **68**, 1959, 525–537.

58. C. E. Shannon, "The Mathematical Theory of Communication." Univ. of Illinois Press, Urbana, Illinois, 1949.

59. C. E. Shannon, Coding theorems for a discrete source with a fidelity criterion, *IRE Nat. Conv. Record* Pt. 4, 1959, 142–163.

60. M. Tasto and P. A. Wintz, Image coding by adaptive block quantization, *IEEE Trans. Comm. Technol.* **COM-19**, 1971, 50–60.

61. M. Tasto and P. A. Wintz, A bound on the rate-distortion function and application to images, *IEEE Trans. Informat. Theory* **IT-18**, 1972, 150–159.

62. N. Wiener, "Extrapolation, Interpolation and Smoothing of Stationary Time Series." MIT Press, Cambridge, Massachusetts, 1949.

63. L. C. Wilkins and P. A. Wintz, A Contour Tracing Algorithm for Data Compression for Two-Dimensional Data, Tech. Rep. No. TR-EE-69-3, School of Elec. Eng., Purdue Univ., West Lafayette, Indiana, Sept. 1970.

64. L. C. Wilkins and P. A. Wintz, Bibliography on data compression, picture properties and picture coding, *IEEE Trans. Informat. Theory* **IT-17**, 1971, 180–199.

65. P. A. Wintz, Transform picture coding, *Proc. IEEE* **60**, 1972, 809–820.

66. *IEEE Proc.* (Special issue on redundancy reduction) **55**, 1967.

67. *IEEE Trans. on Comm. Technol.* (Special issue on signal processing for digital communication) **COM-19**, 1971.

68. *IEEE Proc.* (Special issue on digital picture processing) **60**, 1972.

Chapter 6

Enhancement

Whenever a picture is converted from one form to another, e.g., imaged, copied, scanned, transmitted, or displayed, the "quality" of the output picture may be lower than that of the input. This chapter reviews methods of evaluating picture quality, and of "enhancing" low-quality pictures.

Many enhancement techniques are designed to compensate for the effects of a specific (known or estimated) degradation process. This approach, generally known as *image restoration*, makes extensive use of filtering theory; it will be treated in Chapter 7.

In the present chapter, a more elementary class of image-enhancement methods will be discussed. These include methods of modifying the gray scale (e.g., increasing contrast), deblurring, smoothing or removing noise, and correcting geometrical distortions. In all but the last of these, little or no attempt is made to estimate the actual degradation process that has operated on the picture. These methods, however, do take into account certain general properties of picture degradations. For example, increasing the contrast is a reasonable enhancement operation, since degradation usually attenuates the picture signal; deblurring is reasonable, since degradation usually blurs, and the original picture or object is assumed to have been sharp; and smoothing is reasonable, since degradation usually introduces noise, and the original is assumed to have been smooth.

We may also use these enhancement methods to make the picture more acceptable to, and more effectively usable by, its user. For example, when a picture contains "false contours" due to inadequate quantization (see Section 4.3), their presence may mask important information, which becomes more readily visible when we smooth out the contours. In general, we can use enhancement techniques to suppress selected features of a picture, or to emphasize such features at the expense of other features. From this viewpoint, enhancement can be regarded as selective emphasis and suppresion of information in the picture, with the aim of increasing the picture's usefulness.

The enhancement methods described in this chapter are all quite basic and simple. In practical situations, one must often experiment extensively in order to find an effective method. In the absence of knowledge about how the given picture was actually degraded, it is difficult to predict in advance how effective a particular method will be. It is often necessary to use combinations of methods, or to use "tunable" methods whose parameters vary from place to place in the picture, depending on the local context. The techniques described in this chapter provide elementary examples, and can also be used as building blocks in the design of more complex techniques.

6.1 QUALITY

The "quality" of a picture depends on the purpose for which the picture is intended. The picture may be intended for casual human viewing, as in the case of a TV image, or it may be needed for precise, quantitative measurement of some sort. The types and degrees of degradation that would be objectionable or acceptable might be quite different in these two cases. In this section we will deal primarily with objective quality criteria, not with criteria that involve subjective evaluation.

There are many ways of measuring the "fidelity" of a picture $g(x, y)$ to its original $f(x, y)$. One class of such methods uses simple measures of the similarity or difference between f and g. For example, a widely used difference measure is the mean square deviation $\iint (f-g)^2 \, dx \, dy$. Note that this type of measure cannot distinguish between a few large deviations and many small ones. One can, of course, also use measures such as the mean absolute deviation $\iint |f-g| \, dx \, dy$, the maximum absolute deviation $\max |f-g|$, or various measures of the correlation between f and g (see Section 8.3). If one wishes to evaluate the picture transformation process that took f into g, and not just the individual picture g, one should use averages of such measures, taken over ensembles of input f's. A collection of such measures is reviewed in Levi [17].

The picture transformation process that takes f into g can be quite arbitrary. It certainly need not be linear, and it may not even be deterministic—i.e., it may be "noisy." Over limited ranges of values, however, one can often assume, to a good approximation, that the transformation consists of a linear operation combined additively with signal-independent noise. We shall further assume here that the linear operation is shift invariant (see Section 2.1.2). As we have seen, such an operation is a convolution, i.e., is of the form

$$g(x, y) = \int \int h(x-x', y-y') f(x', y') \, dx' \, dy' = h * f$$

If we also take the noise into account, we have

$$g = h * f + v$$

where $v(x, y)$ is assumed to be uncorrelated with $h * f$.

In Sections 6.1.1–6.1.3, we review some of the standard methods of evaluating a (deterministic) picture transformation, in terms of its effects on various simple f's, such as bright points or lines, sinusoidal bar patterns, etc. In Section 6.1.4, we briefly discuss various types of pictorial noise.

6.1.1 Spread and Transfer Functions

As mentioned in Section 2.1.1, the output of a picture-to-picture transformation $\mathcal{O}$ for a point source input $\delta(x, y)$ (essentially: a bright point centered on a dark background) is called the *point spread function* (PSF) of $\mathcal{O}$. If $\mathcal{O}$ is linear and shift invariant, knowledge of its point spread function determines its effect on any arbitrary input f, since f can be regarded as a linear combination of shifted δ's. As we have seen in Section 2.1, if $\mathcal{O}(f) = h * f$, then the PSF of $\mathcal{O}$ is just $h * \delta = h$.

The PSF can be regarded as a measure of the degradation caused by $\mathcal{O}$, since it describes the blurred output that is obtained when a sharp point is input. In fact, when we convolve h with f, we are simply blurring f by weighted averaging at every point, where the weighting is described by the function h. Note that $\mathcal{O}$ may also involve an overall attenuation (or intensification) of its input. In fact, if the input is a constant c, the output is $h * c = c \int \int h(x, y) \, dx \, dy$. Thus $\int \int h$ is a measure of the overall attenuation produced by $\mathcal{O}$.

Other "spread functions" of $\mathcal{O}$ can also be used to derive quality measures. Two important examples are the *line spread function* (LSF) and the *edge spread function* (ESF); these are the outputs obtained when the input is a bright straight line, or an abrupt step in brightness (a straight "edge"). Note that if $\mathcal{O}$ is not isotropic (i.e., if its point spread function does not have central symmetry), then the LSF and ESF will depend on the orientation of the given line or edge.

More precisely, we can define the LSF as follows: Let $\delta(x)$ be a one-dimensional delta function, defined analogously to $\delta(x, y)$ in Section 2.1.1.

Then $\delta(x)$ has the sifting property $\int g(x)\delta(x-\alpha)\,dx = g(\alpha)$. We can regard $\delta(x)$ as a function of two variables, x and y, which is independent of y. From this viewpoint, $\delta(x)$ can be regarded as a *line source* along the y-axis. The line spread function (for this direction) is then

$$h_l(x, y) \equiv \int\int h(x-x', y-y')\,\delta(x')\,dx'\,dy'$$

Using the sifting property of $\delta(x)$, this becomes

$$\int \left[\int h(x-x', y-y')\,\delta(x')\,dx' \right] dy' = \int h(x, y-y')\,dy'$$

Since we can change the variable of integration from y' to $y-y'$, we see that $h_l(x, y)$ can also be written as $\int h(x, y)\,dy$. Note that this is a function of x alone, and could be denoted by $h_l(x)$. We have also shown that the LSF for a line source in the y-direction is just the integral of the PSF with respect to y, i.e., $h_l = \int h\,dy$. In general, the LSF for a line source in a given direction θ is just the integral of the PSF taken in that direction.

Exercise 1. The function $\delta(x)$ can also be expressed as $\int_{-\infty}^{\infty} \delta(x, y)\,dy$. Use this expression to give another proof that $h_l = \int h\,dy$, based on the sifting property of the two-dimensional delta function. ∎

An edge source (along the y-axis, say) can be defined as a unit step function

$$s(x, y) = \begin{cases} 0, & x < 0 \\ 1, & x \geqslant 0 \end{cases} \quad \text{for all } y$$

The corresponding ESF is then

$$h_e(x, y) \equiv \int\int h(x-x', y-y')\,s(x', y')\,dx'\,dy'$$

It can be seen that $s(x, y) = \int_{-\infty}^{x} \delta(x')\,dx'$, and it follows that the ESF is the indefinite integral, with respect to x, of the LSF; this will be proved in Section 7.1.2. Similarly, the ESF in any direction is the indefinite integral of the corresponding LSF in that direction; or, equivalently, the LSF is the derivative of the ESF.

The Fourier transform $H(u, v)$ of the PSF $h(x, y)$ is called the *optical transfer function* (OTF), or sometimes just the *transfer function*. Its amplitude and phase are called the *modulation transfer function* (MTF) and *phase transfer function* (PTF), respectively. If we denote these functions by $M(u, v)$ and $\Phi(u, v)$, we have

$$H(u, v) = M(u, v)\,e^{j\Phi(u, v)}$$

If the PSF is isotropic, then so is the OTF (Section 2.1.3, Exercise 2). In this case the MTF (and PTF and OTF) become functions of a single variable $\sqrt{u^2+v^2}$, and can be plotted in the form of "MTF curves." We shall see in Section 7.1.2 that the one-dimensional Fourier transform of the LSF, for a line source in a given direction θ, is equal to a cross section of the OTF along a line through the origin with slope $\theta+\frac{1}{2}\pi$; if the OTF is isotropic, all these cross sections are identical.

Just as the PSF, LSF, and ESF are the outputs that result from point, line, and edge sources, respectively, so the OTF specifies the outputs that result from sinusoidal bar pattern inputs (Section 2.1.3) having all possible positions, orientations, and spatial frequencies. To see this, consider the input

$$A + B\cos(2\pi u x')$$

which is a sinusoidal bar pattern with the bars running parallel to the y'-axis. The resulting output is

$$A\int\int h(x-x', y-y')\,dx'\,dy' + B\int\int h(x-x', y-y')\cos(2\pi u x')\,dx'\,dy'$$

in which the first term is a constant times A (see the second paragraph of this section). As for the second term, we can integrate $h(x-x', y-y')$ with respect to y' to obtain the LSF $h_l(x-x')$. The second term thus becomes

$$B\int h_l(x-x')\cos(2\pi u x')\,dx'$$

Letting $x-x'=z$, this can be rewritten as

$$B\int h_l(z)\cos 2\pi u(x-z)\,dz$$

$$= B\left[\cos 2\pi u x\int h_l(z)\cos 2\pi u z\,dz + \sin 2\pi u x\int h_l(z)\sin 2\pi u z\,dz\right]$$

in which the two integrals are, respectively, the real and imaginary parts of the one-dimensional Fourier transform of $h_l(z) = h_l(x-x')$. By the preceding paragraph, this transform is just the cross section of the OTF H along the u-axis, and we can write this cross section as

$$H(u,0) = C(u,0) + jS(u,0) = M(u,0)\,e^{j\Phi(u,0)}$$

where

$$M(u,0) = \sqrt{C^2(u,0)+S^2(u,0)}; \qquad \Phi(u,0) = \tan^{-1}[S(u,0)/C(u,0)]$$

In this notation, the previously mentioned second term becomes

$$B[\cos 2\pi u x\cdot C(u,0) + \sin 2\pi u x\cdot S(u,0)] = BM(u,0)\cos[2\pi u x - \Phi(u,0)]$$

Thus we see that if the input is a sinusoidal bar pattern $A + B\cos(2\pi u x')$, then the output is a sinusoidal bar pattern in the same direction (here: bars parallel to the y-axis), and with the same spatial frequency u, but with amplitude multiplied by $M(u, 0)$ and phase shifted by $\Phi(u, 0)$. In general, for a bar pattern having spatial frequency components (u, v), we would multiply the amplitude by $M(u, v)$ and shift the phase by $\Phi(u, v)$; the details are left as an exercise to the reader. Thus the OTF allows us to determine directly the amplitdues and phases of the outputs that result from arbitrary sinusoidal bar pattern inputs.

6.1.2 An Example: Unweighted Averaging

A basic example of a linear shift-invariant picture degradation is unweighted averaging: the gray level at each point (x, y) of the output picture is the average of the gray levels in the input picture over a neighborhood of (x, y). Let $\mathscr{A}$ denote this neighborhood when the point in question is the origin, and let A be the area of $\mathscr{A}$. Then the unweighted averaging operation has point spread function h given by

$$h(x, y) = \begin{cases} 1/A & \text{for } (x, y) \text{ in } \mathscr{A} \\ 0 & \text{otherwise} \end{cases}$$

Note that $\iint h = \iint_{\mathscr{A}} h = (1/A) \iint_{\mathscr{A}} 1 = 1$, as should be expected since the averaging operation involves no overall attenuation or intensification.

Suppose, in particular, that $\mathscr{A}$ is the square $|x| \leqslant 1/2n$, $|y| \leqslant 1/2n$. Thus h is just the function $n^2 \operatorname{rect}(nx, ny)$ of Section 2.1.1, which has value n^2 inside $\mathscr{A}$ and 0 outside. For this h, the LSF for a line source in the x direction is given by

$$h_l(y) = \int h(x, y)\, dx = \int_{-1/2n}^{1/2n} n^2\, dx = \begin{cases} n & \text{for } |y| \leqslant 1/2n \\ 0 & \text{otherwise} \end{cases}$$

For other directions, the LSF can be computed analogously; note that it is not simply a rotation of this function.

The corresponding cross section of the MTF, for sinusoidal bar patterns in the x-direction, is obtained from the one-dimensional Fourier transform of $h_l(y)$, which is readily (see Section 2.1.3, Exercise 1)

$$F(u, v) = \frac{\sin(\pi v/n)}{\pi v/n} = \operatorname{sinc}\left(\frac{v}{n}\right) \qquad \text{for all } u$$

Since this transform is real, its modulus is just its absolute value, while its phase is identically zero. Thus the MTF is $|\operatorname{sinc}(v/n)|$, so that the output of our unweighted averaging operation for the input $A + B \cos(2\pi v y)$ is $A + B|\operatorname{sinc}(v/n)| \cos(2\pi v y)$. Analogous results can be obtained for other directions.

It should be pointed out that $|\mathrm{sinc}(v/n)|$ does not decrease monotonically with increasing spatial frequency v. In fact, this function does decrease to zero as v increases to n, but it then increases again; it has zeros at integer multiples of n, separated by peaks which become smaller and smaller as v increases. This corresponds to the fact that when the sinusoidal bar frequency is an exact multiple of n, the averaging neighborhood covers an integer number of periods of the bar pattern, thus yielding a constant value of the average at any point. When the frequency is higher or lower, however, the average fluctuates from point to point, and the sinusoidal modulation is still visible. (On the fact that the modulation is visible even when the frequency gets higher, see the discussion of pseudoresolution in Section 6.1.3.)

Unweighted averaging over a circular, rather than square, neighborhood will produce isotropic spread and transfer functions, but the MTF will still oscillate, rather than decreasing monotonically. One can obtain a monotonically decreasing MTF by performing suitably weighted averaging. For example, if the averaging in the y-direction is weighted as $\mathrm{sinc}(y/n)$, the MTF for sinusoidal bars parallel to the x-axis will be (see Section 2.1.3) $n \, \mathrm{rect}(nv)$, which is "flat" for $|v| \leqslant n/2$ and then cuts off sharply to zero. Note that this averaging operation may produce negative values in the output "picture," since it involves both positive and negative weights. Such an MTF could be achieved approximately, but not exactly, if a finite averaging neighborhood were used. On the other hand, such an MTF can be very easily obtained by operating in the spatial frequency domain rather than in the space domain— namely, by taking the Fourier transform F of the given picture f; suppressing all but the low spatial frequencies from F [i.e., multiplying F by $n \, \mathrm{rect}(nv)$]; and then taking the inverse transform.

6.1.3 Resolution and Acutance

Classically, image quality has often been measured in terms of the output resulting from simple input patterns such as steps and bars. One such measure is *resolution* or *resolving power*, which describes the distinguishability of small, close objects, such as the sets of bars on a resolution chart, in the output. If the bar width is b units, and the spaces are equal to the widths, then we have $1/2b$ "line pairs" (bar + space = pair) per unit distance. The resolution, informally, is the greatest number of input line pairs per unit distance such that, on the output, we can still count the bars correctly.

To illustrate the meaning of this definition, consider the unweighted averaging operation discussed in Section 6.1.2, where we assume that the bars are parallel to (say) the x-axis and are much longer than they are wide. If there are fewer than n line pairs per unit distance (i.e., $1/2b < n$), averaging over a square of side $1/n$ does not totally average out the bars, and correct counting should still be possible. In fact, it is easily verified that after averaging, the cross section of the bars still has peaks at the centers of the bars and troughs at the centers of the spaces. On the other hand, if $1/2b = n$, so that the averaging neighborhood size is exactly equal to the period of the bar pattern, then for any position of the neighborhood, the value of the average is constant, and the bars can no longer be "resolved."

It is important to observe that if the averaging neighborhood is still bigger, i.e., if $1/2b > n$, the averaged image once again contains a bar pattern. For example, suppose that $1/n = 3b$, so that the neighborhood is as wide as two bars and a space, or two spaces and a bar. In this case, when the neighborhood is centered at the center of a bar, we are averaging one bar and two spaces, giving an average gray level of $\frac{1}{3}$ (assuming the bars to be 1 and the spaces 0). On the other hand, when it is centered at the center of a space, we have the average of one space and two bars, or $\frac{2}{3}$. Thus here again we see bars in the output picture; but they are in the wrong places! Moreover, we will not be able to count them correctly; if the original pattern contained k bars, we will obtain peak gray levels at the centers of the $k-1$ spaces between these bars, so that we will count $k-1$ instead of k.

The phenomenon described in the preceding paragraph is known as *spurious resolution* or *pseudoresolution*. It is illustrated in Fig. 1, where a bar pattern (Fig. 1a) has been averaged using square neighborhoods of sides 3, 5, 7, 9, and 11; the resulting outputs are shown in Figs. 1b–f, respectively. Note that in (e) and (f) the bar counts are no longer reliable.

Exercise 2. Explicitly calculate the cross section of the averaged bar pattern as a function of bar size, averaging window size, and position. ▮

Centrally weighted averaging preserves resolution over a much greater

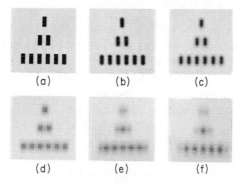

Fig. 1 Unweighted averaging. (a) Bar pattern (the bar size is 4×10 points). (b)–(f) Results of averaging (a) using square neighborhoods of sides 3, 5, 7, 9, and 11. Note the spurious resolution in (e) and (f).

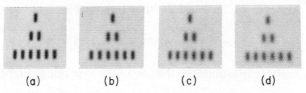

$$(a) \qquad (b) \qquad (c) \qquad (d)$$

Fig. 2   Weighted averaging by iterated unweighted averaging. (a) Result of averaging Fig. 1b, using a square neighborhood of side 3; in other words, 3 × 3 averaging has now been performed twice in succession on Fig. 1a. (b)–(d) Results of averaging (a)–(c) in the same way—i.e., Fig. 1a has now been averaged 3, 4, and 5 times.

range of neighborhood sizes than does unweighted averaging. An example is provided by Fig. 2, which shows the results of *iterated* 3 × 3 averaging applied to the same bar pattern used in Fig. 1.

Exercise 3.   Show that if we perform unweighted averaging m times, each time using a $(2k+1) \times (2k+1)$ neighborhood, the result is a weighted average over a $(2mk+1) \times (2mk+1)$ neighborhood, with the weights decreasing exponentially from the center. The weights can be explicitly expressed in terms of binomial coefficients. ∎

Another useful class of image quality measures relates to the average steepness of an edge in the output picture that results from a perfect step edge in the input. We recall that the output edge cross section is described by the ESF, which is the indefinite integral of the LSF, as indicated in Section 6.1.1. If we denote this function by $h_e(y)$ (for an edge along the x-axis), then we can measure its average steepness by an expression of the form

$$\frac{1}{h_e(b) - h_e(a)} \int_a^b \left(\frac{dh_e}{dy}\right)^2 dy \qquad \text{or} \qquad \int_a^b h_l(y)^2 \, dy \Big/ \int_a^b h_l(y) \, dy$$

where $h_l(y)$ is the LSF, and where a, b are points at or near the top and bottom of the output edge, respectively. This expression is known as *acutance*; it has been shown experimentally to correlate well with subjective judgments of edge sharpness.

As a simple illustration of this concept, we again consider the case of unweighted averaging. For an edge along the x-axis, say, having value 0 for $y \leqslant 0$ and 1 for $y > 0$, the ESF corresponding to an averaging neighborhood of size $1/n$ is readily

$$h_e(y) = \begin{cases} 0 & \text{for} \quad y \leqslant -1/2n \\ ny + \tfrac{1}{2} & \text{for} \quad -1/2n \leqslant y \leqslant 1/2n \\ 1 & \text{for} \quad y \geqslant 1/2n \end{cases}$$

Thus $dh_e/dy = h_l(y) = n$ for $|y| \leqslant 1/2n$, and 0 elsewhere, as found in Section

6.1.2. If we take $a = -1/2n$, $b = 1/2n$, the acutance becomes

$$1/(1-0) \int_{-1/2n}^{1/2n} n^2 \, dy = n^2 \cdot (1/n) = n$$

which is inversely proportional to the size of the averaging neighborhood.

Exercise 4. Compute the acutance for the case of a circular averaging neighborhood of diameter $1/n$. ▮

In the unweighted averaging case, acutance and resolution are numerically equal; an averaging neighborhood size of $1/n$ yields a resolution of (up to) n line pairs per unit length, and an acutance of n. In general, however, acutance and resolution can have very different values. In fact, acutance can be low while resolution is high, and vice versa. Suppose, for example, that the point spread function, in cross section, consists of a broad plateau surmounted by a sharp peak. Because of the plateau, edges become very blurred, so that the acutance will be low. On the other hand, bars should remain resolvable as long as their period is larger than the width of the peak.

Exercise 5. Treat the case of "pyramidally" weighted averaging, in which the weights decline linearly from the center of the square averaging neighborhood to its edges. In particular, determine the spread and transfer functions and the acutance. ▮

6.1.4 Noise

Pictures are subject to many different types of noise. Some of these are independent of the picture signal, but others are not; some are uncorrelated from point to point, while others are "coherent." In the following paragraphs we give a few examples of commonly encountered kinds of noise.

When a picture is transmitted, *channel noise* is introduced whose value is generally independent of the strength of the picture signal. The situation is similar when a picture is scanned by a vidicon television camera. In these cases, we can write $g = f + v$, where the noise v and input picture f are uncorrelated.

A simple illustration of the effects of picture-independent noise is shown in Fig. 3. Here the bars have constant gray level, say s, and the spaces have gray level r. In Fig. 3, each point has its gray level incremented or decremented by an amount z, randomly chosen to lie in the range $|z| \leqslant \theta(s-r)$, for various

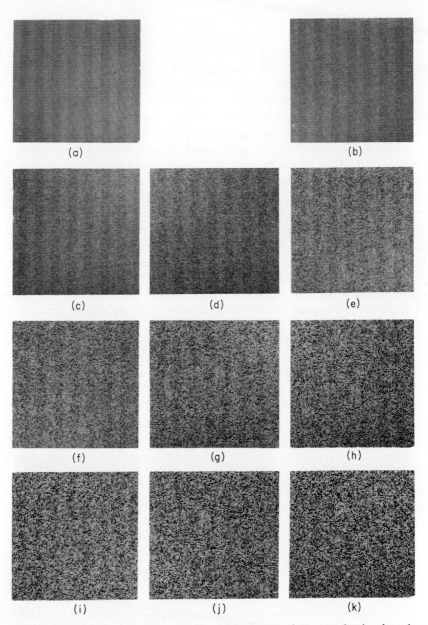

Fig. 3   Signal-independent noise. (a) Bar pattern; spaces have gray level r, bars have gray level s. (b)–(k) Results of incrementing or decrementing the gray level of each point in (a) by an amount z, randomly chosen in the range $|z| \leqslant \theta(s-r)$, for $\theta = 1, 2, ..., 10$.

values of θ. Note that even when the range of noise levels exceeds the difference between the bars and the background, the bars are still visible.

The problem of estimating the true gray levels of a picture when noise is present will be discussed in Section 7.5. Such estimates require some degree of knowledge about the statistics of the picture and of the noise. If we know that the nonnoisy picture contains a large region of constant gray level, we can get some insight into the noise statistics by analyzing the gray level fluctuations in the corresponding region of the noisy picture. For example, if we assume the noise to have a Gaussian distribution with zero mean, its standard deviation is just the standard deviation of these fluctuations. We can also estimate the noise autocorrelation and power spectrum (assuming ergodicity) by measuring the autocorrelation and power spectrum of the noisy picture over the given region.

In many cases the noise level does depend on that of the picture signal; this is true, for example, when a picture is scanned by a flying-spot scanner. If the noise is proportional to the signal, i.e., $g = f + v_1 f$, we have $g = f(1 + v_1) = fv$ (say), so that we can regard this situation as one where we have uncorrelated noise that is multiplicative rather than additive. A simple example is the coherent "noise" in a television picture due to the presence of the TV raster lines; here v is a bar pattern that has maxima on the raster lines and zeros between them. (For methods of removing such noise patterns, see Section 6.5.1.) A less trivial example is *photographic graininess*, which arises from the fact that photographic images are formed from clumps of developed "grains" in the emulsion.

Noise proportional to the signal is illustrated in Fig. 4, which is analogous to Fig. 3 except that the noise ranges for points on the bars and on the spaces

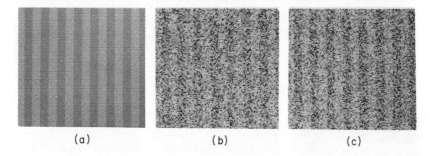

(a)   (b)   (c)

Fig. 4 Signal-dependent noise. (a) Bar pattern having two gray levels, r and s. (b) Result of adding noise to (a) as in Fig. 3, except that the noise ranges for the spaces and bars are proportional to r and s, respectively. (c) Result of using the average of the two noise ranges of (b) at every point of (a); here the noise is no longer signal dependent. The bars are more clearly visible, because the noise range for the spaces is not a subinterval of the range for the bars, as it was in (b).

are proportional to s and r, respectively. Note that the bars are somewhat less visible in this case.

An important type of noise in digital pictures is *quantization noise* (or quantization error), which is the difference between a quantized picture and its original. In Section 4.3.1 we saw how, for a given probability density of gray levels, this error could be minimized by suitable choice of the quantization levels.

One often wants to convert a picture that contains shades of gray into a "black and white" picture by thresholding it (see Section 8.1). The resulting output picture can be regarded as consisting of black "objects" on a white "background," or vice versa. If the original objects were too noisy, the black regions may contain scattered white points, and the white regions scattered black points. This condition is known as *salt-and-pepper noise*. More generally, we can apply this term to any situation in which scattered points of a picture are markedly darker or lighter than their immediate surroundings.

Two simple examples of salt-and-pepper noise are given in Figs. 5 and 6. Figure 5 shows a black-and-white bar pattern in which a fraction p of the black points have been changed to white and vice versa, where

$$p = 0, 0.1, 0.2, 0.3, 0.4, 0.5$$

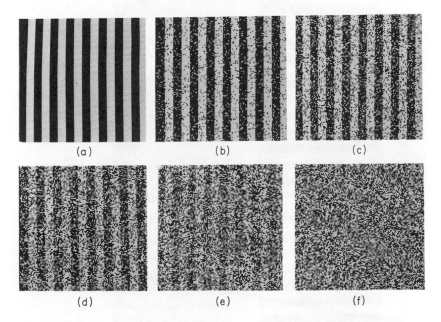

(a)   (b)   (c)

(d)   (e)   (f)

Fig. 5   Salt-and-pepper noise. (a) Black-and-white bar pattern. (b)–(f) Results of changing fraction p of the black points of (a) to white, and vice versa, for $p = 0.1, 0.2, 0.3, 0.4$, and 0.5.

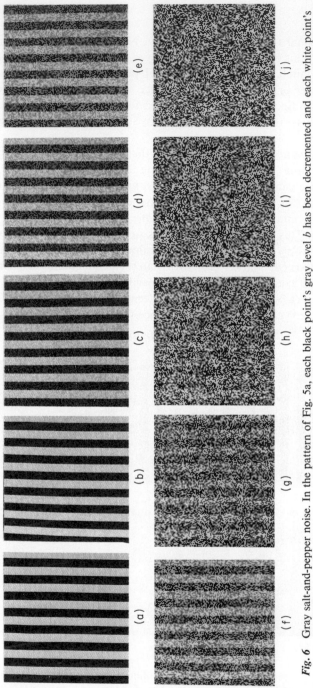

Fig. 6 Gray salt-and-pepper noise. In the pattern of Fig. 5a, each black point's gray level b has been decremented and each white point's level w has been incremented by a random amount z in the range $0 \leq z \leq \theta(b-w)$, where $\theta = 0.1, 0.2, \ldots, 0.9, 1.0$ in parts (a)–(j), respectively.

in Figs. 5a–f, respectively. In Fig. 6, each black point has had its gray level decremented, and each white point has had its level incremented, by a random amount z in the range $0 \leqslant z \leqslant \theta(b-w)$, where $b = $ black, $w = $ white, and $\theta = 0.1, 0.2, \ldots, 0.9, 1$ in Figs. 6a–j, respectively. A method of removing salt-and-pepper noise is described in Section 6.5.1.

Noise may depend not only on the gray level of the picture at the given point, but also on the levels at nearby points. This is so for many of the examples previously mentioned—e.g., for the flying-spot scanner, since the spot size is finite; and for photographic grain, since the clumping is influenced by local conditions. An interesting artificial example is *random walk noise*, where the picture is degraded by interchanging the gray levels of randomly chosen pairs of nearby points.

As we have seen in Chapters 4 and 5, a picture is often processed by applying an invertible transformation to it (e.g., taking its Fourier transform), operating on the transform, and then applying the inverse transformation to obtain the desired processed picture as output. Errors (or noise) may be introduced by the operations in the transform domain, and it is important to analyze the effects of these errors on the output picture.

If picture-independent noise is added to the Fourier transform of a picture, the reconstructed picture will also contain independent noise, which by Parseval's theorem (Section 2.1.4) will have the same power. A similar result holds if noise is added to the power spectrum of the picture. Noise in the phase of the Fourier transform has more complicated effects. One can estimate the output noise autocorrelation in this case in terms of that of the ideal output picture; for the details, the reader is referred to Anderson and Huang. [2], where the case of multiplicative noise in the Fourier domain is also treated.

Quantization noise in the Fourier domain is a case of special interest, since the range of values in the Fourier transform of a picture is generally much greater than that in the picture itself, with most of the power being found at low spatial frequencies. This makes it especially important to use unequally spaced quantization levels (see Section 4.3.1) when quantizing a Fourier transform. Errors in quantizing the phase of the transform tend to have much worse effects on the reconstructed picture than do errors in quantizing the amplitude.

6.2   GRAY SCALE MODIFICATION

A simple but surprisingly powerful class of enhancement operations involves modifying the gray scale of the given picture. Two types of such operations will be discussed here. The first is *gray level correction*, which modifies the gray levels of the individual picture points so as to compensate for uneven "exposure" when the picture was originally recorded. The second type is *gray scale transformation*, which has the aim of changing the gray scale in a uniform way throughout the picture, or perhaps throughout some region of the picture, usually in order to increase contrast and thereby make details of the picture more easily visible. An important special case of this is *histogram modification*, in which a gray scale transformation is used to give the picture a specified distribution of gray levels.

6.2.1 Gray Level Correction

A picture recording system should map object brightness into picture gray level in a monotonic fashion, and this mapping should ideally be the same at every point of the picture. In practice, however, the mapping often varies from point to point. For example, light passing along the axis of an optical system is generally attenuated less than light that passes through the system obliquely. Thus when an image is formed by the system, the parts of the image far from the axis will be attenuated relative to the parts near the axis; this phenomenon is called "vignetting." As another example, the photocathode of a vidicon may not be equally sensitive at all points; thus in a picture obtained using the vidicon, equal gray levels may not correspond to equally bright points in the scene.

If we can determine the nonuniform "exposure mapping" that produced a given picture, it is then straightforward to correct the picture, by changing the gray level of each point, to achieve the effect of a uniform mapping. Specifically, suppose that we describe the nonuniform mapping by an expression of the form

$$g(x, y) = e(x, y) f(x, y)$$

where $f(x, y)$ is the ideal gray level that should have resulted at picture point (x, y) had the "exposure" been uniform [i.e., $f(x, y)$ is the desired monotonic function of the object brightness at the object point corresponding to (x, y)], and $g(x, y)$ is the actual gray level at point (x, y), due to the nonuniform mapping. To determine the function $e(x, y)$, we can calibrate the picture recording system by taking a picture of a uniform field of known brightness. For such a field, f is a known constant, call it c. If $g_c(x, y)$ is the picture of the uniform field, we have $e(x, y) = g_c(x, y)/c$. Once we know $e(x, y)$, we can correct any picture $g(x, y)$ obtained by the system (as long as the calibration does not change!), since

$$f(x, y) = g(x, y)/e(x, y) = cg(x, y)/g_c(x, y).$$

In any practical gray scale, the range $[z_1, z_K]$ of gray levels used is limited

by the dynamic ranges of the available display devices. The correction operation just described may give rise to gray levels outside the allowable range.
One way of handling this situation is simply to change any gray level less than
z_1 to z_1, and any level greater than z_K to z_K. Another possibility is to shrink
and/or shift the scale of the corrected gray levels until it falls within the
allowable range. (Methods of doing this will be described in Section 6.2.2.)
It should also be noted that even if the uncorrected gray levels were quantized
to a discrete set of values $z_1, ..., z_K$, the corrected levels—even though they
lie in the range $[z_1, z_K]$—need not have these discrete values. Thus gray level
correction of a quantized picture must be followed by requantization.

6.2.2   Gray Scale Transformation

We next consider gray scale transformations that are the same at every
point of the picture (or a region), rather than varying from point to point.
Such a transformation can be expressed as a mapping from the given gray
scale z into a transformed gray scale z', i.e.,

$$z' = t(z)$$

We shall assume that in both the old and new gray scales, the allowable range
of gray levels is the same, i.e., $z_1 \leqslant z \leqslant z_K$ and $z_1 \leqslant z' \leqslant z_K$. We can now discuss how to design gray scale transformations that have enhancing effects,
e.g., that increase contrast.

It is easy to increase the contrast of a picture if the picture does not occupy
its full allowable gray level range. (This can happen if the picture recording
device used had a smaller dynamic range than $[z_1, z_K]$, or if the picture was
originally "underexposed.") Suppose that, in the given picture f, we have
$a \leqslant z = f(x, y) \leqslant b$ for all x, y, where $[a, b]$ is a subinterval of $[z_1, z_K]$. Let

$$z' = \frac{z_K - z_1}{b-a}(z-a) + z_1 = \frac{z_K - z_1}{b-a}z + \frac{z_1 b - z_K a}{b-a}$$

This simple linear gray scale transformation stretches and shifts the gray
scale to occupy the full range $[z_1, z_K]$.

A similar approach can be used if *most* of the gray levels of the given picture
lie in the subrange $[a, b]$. In this case, we can use the transformation

$$z' = \begin{cases} \dfrac{z_K - z_1}{b-a}(z-a) + z_1 & \text{for} \quad a \leqslant z \leqslant b \\[2mm] z_1 & \text{for} \quad z < a \\[2mm] z_K & \text{for} \quad z > b \end{cases}$$

This piecewise linear transformation stretches the $[a, b]$ interval of the original gray scale; but it *compresses* the intervals $[z_1, a]$ and $[b, z_K]$—in fact, it collapses them to single points. This may be tolerable, however, if very few points have gray levels in these intervals, so that little information is lost when we compress.

More generally, we can stretch selected regions of the gray scale, at the cost of compressing other regions, if we want to bring out detail in the stretched regions, and we do not care about loss of information in the compressed regions. As a simple example, suppose that the gray scale range is [0, 30]; then the transformation

$$z' = \begin{cases} z/2 & \text{for} \quad z \leqslant 10 \\ 2z - 15 & \text{for} \quad 10 \leqslant z \leqslant 20 \\ (z/2) + 15 & \text{for} \quad 20 \leqslant z \leqslant 30 \end{cases}$$

compresses the gray scale by a factor of 2 in the ranges [0, 10] and [20, 30] (on the original), while expanding it by a factor of 2 in the range [10, 20].

Exercise 6.   Give the equations for the transformation that stretches gray scale range [0, 10] into [0, 15], shifts range [10, 20] to [15, 25], and compresses range [20, 30] into [25, 30].   ∎

The gray scale transformations that we use can, of course, be smooth rather than piecewise linear. We can implement any desired mathematical transformation $z' = t(z)$—quadratic, logarithmic, or completely arbitrary— subject only to the restriction that the results lie in the allowable range $[z_1, z_K]$—i.e., that $z_1 \leqslant t(z) \leqslant z_K$ for all $z_1 \leqslant z \leqslant z_K$. This can always be achieved by incorporating a suitable shift and scale factor into the transformation. We can do this as follows: given any $t(z)$, let $t_1 = \min t(z)$ and $t_K = \max t(z)$ for $z_1 \leqslant z \leqslant z_K$. Then the modified transformation t' defined by

$$t'(z) = \frac{z_K - z_1}{t_K - t_1} [t(z) - t_1] + z_1$$

satisfies $z_1 \leqslant t'(z) \leqslant z_K$ for all $z_1 \leqslant z \leqslant z_K$.

Exercise 7.   Give the equations for a transformation t for which $t(z)$ is a linear function of $\log z$ over the range $10 \leqslant z \leqslant 100$.   ∎

Two simple examples of contrast-stretching gray scale transformations are shown in Figs. 7 and 8. The graph of such a transformation, $t(z)$ as a function of z, is a useful aid in visualizing the effect of t. In ranges where this graph has slope < 1, contrast is compressed, while slope > 1 implies that contrast is stretched.

The examples of transformations considered up to now have all been monotonic [i.e., $z' \leqslant z''$ implies $t(z') \leqslant t(z'')$] and continuous. It is sometimes

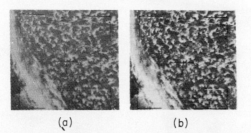

(a)   (b)

Fig. 7   Contrast stretching. (a) "Underexposed" picture (all gray levels in middle half of range). (b) Result of stretching the gray scale of (a) by a factor of 2.

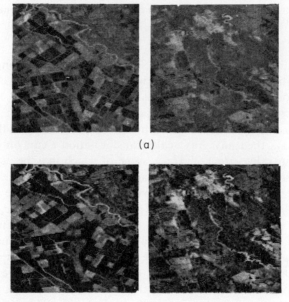

(a)

(b)

Fig. 8   Partial contrast stretching. (a) Input pictures. (b) Result of stretching the middle third of part (a)'s gray scales by a factor of 2 while compressing the upper and lower thirds by a factor of 2.

of interest, however, to consider nonmonotonic discontinuous transformations in which different ranges of input gray levels are mapped into the same range of output levels. As a simple example, if the input range is $[0, 30]$, the discontinuous transformation

$$t(z) = \begin{cases} 3z, & 0 \leqslant z \leqslant 10 \\ 3(z-10), & 10 < z \leqslant 20 \\ 3(z-20), & 20 < z \leqslant 30 \end{cases}$$

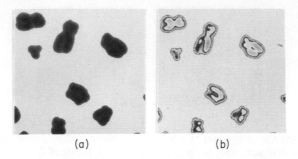

(a)   (b)

Fig. 9 Discontinuous, nonmonotonic contrast stretching. (a) Input picture (same as Fig. 2b of Chapter 8). (b) Result of stretching each third of part (a)'s nonzero gray levels to cover the full gray level range.

maps each third of this range into $[0, 30]$. Within these subranges, we have threefold contrast stretching; but as we cross from one range to another, strong false contours are introduced, as illustrated in Fig. 9.[†]

This discussion has treated the input gray scale as though it were continuous rather than quantized. If it is, in fact, quantized, say to the K discrete values $z_1, ..., z_K$, then any gray scale transformation t can only produce K discrete output values $t(z_1), ..., t(z_K)$, some of which may be equal.[‡] Nevertheless, contrast stretching can be advantageous, since it enables us to space (some of) the output values farther apart, so that they become more clearly distinguishable. In a well-designed gray scale, consecutive levels z_i, z_{i+1} should not be easily distinguishable; in fact, if they were, we could never get the impression of continuously varying gray levels in a picture. Thus detail that is not sharply delineated in the input is enhanced in the output when we spread the gray levels. Note, however, that once the levels have been spread far enough apart to become easily distinguishable, there is little to be gained in spreading them still farther apart, since all the available information (in the gray level range that has been stretched) has already been brought out. Note also that in ranges where the gray scale is compressed, information is lost, since when we requantize, several input levels may be mapped into the same output level.

If a picture's gray scale has been distorted by a known gray scale transformation $z' = t(z)$, the distortion can in principle be corrected by applying

---

[†] Another important "contrast-stretching" technique is to map the gray levels of a picture into *colors*—e.g., low levels into shades of red; higher ones into orange shades; still higher ones into yellow, and so on. This can greatly enhance the visibility of detail in the picture, because the eye is more sensitive to differences in hue than it is to differences in brightness.

[‡] See, however, Section 6.2.3, where we discuss how to map a single input gray level into more than one output level.

the inverse transformation $z = t^{-1}(z')$. For example, this approach can be used to correct for the effects of nonlinearities in the gray scales of display devices, or of recording media such as photographic film. In practice, of course, implementation of t^{-1} may be subject to the difficulties previously discussed.

6.2.3   Histogram Modification

Given a picture f, let $p_f(z)$ denote the relative frequency with which gray level z occurs in f, for all z in the gray level range $[z_1, z_K]$ of f. The graph of $p_f(z)$ as a function of z, normalized so that $\int_{z_1}^{z_K} p_f(z)\, dz$ is equal to the area of f, is called the *histogram* of f. If f is quantized, and has gray levels $z_1, \ldots, z_K$, its histogram can be represented as a bar graph having K bars. As we shall see in Chapters 8 and 10, the histogram of f can provide useful information about how to segment f into parts, and it also serves as a basis for measuring certain textural properties of f. Note that for any $z_1 \leqslant a \leqslant b \leqslant z_K$, the integral $\int_a^b p_f(z)$ measures how heavily the gray scale range $[a, b]$ is populated, i.e., what fraction of f's points have their gray levels in that range.

In this section we describe how to transform a picture's gray scale so as to give the picture a specified histogram $q(z)$. The following are some cases where this type of transformation might be required:

(a)   If we want to quantize a picture f to K discrete levels in such a way as to minimize quantization error (Section 4.3.1), the K levels should be spaced close together in heavily populated regions of f's gray scale, but they can be farther apart in sparse regions. One way of doing this is to pick the levels at the midpoints between the successive K-tiles of f's histogram; this means that when we quantize, just one Kth of the points of f will be quantized to each of the levels. This "tapered" quantization scheme thus transforms f's histogram into a "flat" bar graph in which all the bars have equal height.

(b)   Suppose that we want to compare two pictures f_1 and f_2 in order to (say) detect differences between them. If the pictures were obtained under different lighting conditions, we must somehow compensate for this, since otherwise they will have different gray levels at every point even if they are pictures of the same scene. One way to carry out this compensation might be to transform the gray scale of f_1 so that its histogram matches that of f_2, or to transform both pictures so that they have some standard histogram. (On picture matching see Section 8.3.)

(c)   We often want to measure certain properties of a picture f in order to classify or describe it. If these properties depend on the gray levels that are present in f, their values will be sensitive to the lighting conditions under which f was obtained. This sensitivity can be reduced by "normalizing" f so that it has some standard histogram. (This topic is discussed further in Section 10.1.2.)

Our treatment of the histogram modification problem will deal only with the case of a *digital* picture f, say, having MN points, and quantized to K levels $z_1, ..., z_K$. Let level z_i have p_i points; thus the height of the ith bar in f's histogram is proportional to p_i, and $\sum_{i=1}^{K} p_i = MN$. Let the number of points having level z_i in the desired histogram be q_i, where $\sum_{i=1}^{K} q_i = MN$.

In order to transform the p histogram into the q histogram, we proceed as follows: Suppose that

$$\sum_{i=1}^{k_1-1} p_i \leqslant q_1 < \sum_{i=1}^{k_1} p_i$$

This means that the number of points in f that have gray levels $z_1, ..., z_{k_1-1}$ is at most q_1. Evidently, all of these points should be given gray level z_1 in the transformed picture f'. In addition, if there are fewer than q_1 of these points, some of the points of f that have gray level z_{k_1} should also be given gray level z_1 in f'. (One can choose these points randomly, or one can make their choice depend on the gray levels of their neighboring points—e.g., use points having low average neighborhood gray levels first.)

Next, let

$$\sum_{i=1}^{k_2-1} p_i \leqslant q_1 + q_2 < \sum_{i=1}^{k_2} p_i$$

This tells us which points of f should be given gray level z_2 in f', namely,

(a)   all points of gray level z_{k_1} that were not given level z_1 in f' (provided that $k_2 > k_1$; if $k_2 = k_1$, we must further subdivide the points that have level z_{k_1});

(b)   all points of gray levels $z_{k_1+1}, ..., z_{k_2-1}$;

(c)   if the left-hand inequality is strict, also some points of gray level z_{k_2} (which must be chosen as in the preceding paragraph).

The procedure is similar for the succeeding output gray levels $z_3, ..., z_K$. The general case for output level z_h is

$$\sum_{i=1}^{k_h-1} p_i \leqslant \sum_{i=1}^{h} q_i < \sum_{i=1}^{k_h} p_i.$$

Note that for the last output level z_K we have $\sum_{i=1}^{K} p_i = \sum_{i=1}^{K} q_i = MN$, so that we always end with no points left over.

As a simple example, suppose $K = 8$ and that the p's and q's are given by

i	1	2	3	4	5	6	7	8
p_i	1	7	21	35	35	21	7	1
q_i	16	16	16	16	16	16	16	16

The first few steps of the computation are as follows:

(1) $p_1 + p_2 = 8 < q_1 = 16 < p_1 + p_2 + p_3 = 29$—i.e., $k_1 = 3$. Thus all points of f having levels z_1 and z_2, as well as 8 of the 21 points having level z_3, get level z_1 in f'.

(2) $p_1 + p_2 + p_3 = 29 < q_1 + q_2 = 32 < p_1 + p_2 + p_3 + p_4 = 64$. Here $k_2 = 4$; thus the remaining 13 points of f that had level z_3, as well as 3 of the 35 points having level z_4, get level z_2 in f'.

(3) $p_1 + p_2 + p_3 = 29 < q_1 + q_2 + q_3 = 48 < p_1 + p_2 + p_3 + p_4 = 64$. Here $k_3 = k_2 = 4$, and we must further subdivide level z_4 of f—namely, 16 of its remaining points get level z_3 in f'.

Exercise 8.   Finish this example.   ∎

Two variations on the histogram modification process are illustrated in Fig. 10; the output histogram is flat in both cases. In one version, when a gray level of f has to be subdivided, the points are chosen randomly. In the second version, the choices depend on average neighborhood gray level, as previously described. (In case of a tie in average neighborhood gray level, random choice is used; but this happens much less frequently than in the first variation.) Not surprisingly, the first method appears to give somewhat noisier results. In general, the results will be noisy when we subdivide a gray level z such that f contains large regions having constant level z. (Imagine what would happen if f were half black and half white, and we tried to force all levels to occur equally often!)

As Fig. 10 shows, when a picture's histogram is transformed so that all gray levels occur equally often, the result tends to have higher contrast. In other words, performing this histogram "flattening" or "equalization" transformation not only "normalizes" the picture, as discussed earlier, but also tends to enhance it. This is because when we flatten the histogram, the points in densely populated regions of the gray scale are forced to occupy a larger number of gray levels, so that these regions of the gray scale are stretched. At the same time, points in sparse regions of the scale are forced to occupy fewer levels, so that these regions are compressed; but the stretched regions are more populous than the compressed regions, so that the overall effect is one of contrast enhancement. It may also be noted that, of all pictures having a given size and a given number of gray levels, pictures having flat histograms contain the greatest amount of information in the sense of Section 5.4.1.

6.3   GEOMETRIC CORRECTION

Another important class of image-enhancement techniques involves the correction of geometrical distortions that may be present in a picture. Some

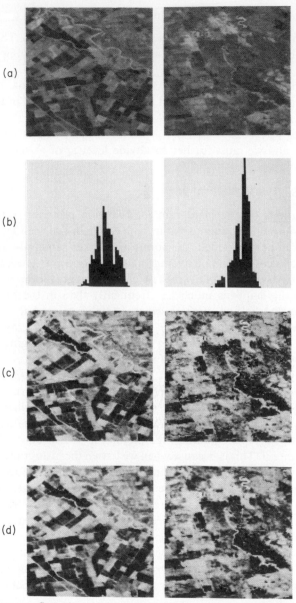

Fig. 10 Histogram flattening. (a) Input pictures. (b) Histograms for (a). (c) Results of transforming (a)'s gray scales so that they have flat histograms. When a gray level is subdivided, the decisions as to which points get which new levels are made randomly. (d) Same as (c), except that when a gray level is subdivided, the new level assigned to a point depends on the average gray level of the point's four horizontal and vertical neighbors (see text). The results are less noisy than in (c).

common examples of such distortions include perspective distortion, resulting when a picture is taken from an oblique viewing angle; and pincushion or barrel distortion, due to limitations of optical imaging or electronic scanning systems.

An arbitrary geometrical distortion is defined by equations that specify the transformation from the undistorted coordinate system (x, y) to the distorted system (x', y'). These equations are of the general form

$$x' = h_1(x, y), \qquad y' = h_2(x, y).$$

In the case of perspective distortion, the transformation is linear, i.e., of the form

$$x' = ax + by + c, \qquad y' = dx + ey + f.$$

Let f be an ideal, undistorted picture, and let g be the result of distorting f by a known geometrical transformation, defined by the functions h_1 and h_2. Thus $g(x', y') = f(x, y)$, i.e., the gray level that should appear at (x, y) actually appears at (x', y'). We shall now describe how to restore f, given g, h_1, and h_2, where we assume that f and g are both digital pictures.

(1) For each point (x_0, y_0) in the desired picture f, find the corresponding location $(\alpha, \beta) = (h_1(x_0, y_0), h_2(x_0, y_0))$ in the distorted picture g. Note that (α, β) will not, in general, coincide with any of the digital points in g, since α and β need not be integers.

(2a) Find the digital point (x_1, y_1) of g that lies nearest to (α, β), and let $f(x_0, y_0) = g(x_1, y_1)$, i.e., give (x_0, y_0) in f the gray level that (x_1, y_1) has in g.

(2b) Alternatively, suppose that (α, β) is surrounded by the four digital points

$$(x_1, y_1), \qquad (x_1 + 1, y_1), \qquad (x_1, y_1 + 1), \qquad (x_1 + 1, y_1 + 1)$$

of g, i.e., that $x_1 \leqslant \alpha < x_1 + 1$ and $y_1 \leqslant \beta < y_1 + 1$. Determine the gray level of (x_0, y_0) in f by some form of interpolation between the gray levels of these four points in g. (If desired, more than just the four points can be used.) For example, we can use linear interpolation, i.e., we give (x_0, y_0) the gray level

$$f(x_0, y_0) = (1 - \alpha')(1 - \beta') g(x_1, y_1) + \alpha'(1 - \beta') g(x_1 + 1, y_1)$$
$$+ (1 - \alpha') \beta' g(x_1, y_1 + 1) + \alpha' \beta' g(x_1 + 1, y_1 + 1)$$

where $\alpha' = \alpha - x_1$ and $\beta' = \beta - y_1$.

It should be pointed out that (α, β) may lie outside the picture g, so that we may not be able to assign it a gray level by this method. In other words, it should be realized that the corrected picture f will not, in general, be rectangular. If the transformation is discontinuous, f may not even be connected!

Exercise 9. If $h_1(x, y) = x/2$, $h_2(x, y) = y/2$, and g is

$$
\begin{array}{ccc}
1 & 1 & 1 \\
1 & 1 & 1 \\
1 & 1 & 1
\end{array}
$$

(surrounded by 0's), construct f, using both alternatives (2a) and (2b). Note that in the second case, f has more than just the two gray levels 0 and 1. ∎

Exercise 10. If $h_1(x, y) = (x-y)/\sqrt{2}$, $h_2(x, y) = (x+y)/\sqrt{2}$ (i.e., the transformation is a 45° rotation), and g is

$$
1\ 1\ 1\ 1\ 1\ 1\ 1
$$

(surrounded by 0's), construct f. What if g is

$$
\begin{array}{ccc}
& 1 & \\
1 & & \\
1 & & ? \\
1 & & \\
1 & &
\end{array}
$$

Note that "digital rotation" is not a one-to-one transformation! ∎

If h_1 and h_2 are not known, but a distorted picture g of a known pattern, such as a regular grid, is available, it is possible to determine an approximation to the distortion transformation by measuring the positions of grid points in g. For example, suppose that we are given three neighboring grid points that form a small triangle; let their ideal coordinates (in the regular grid) be (r_1, s_1), (r_2, s_2), (r_3, s_3), and let their positions in g be (u_1, v_1), (u_2, v_2), (u_3, v_3), respectively. We can find a linear transformation

$$ x' = ax + by + c, \qquad y' = dx + ey + f $$

which maps the three points into their distorted positions by solving the six equations

$$
\begin{array}{ll}
u_1 = ar_1 + bs_1 + c, & v_1 = dr_1 + es_1 + f \\
u_2 = ar_2 + bs_2 + c, & v_2 = dr_2 + es_2 + f \\
u_3 = ar_3 + bs_3 + c, & v_3 = dr_3 + es_3 + f
\end{array}
$$

for a, b, c, d, e, and f. This transformation can then be used to correct the distortion of the triangular portion of g surrounded by the lines joining the three points. The process can then be repeated for other triples of grid points. In effect, this method constructs a piecewise linear approximation to the unknown distortion. Of course, higher-order approximations can also be used if desired. An example of a distorted picture that has been corrected by piecewise approximation of the distortion derived from the distorted image of a grid is shown in Fig. 11.

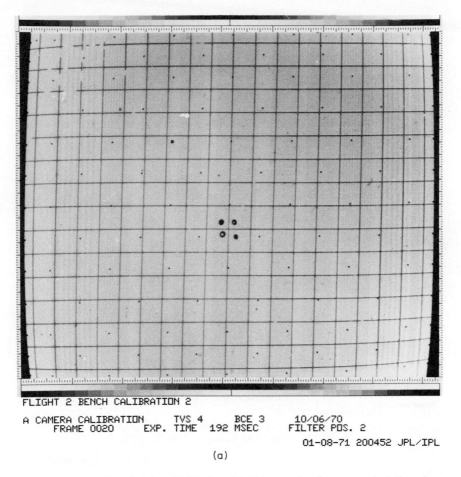

FLIGHT 2 BENCH CALIBRATION 2

A CAMERA CALIBRATION   TVS 4   BCE 3   10/06/70
      FRAME 0020   EXP. TIME 192 MSEC   FILTER POS. 2
                                                    01-08-71 200452 JPL/IPL
                              (a)

Fig. 11   Geometric correction. (a) Mariner 9 grid target, showing geometrical distortion.

6.4 SHARPENING

As pointed out in Section 6.1.1, picture degradation generally involves blurring, the extent of which is described by the spread functions of the degradation operation. In this section we discuss some simple methods for counteracting blur by "crispening" or "sharpening" the picture. Methods of deblurring based on estimation of the blur will be treated in Chapter 7.

Blurring is an averaging, or integration, operation; this suggests that we may be able to sharpen by performing differentiation operations. Blurring also weakens high spatial frequencies more than low ones; this suggests that

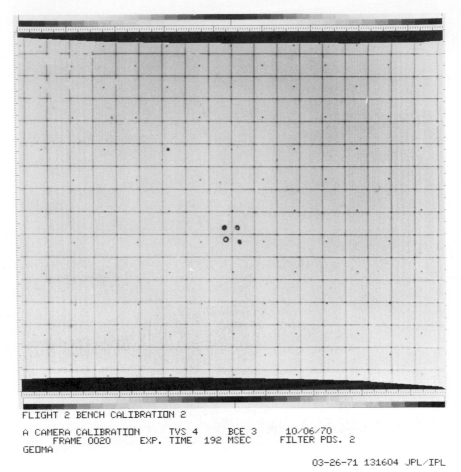

FLIGHT 2 BENCH CALIBRATION 2

A CAMERA CALIBRATION   TVS 4   BCE 3   10/06/70
        FRAME 0020   EXP. TIME 192 MSEC   FILTER POS. 2
GEOMA

                                           03-26-71 131604 JPL/IPL

(b)

Fig. 11 Geometric Correction. (b) Results of distortion removal. (From D. A. O'Handley and W. B. Green, Recent developments in digital image processing at the Image Processing Laboratory of the Jet Propulsion Laboratory, *Proc. IEEE* **60**, 1972, 821–828, Fig. 1.)

pictures can be sharpened by emphasizing their high spatial frequencies. Differentiation operations are discussed in Section 6.4.1, and a particularly useful class of these operations, the "Laplacians," are treated in Section 6.4.2. High-emphasis spatial frequency filtering is treated in Section 6.4.3.

When a picture is noisy as well as blurred, differentiation and high-emphasis filtering cannot be used indiscriminately to sharpen it, since the noise generally involves high rates of change of gray level, and it usually becomes stronger than the picture signal at high frequencies. These methods should be restricted, if possible, to frequency ranges where the picture is stronger than the noise. Alternatively, one should attempt to remove or reduce the noise before attempting to sharpen the picture. Methods of noise removal and reduction are treated in Section 6.5.

6.4.1 Differentiation

Any partial derivative operator $D = \partial^n/\partial x^k\, \partial y^{n-k}$ is a linear operator; it follows that any linear combination of D's is also a linear operator. Any arbitrary combination of D's is a local operator, since its value for a picture f at a point (x, y) depends only on the values of f in any small neighborhood of (x, y). This implies that such operators are all shift invariant. We assume in this section that all the derivatives of f exist and are continuous.

It is of particular interest to construct derivative operators that are *isotropic*, i.e., rotation invariant (in the sense that rotating f and then applying the operator gives the same result as applying the operator to f and rotating the output). We want our operators to be isotropic because we want to sharpen blurred features, such as edges and lines, that run in any direction. It can be shown that

(a)   An isotropic linear derivative operator can involve only derivatives of even orders.

(b)   In an arbitrary isotropic derivative operator, derivatives of odd orders can occur only raised to even powers.

We shall now illustrate, by a simple example, how isotropic derivative operators can be constructed. A rotation is defined by the coordinate transformation equations

$$x = x' \cos\theta - y' \sin\theta, \qquad y = x' \sin\theta + y' \cos\theta$$

where (x, y) are the unrotated and (x', y') are the rotated coordinates. The first partial derivatives of a picture f in the (x', y') system are given, in terms of its first partials in the (x, y) system, by

$$\frac{\partial f}{\partial x'} = \frac{\partial f}{\partial x}\frac{\partial x}{\partial x'} + \frac{\partial f}{\partial y}\frac{\partial y}{\partial x'} = \frac{\partial f}{\partial x}\cos\theta + \frac{\partial f}{\partial y}\sin\theta,$$

$$\frac{\partial f}{\partial y'} = \frac{\partial f}{\partial x}\frac{\partial x}{\partial y'} + \frac{\partial f}{\partial y}\frac{\partial y}{\partial y'} = -\frac{\partial f}{\partial x}\sin\theta + \frac{\partial f}{\partial y}\cos\theta$$

These partial derivatives themselves are thus not rotation invariant; but the sum of their squares in invariant, since we readily have

$$\left(\frac{\partial f}{\partial x'}\right)^2 + \left(\frac{\partial f}{\partial y'}\right)^2 = \left(\frac{\partial f}{\partial x}\right)^2 + \left(\frac{\partial f}{\partial y}\right)^2$$

Assertion (b) suggests that this sum of squares is of the simplest possible form for an isotropic derivative operator (ignoring degenerate operators that involve no derivatives at all!), since it uses only first-order derivatives raised to the smallest possible even power.

Exercise 11. Prove that the *Laplacian* operator $(\partial^2/\partial x^2)+(\partial^2/\partial y^2)$ is rotation invariant. (By assertion (a) this has the simplest possible form for a nondegenerate linear isotropic derivative operator, since it uses only derivatives of the smallest possible even order.) ▌

As the previous equations show, the partial derivative of a picture f in an arbitrary direction (e.g., $\partial f/\partial x'$) is a linear combination of its partial derivatives in the x- and y-directions. In fact, given the derivatives in any two noncollinear directions, not necessarily perpendicular to each other, we can express the derivative in any other direction as a linear combination of them.

We can find the direction in which the partial derivative of f has maximum value by differentiating $\partial f/\partial x'$ with respect to θ, setting the result equal to zero, and solving for θ. This gives us

$$-\frac{\partial f}{\partial x}\sin\theta + \frac{\partial f}{\partial y}\cos\theta = 0$$

so that the desired direction is

$$\theta_n = \tan^{-1}\left(\frac{\partial f}{\partial y}\bigg/\frac{\partial f}{\partial x}\right)$$

Note that $\theta_n+\pi$ is also a solution; in one of these directions, $\partial f/\partial x'$ is a maximum, while in the other direction it is a minimum, since it has equal magnitude but opposite sign, i.e., $\partial f/\partial(-x') = -\partial f/\partial x'$. For this θ_n we have

$$\cos\theta_n = \frac{1}{\sqrt{1+\tan^2\theta_n}} = \frac{\partial f}{\partial x}\bigg/\sqrt{\left(\frac{\partial f}{\partial x}\right)^2+\left(\frac{\partial f}{\partial y}\right)^2}$$

$$\sin\theta_n = \sqrt{1-\cos^2\theta_n} = \frac{\partial f}{\partial y}\bigg/\sqrt{\left(\frac{\partial f}{\partial x}\right)^2+\left(\frac{\partial f}{\partial y}\right)^2}$$

Thus the maximum value of $\partial f/\partial x'$ is given by

$$\frac{\partial f}{\partial x}\cos\theta_n + \frac{\partial f}{\partial y}\sin\theta_n = +\sqrt{\left(\frac{\partial f}{\partial x}\right)^2+\left(\frac{\partial f}{\partial y}\right)^2}$$

The vector whose magnitude is $\sqrt{(\partial f/\partial x)^2+(\partial f/\partial y)^2}$, and whose direction is θ_n (or $\theta_n+\pi$, whichever gives the positive partial derivative) is called the *gradient* of f. As we saw previously the magnitude of the gradient is a rotation-invariant derivative operator. The gradient, and some of its generalizations, are discussed further in Section 8.2.1, where we deal with edge detection operations.

Exercise 12.   Prove that the partial derivative of f is zero in the direction θ_t perpendicular to θ_n.   ∎

Exercise 13.   Prove that the magnitude of the gradient is equal to $\sqrt{(\partial f/\partial \alpha)^2 + (\partial f/\partial \beta)^2}$, where α and β are any two perpendicular directions.   ∎

For digital pictures, we must use differences rather than derivatives. The first differences in the x- and y-directions are

$$\Delta_x f(i,j) \equiv f(i,j) - f(i-1,j)$$
$$\Delta_y f(i,j) \equiv f(i,j) - f(i,j-1)$$

First differences in other directions can be defined as linear combinations of the x- and y-differences, e.g., as

$$\Delta_\theta f(i,j) = \Delta_x f(i,j) \cos\theta + \Delta_y f(i,j) \sin\theta$$

Using this definition, the maximum directional difference, which we call the *digital gradient*, has magnitude and direction

$$\sqrt{(\Delta_x f)^2 + (\Delta_y f)^2} \qquad \text{and} \qquad \tan^{-1}(\Delta_y f/\Delta_x f)$$

respectively. (Several other definitions of the digital gradient will be given in Section 8.2.1.)

Exercise 14.   What happens to the magnitude of the digital gradient under rotation of the digital picture (see Section 6.3)?   ∎

The higher difference operators can be defined by repeated first differencing, e.g.,

$$\Delta_x^2 f(i,j) \equiv \Delta_x f(i+1,j) - \Delta_x f(i,j)$$
$$= [f(i+1,j) - f(i,j)] - [f(i,j) - f(i-1,j)]$$
$$= f(i+1,j) + f(i-1,j) - 2f(i,j)$$

and similarly

$$\Delta_y^2 f(i,j) = f(i,j+1) + f(i,j-1) - 2f(i,j)$$

We have used these definitions [rather than, e.g., $\Delta_x f(i,j) - \Delta_x f(i-1,j)$] because they are symmetric with respect to (i,j). We could have also used symmetrical first differences, i.e.,

$$\Delta_x f(i,j) = f(i+1,j) - f(i-1,j)$$

and

$$\Delta_y f(i,j) = f(i,j+1) - f(i,j-1)$$

but we did not do so because they ignore the gray level at (i,j) itself.

6.4.2   The Laplacian

The *Laplacian* is the linear derivative operator

$$\nabla^2 f \equiv \frac{\partial^2 f}{\partial x^2} + \frac{\partial^2 f}{\partial y^2}$$

As indicated in Exercise 11, it is rotation invariant.

To see the relevance of the Laplacian to picture sharpening, suppose that the blur in the picture is a result of a diffusion process that satisfies the well-known partial differential equation

$$\partial g / \partial t = k \, \nabla^2 g$$

where g is a function of x, y, and t (time), and $k > 0$ is a constant. At $t = 0$, $g(x, y, 0)$ is the unblurred picture $f(x, y)$; at some $t = \tau > 0$, we have the observed blurred picture $g(x, y, \tau)$. If we expand $g(x, y, t)$ in a Taylor series around $t = \tau$, we have

$$g(x, y, 0) = g(x, y, \tau) - \tau \frac{\partial g}{\partial t}(x, y, \tau) + \frac{\tau^2}{2} \frac{\partial^2 g}{\partial t^2}(x, y, \tau) - \cdots.$$

If we ignore the quadratic and higher-order terms, and substitute f for $g(x, y, 0)$ and $k \, \nabla^2 g$ for $\partial g / \partial t$, this gives us

$$f = g - k\tau \, \nabla^2 g.$$

Thus, to a first approximation, we can restore the unblurred picture f by subtracting from g a positive multiple of its Laplacian. (Higher-order approximations based on the Taylor series expansion could also be used, if desired.)

Diffusion is not necessarily an appropriate model for picture blur; but the foregoing at least makes it plausible that we can achieve sharpening by a subtractive combination of the picture and its Laplacian. Other rationales for the relevance of the Laplacian to picture sharpening will be given later. It should be mentioned that according to the diffusion model, a point source blurs into a spot with a Gaussian distribution of brightness whose variance is proportional to $k\tau$; we can thus estimate $k\tau$ by fitting a Gaussian to the point spread function.

Several modifications to this Laplacian method have been proposed, with the aim of reducing noise sensitivity. One suggestion is to use, instead of the Laplacian, the second partial derivative in the gradient direction θ_n. A further refinement is to use a linear combination of the second partial derivatives in the two perpendicular directions θ_n and θ_t. This last scheme can be designed to smooth the picture in the direction along an edge at the same time that it deblurs the edge (see also Section 6.5.2).

For a digital picture, the discrete analog of the Laplacian is

$$\nabla^2 f(i, j) \equiv \Delta_x^2 f(i, j) + \Delta_y^2 f(i, j)$$
$$= [f(i+1, j) + f(i-1, j) + f(i, j+1) + f(i, j-1)] - 4f(i, j)$$

Note that this is proportional, by the factor $-\frac{1}{5}$, to

$$f(i, j) - \tfrac{1}{5}[f(i+1, j) + f(i-1, j) + f(i, j) + f(i, j+1) + f(i, j-1)]$$

which is the difference between the gray level $f(i, j)$ and the average gray level in a neighborhood of (i, j), where the neighborhood consists of (i, j) and its four horizontal and vertical neighbors. Thus we see that the digital Laplacian of a picture f is obtained, up to a constant factor, by subtracting a blurred (i.e., averaged) version of f from f itself.

Alternative digital "Laplacians" can be defined by using different neighborhoods (e.g., the 3×3 neighborhood consisting of (i, j) and its 8 horizontal, vertical, and diagonal neighbors), or by using a weighted average over the neighborhood. One could also combine the original and blurred pictures nonlinearly, e.g., divide by the blurred picture instead of subtracting it. The operation of subtracting a blurred version of a picture from the original picture (e.g., adding a blurred negative to a sharp positive) is a well-known one in photography, where it is called "unsharp masking."

The effect of the digital Laplacian on a picture can be understood by considering a simple one-dimensional example. Suppose that the $1 \times n$ digital "picture" f contains the sequence of gray levels

$$..., 0, \quad 0, \quad 0, \quad 1, \quad 2, \quad 3, \quad 4, \quad 5, \quad 5, \quad 5, \quad 5, \quad 5,$$
$$5, \quad 6, \quad 6, \quad 6, \quad 6, \quad 6, \quad 6, \quad 3, \quad 3, \quad 3, \quad 3, ...$$

It is easily verified that $\Delta_x^2 f$ at these points has the values

$$..., 0, \quad 0, \quad 1, \quad 0, \quad 0, \quad 0, \quad 0, \quad -1, \quad 0, \quad 0, \quad 0, \quad 0,$$
$$1, \quad -1, \quad 0, \quad 0, \quad 0, \quad 0, \quad -3, \quad 3, \quad 0, \quad 0, \quad 0, ...$$

Note that this is zero on flat intervals or ramps, as is appropriate for a second difference operator; and it is nonzero at the top and bottom of a ramp, and just on each side of a step. (The use of the Laplacian as an edge detector will be discussed in Section 8.2.) If we subtract $\Delta_x^2 f$ from f, point by point, we obtain

$$..., 0, \quad 0, \quad -1, \quad 1, \quad 2, \quad 3, \quad 4, \quad 6, \quad 5, \quad 5, \quad 5, \quad 5,$$
$$4, \quad 7, \quad 6, \quad 6, \quad 6, \quad 6, \quad 9, \quad 0, \quad 3, \quad 3, \quad 3, ...$$

We have now created gray level "undershoots" at the bottom of the ramp and on the low sides of the edges, and "overshoots" at the ramp top and on the high sides of the edges. This has the effect of increasing the average ramp steepness, and of increasing the contrast just at the edges. Thus we have confirmed, in this case, that subtracting the Laplacian from the picture should have a deblurring effect.

In two dimensions, $f(i,j) - \nabla^2 f(i,j)$ is

$$5f(i,j) - [f(i+1,j) + f(i-1,j) + f(i,j+1) + f(i,j-1)]$$

Readily, we have $f - \nabla^2 f = f$ when (i,j) is in the middle of a ramp or flat region. If (i,j) is just at the bottom of a ramp, or on the low side of an edge, so that some of the neighbors' gray levels are higher than $f(i,j)$, while none are lower, we have $(f - \nabla^2 f) < f$ at (i,j), which produces an undershoot. Similarly, at the top of a ramp or on the high side of an edge, none of the neighbors' gray levels are higher than $f(i,j)$, and some are lower, so that $(f - \nabla^2 f) > f$ at (i,j), producing an overshoot. A simple two-dimensional example illustrating these phenomena is shown in Fig. 12; some real cases are shown in Fig. 13.

We can obtain further insight into why subtracting a digital picture's Laplacian from the picture should have a deblurring effect by considering one last example. Suppose that the original picture f was blurred by adding to the

(a)   (b)

Fig. 12 Results of subtracting the Laplacian from the original picture in some simple cases. (a) Original pictures (ramp and bar). (b) Pictures minus Laplacians.

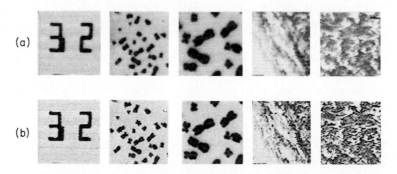

Fig. 13 Results of subtracting the Laplacian from the original picture in some real cases. (a) Original pictures. (b) Pictures minus Laplacians (negative values have been set to 0, and values beyond the end of the gray scale have been set to the highest gray level).

gray level of each point a small fraction ε of the sum of its four neighbors' levels, yielding the blurred picture g in which

$$g(i,j) = f(i,j) + \varepsilon s(i,j)$$

where $s(i,j) = f(i+1,j) + f(i-1,j) + f(i,j+1) + f(i,j-1)$. Let us subtract from g a multiple (by θ, say) of its Laplacian, obtaining

$$(1+4\theta)\,g(i,j) - \theta\,[g(i+1,j) + g(i-1,j) + g(i,j+1) + g(i,j-1)]$$

If we substitute for the g's in this expression, we obtain

$$(1+4\theta)\,f(i,j) + (1+4\theta)\,\varepsilon s(i,j)$$

$$- \theta s(i,j) - \theta\varepsilon\,[s(i+1,j) + s(i-1,j) + s(i,j+1) + s(i,j-1)]$$

Let us choose θ so that the coefficient of $s(i,j)$, which is $(1+4\theta)\varepsilon - \theta$, vanishes; this requires $\theta = \varepsilon/(1-4\varepsilon)$. If ε is sufficiently small, we can ignore terms of order ε^2; in particular, we can ignore the last term, which is a multiple of $\theta\varepsilon \doteq \varepsilon^2$. Hence

$$g(i,j) - \frac{\varepsilon\,\nabla^2 g(i,j)}{1-4\varepsilon} \doteq \left(1 + \frac{4\varepsilon}{1-4\varepsilon}\right) f(i,j)$$

Thus by subtracting a suitable multiple of g's Laplacian from g, we have approximately restored the unblurred f.

Exercise 15. In optical microscopy, let f_k be the cross section of the specimen at depth k. When we focus on that cross section, the image g_k that we obtain is degraded by the presence of blurred images of neighboring cross sections, say $g_k \doteq f_k + \varepsilon(\bar{f}_{k+1} + \bar{f}_{k-1})$, where the overbars denote local averaging. Show how f_k can be restored, to a first approximation, by subtracting a suitable multiple of $(\bar{g}_{k+1} + \bar{g}_{k-1})$ from g_k. (On this method see M. Weinstein and K. R. Castleman, Reconstructing 3-D specimens from 2-D section images, *in* "Quantitative Imagery in the Biomedical Sciences," pp. 131–137. R. E. Herron (ed.), S.P.I.E., Redondo Beach, California, 1972.) ∎

6.4.3 High-Emphasis Filtering

The effects of derivative operators in sharpening a picture can also be interpreted from a spatial frequency standpoint. Since the derivative of $\sin nx$ is $n \cos nx$, we see that the higher the spatial frequency of a sinusoidal pattern, the higher the amplitude of its derivative is. Thus differentiation strengthens high spatial frequencies more than it does low ones. (Conversely, the integral of $\cos nx$ is $(1/n) \sin nx$, so that integration weakens high frequencies more than it does low ones.)

The effect of subtracting a Laplacian from a picture can be explained along similar lines. As we saw in Section 6.4.2,

$$f(i,j) - \nabla^2 f(i,j) = f(i,j) + 5\{f(i,j) - \tfrac{1}{5}[f(i+1,j) + f(i-1,j)$$
$$+ f(i,j+1) + f(i,j-1)]\}$$

Here the second term is proportional to the difference between the original picture f and a blurred version of f, call it $\bar{f}$. Now in $\bar{f}$, high spatial frequencies have been weakened more than low ones. Hence when we subtract $\bar{f}$ from f, since the Fourier transform is a linear operator, the low frequencies in f are more or less cancelled out, while the high frequencies remain relatively intact. Thus, when we add a multiple of $f - \bar{f}$ to f, we are boosting the high frequencies, while leaving the low ones relatively unaffected.

These qualitative remarks can help to justify the use of Laplacians and other derivative operators to sharpen pictures. These methods should, however, be considered only in the absence of knowledge about the blur that is to be corrected. When such knowledge is available, restoration techniques such as inverse filtering (Section 7.2) should be used instead. In such techniques, the degree of emphasis on the high frequencies, as a function of frequency, is chosen so as to just compensate for the weakening caused by the blurring. If noise is present in addition to the blur, however, frequencies at which the noise is strong should not be emphasized. Typically, a high-emphasis filtering operation designed for blur compensation has a transfer function which increases with increasing spatial frequency, up to the point where noise would begin to dominate; it then drops off, more or less abruptly, to zero. The form of the dropoff can be designed to simulate some familiar type of transfer function; this helps avoid artifacts in the sharpened picture. A good example of the type of enhancement that can be achieved by this type of filtering is shown in Fig. 14.

By the Convolution theorem (Section 2.1.4), convolving a picture g with some function h gives the same result as multiplying the Fourier transform G of g by the Fourier transform H of h, and then taking the inverse transform. We also saw in Section 2.1.2 that any shift-invariant linear operation is a convolution operation. Analogous results hold for discrete convolutions, using the discrete Fourier transform (Section 2.2.1). Now the digital Laplacian ∇^2 is a shift-invariant linear operation; in fact, applying it to a picture g is the same as convolving g with the array h_L defined by

$$\begin{array}{ccc} & 1 & \\ 1 & -4 & 1 \\ & 1 & \end{array}$$

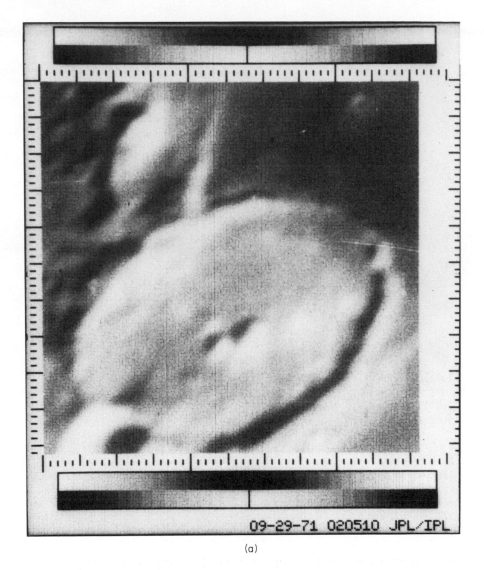

(a)

Fig. 14 High-emphasis spatial frequency filtering. (a) The lunar crater Gassendi, blurred by atmospheric turbulence.

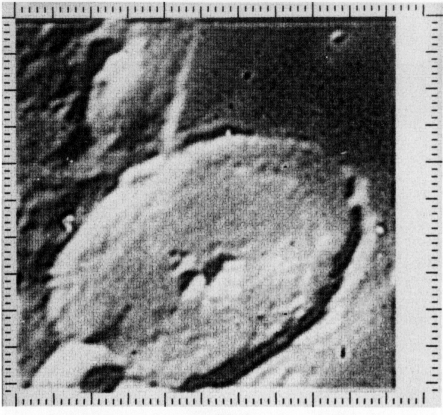

(b)

Fig. 14 High-emphasis spatial frequency filtering. (b) Results of high-emphasis filtering. (From D. A. O'Handley and W. B. Green, Recent developments in digital image processing at the Image Processing Laboratory at the Jet Propulsion Laboratory, *Proc. IEEE* **60**, 1972, 821–828, Fig. 6.)

in which all points not shown have value 0. Similarly, computing $g - \nabla^2 g$ is the same as convolving g with h_{D}, defined by

$$
\begin{array}{ccc}
 & -1 & \\
-1 & 5 & -1. \\
 & -1 &
\end{array}
$$

These space-domain convolution operations can also be performed in the frequency domain, by multiplying the discrete Fourier transform H_L (or H_D) by G and then inverse transforming.

Exercise 16. Compute the discrete Fourier transform of the array h_D (which you may assume to be $N \times N$). ▮

Conversely, given any frequency-domain high-emphasis filter H, we can apply it to a picture g in two ways: multiply g's Fourier transform G by H and inverse transform; or convolve g with the inverse Fourier transform h of H. The h's obtained in this way will generally have both positive and negative values, but $|h|$ will drop off rapidly from a central peak; thus a good approximation to the desired filtering can be obtained using only a small piece of h surrounding the peak. In the digital case, we can still obtain reasonable results if we take this piece to be as small as 3×3; the resulting truncated h generally resembles the $g - \nabla^2 g$ operator, having a positive peak symmetrically surrounded by smaller negative values, whose magnitudes add up to about 1 less than the value at the peak.

Exercise 17. Compute the h that corresponds to the one-dimensional high-emphasis filter H for which $H(n) = |n| + 1$, $-3 \leqslant n \leqslant 4$ (i.e., $N = 8$). ▮

It is sometimes advantageous to combine spatial frequency filtering with a nonlinear transformation of the gray scale. As an important example of this, suppose that we want to remove shading effects, due to uneven illumination, from a picture, without degrading the detail in the picture. Now the illumination incident on an object usually varies more slowly than does the object's reflectance. Thus uneven illumination has more significant effects at low than at high spatial frequencies, and its effects can be reduced by high-emphasis filtering. In this connection, however, it is important that, in the picture being filtered, brightness be represented on a logarithmic scale. (An example of such a scale is photographic density, which is proportional to the logarithm of transmittance or reflectance of the photograph, and is therefore usually a logarithmic function of the exposure.) Indeed, suppose that the illumination and reflectance are $i(x, y)$ and $r(x, y)$, respectively; then the resultant object brightness is given by $b(x, y) = i(x, y) r(x, y)$. It follows that the logarithm of brightness is

$$\log b(x, y) = \log i(x, y) + \log r(x, y)$$

in which i and r enter additively, rather than multiplicatively. Since the Fourier transform is a linear operation, it is now possible to deemphasize selectively the effects of $i(x, y)$ by high-emphasis filtering. If desired, the results can be redisplayed on a linear brightness scale by performing a pointwise exponentiation. An example of a picture which has been processed in this way is shown in Fig. 15.

(a)

Fig. 15 High-emphasis filtering using a logarithmic transformation. (a) Original picture.

6.5 SMOOTHING

As we saw in Section 6.1.4, many types of noise can be present in a picture. In this section we describe some simple methods of noise removal, or more generally, of making a picture "smoother." Optimal estimation of a picture that has been degraded by noise will be treated in Section 7.5.

The basic difficulty with noise removal and smoothing techniques is that, if applied indiscriminately, they tend to blur the picture, which is usually objectionable. In particular, one usually wants to avoid blurring sharp edges

(b)

Fig. 15   High-emphasis filtering using a logarithmic transformation. (b) Results of filtering. (From T. G. Stockham, Jr., Image processing in the context of a visual model, *Proc. IEEE* **60**, 1972, 828–842, Fig. 11.)

or lines that occur in the picture. Our main concern in this section will be with methods which permit smoothing without introducing undesirable blurring.

A general approach to picture smoothing would be to define a cost function φ for evaluating the various possible smoothings $\hat{f}$ of a given noisy picture g. This φ should depend both on the irregularity of $\hat{f}$ and on the discrepancy between $\hat{f}$ and g. For example, φ could be a weighted integral over the picture of $(\partial \hat{f}/\partial x)^2 + (\partial \hat{f}/\partial y)^2 + (\hat{f} - g)^2$; the first two terms give the mean magnitude of the gradient of $\hat{f}$, while the third term is the mean squared difference between

$\hat{f}$ and g. The weighting function in the integral should be defined so that φ is sensitive to noisy fluctuations in $\hat{f}$, but not to high values of $\hat{f}$'s gradient that are due to the presence of sharp long edges or lines, which are presumably not noise. Given φ, one can attempt to find a smoothed f that minimizes it; if we used a discrete version of φ, defined as a sum over the points of the digital pictures $\hat{f}$ and g, this minimization can be done, in principle, using mathematical programming techniques. In practice, however, since the number of variables in this global optimization problem is very large, it may be too difficult to find the optimum $\hat{f}$ in this way unless the picture g is small. The nonoptimal techniques described in this section involve only local processing of g, and should thus be much simpler computationally.

6.5.1 Noise Removal

If the noise occurs in known positions in the picture, or if we are able to distinguish the noise from the rest of the picture, it becomes relatively easy to remove the noise without bad effects on other parts of the picture, since we can operate only on the noise, leaving the rest of the picture intact.

As a simple example, suppose that a periodic line pattern v has been added to the picture f, as in Fig. 16a. We can remove the lines, without affecting the rest of the picture, by changing the gray level of each line point to the average of the levels of neighboring picture points. The results of line removal by this type of "interpolation" are shown in Fig. 16b. Alternatively, we can remove

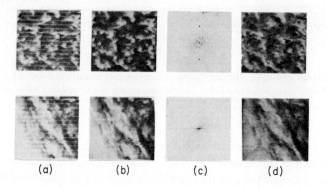

(a)   (b)   (c)   (d)

Fig. 16 Suppression of periodic noise by spatial frequency filtering. (a) Original pictures. (b) Results of removing the lines by interpolation (each point on a line is replaced by the average of the points above and below). (c) Power spectra; note high-value points corresponding to spectrum of lines. (d) Results of removing small neighborhoods of these high-value points and reconstructing the pictures.

the lines by operating in the Fourier domain. If we take the Fourier transform of the noisy picture $f + v$, we obtain the sum $F + N$ of the transforms of f and v (see Fig. 16c). Now the Fourier transform N of the line pattern v has all its energy concentrated at a set of small spots along a line perpendicular to the direction of the lines. If we suppress these spots from the transform (zero them out, or remove them by interpolation), and then take the inverse transform, we obtain a smoothed picture in which v has been deleted, while f is relatively unaffected, as shown in Fig. 16d.

Exercise 18. Describe how to remove a line pattern which has been combined multiplicatively, rather than additively, with a picture (i.e., fv rather than $f + v$). ▌

It should be pointed out that interpolation to replace the removed noise is easy when the noise is fine (isolated points, thin lines, etc.), so that each noise point has nonnoise neighboring points. When the noise is coarse (large artifacts, wide bars, etc.) it becomes much harder to replace it inconspicuously.

Another case in which noise removal is relatively easy is when the noise consists of isolated points that contrast with their neighbors (e.g., "salt-and-pepper" noise). Here we can attempt to detect noise points by comparing each point's gray level z with the levels z_i of its neighbors. If z is substantially larger (or smaller) than all, or nearly all, of the z_i, we can classify it as a noise point, and remove it by interpolation, i.e., replace z by the average of the z_i. An example of noise removal by this method is shown in Fig. 17.

The method just described has several parameters that can be adjusted to suit the characteristics of the noise that is to be detected. These parameters include the neighborhood size, the threshold amount Δ by which z and the z_i must differ ("substantially"), and the number k of z_i from which z must differ ("nearly all"). The parameter Δ might be defined as some multiple of the estimated standard deviation σ of the noise; as pointed out in Section

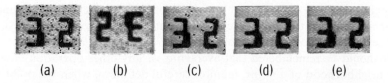

(a)   (b)   (c)   (d)   (e)

Fig. 17 Suppression of "salt-and-pepper" noise by selective local averaging. (a) Noisy picture. (b) Results of blurring (a)—each point replaced by the average of its eight neighbors. (c)–(e) Results of doing the replacement only for points that differ by at least three gray levels from k of their neighbors, for $k = 8, 7, 6$.

6.1.4, we can estimate σ by measuring the standard deviation of gray level over a region that we believe to be constant in the nonnoisy picture. (We assume here that the noise has zero mean.) The choice of parameter k is important in discriminating between noise points and points that lie on edges or lines; the latter type of point will differ from some of its neighbors, but should not differ from "nearly all" of them.

A simplified version of this method, which is often proposed, is to compare z to the average of the z_i (i.e., to examine the value of the digital Laplacian at the given point), rather than to the z_i individually. This approach, however, would be less able to distinguish isolated points from points on edges or lines, since the value of the Laplacian at a low-contrast isolated point can be the same as its value on a medium-contrast line or a high-contrast edge, as we will see in Section 8.2.1. Simplification of the method is possible, however, when the given picture is two-valued ("black and white"). Here we can detect noise points by counting the number of their neighbors from which they differ. A black point that has too many ("nearly all") white neighbors can be changed from black to white, and vice versa. (Replacement by averaging would not be appropriate here, since we want the cleaned picture to remain two-valued.)

As just described the method makes a forced-choice decision as to whether or not the given point is a noise point. If we decide that the point is noise, we change z to the average of the z_i; in other words, in choosing the new gray level for the point, we pay no attention to its old level. On the other hand, if we decide that the point is not noise, we do not change z at all, i.e., we pay no further attention to its neighbors' levels. A more general approach would be to make a "fuzzy" decision about the point, say that it has probability p of being a noise point. We could then give the point a new gray level which would be a weighted average of z and the z_i's, e.g., $(1-p)z + p\sum z_i/k$. If p is high, so that the point is highly likely to be a noise point, this formula gives high weight to the neighborhood average and low weight to the point's old gray level; while if p is low, the opposite is true.

If the average over a large neighborhood is used to replace the gray level of a suspected noise point, more weight should normally be given to nearby neighbors than to distant neighbors. Alternatively, one could determine the proper gray level to interpolate at the point by fitting some type of surface to the gray levels of the nieghboring points.

It should be remembered that averaging, or any other operation based on the neighborhood of a point, requires careful definition when the point is on or near the border of the picture. This problem is especially troublesome when large neighborhoods are used. One can, of course, give the operation a complex definition to cover the cases of points near the picture border. As a simpler alternative, one can define the operation uniformly throughout the picture, treating neighborhood points that lie outside the picture as having value zero (say), and then ignoring the results near the border as not meaningful.

6.5.2 Averaging

The smoothing methods described in the preceding section depend on first distinguishing noise from nonnoise in the picture, then removing the noise and "mending" the picture by interpolation. In this section, we discuss another class of smoothing methods which do not depend on identification and removal of the noise. Instead, these methods weaken the noise by applying some type of averaging to the picture. Since averaging blurs the picture, these methods too must be applied with care, to avoid degrading sharp detail such as lines or edges.

To see why averaging can weaken noise, suppose that the noise values at each point are independent samples chosen from a distribution having mean 0 and standard deviation σ. Suppose also that we are averaging over a set of picture points whose nonnoisy gray levels are $z_1, ..., z_n$, and let the noise values at these points be $w_1, ..., w_n$. Then in the average, $(z_1 + \cdots + z_n)/n + (w_1 + \cdots + w_n)/n$, the second term can be regarded (see Section 2.3.5, Exercise 6) as a sample of a random variable with mean 0 and standard deviation $\sigma/\sqrt{n}$. Thus by averaging, we have reduced the amplitude of the noise fluctuations.

We can perform the averaging without any danger of blurring when we are given several independently noisy copies $f_i(x, y)$ of a picture, e.g., several grainy photographs, or several "snowy" TV frames, of the same scene, where the noise values of the copies at a given point are all independent. In this case, we can reduce the noise by pointwise averaging the copies, obtaining the new picture f defined by

$$f(x, y) = (1/n) \sum_{i=1}^{n} f_i(x, y).$$

An example of smoothing by averaging of independent copies is shown in Fig. 18.

The method of pointwise averaging can also be used when the picture to be smoothed is symmetric or periodic (except for the noise). Specifically, let P be a pattern or region that occurs several times on the picture; let $P_1, ..., P_n$ be the noisy instances of P. If we average these instances pointwise, we obtain a less noisy version $\bar{P}$ of P, and we can replace each of the P_i's by $\bar{P}$ to obtain a smoother picture.

If we only have a single noisy picture available, we can attempt to reduce the

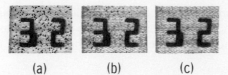

(a)   (b)   (c)

Fig. 18 Smoothing by averaging over different instances of the noise. (a)–(c) Averages of 2, 4, and 8 independently noisy versions of Fig. 17a.

noise level by *local* averaging, i.e., we give each point a new gray level which is the average of the original gray levels in some neighborhood of the point (including the point itself). This is relatively straightforward if the noise is finer grained than the smallest picture detail of interest; it should then be possible, by averaging over a small neighborhood of every point, to reduce the noise while keeping the blur at a negligible, or at least tolerable, level. Alternatively, we may be able to remove the noise by suppressing or deemphasizing high spatial frequencies—or equivalently, emphasizing low spatial frequencies—in the picture's Fourier transform. This last approach has been successfully used to remove film-grain noise from photographs.

If the given picture is two-valued, noise that is smaller than the picture detail can be removed by a process of shrinking and reexpanding. For example, suppose that we change all black points to white if they have any white neighbors, and then change all white points to black if they have any black neighbors. The first step shrinks all black regions, while the second step reexpands them; but a black object that is two points wide or less will disappear completely at the first step, so that the second step cannot restore it. Thus this process deletes not only isolated points, but also thin lines. Other uses of shrinking and expanding operations will be discussed in Sections 9.2.4 and 9.2.5.

Local averaging and low-emphasis filtering cannot be used indiscriminately if the picture contains sharp edges or fine lines, since these involve high spatial frequencies. One can, however, use some sort of *selective* local averaging in which the presence of edges and lines is detected and the averaging is done only at points where they are not present. (Edge and line detection techniques are discussed in Sections 8.2 and 8.3.) As a refinement of this idea, when an edge (or line) is present at a point, one can take a directional average at that point, involving only those neighbors of the point that lie in the direction along the edge; while if no edge is present, one can average isotropically, using neighbors on all sides of the point.

The size of the neighborhood that should be used in local averaging depends on the degree of noise reduction that is desired; as pointed out earlier, averaging n points should reduce the standard deviation of the noise by a factor of $\sqrt{n}$. If the noise is signal dependent, one may want to use a different neighborhood size at each point, depending on the average picture gray level in the vicinity of that point, in order to obtain a uniform reduced noise level throughout the picture. One way of obtaining such variable-sized neighborhoods is to calculate the sums of the gray levels in a series of expanding neighborhoods of a point, and stop expanding when the sum reaches a preset threshold. The resulting neighborhood will be large if the gray levels in the vicinity are generally low, and vice versa. The average gray level over the neighborhood is then inversely proportional to the neighborhood's area, since the sum of gray levels over the neighborhood is a constant. A variable-neighborhood averaging method of this type has been used to smooth "quantum-limited" pictures, which consist of clusters of dots; here the neighborhood size at a point was determined by expanding the neighborhood until it contained a preset number of dots.

Whether the neighborhood size used is fixed or variable, it would be desirable for the neighborhood not to cross over edges in the picture, and not to contain lines, since we want the neighborhood average to represent a "uniform" piece of the picture around the given point. We can "grow" such a neighborhood around the point by accepting neighboring points if their gray levels do not differ too greatly from that of the point or of each other; then accepting neighbors of these neighbors if their levels are not too different from the average level of the already accepted points; and so on (see Section 8.4.3 on such methods of "region growing"). Here again, "too different" can be defined in terms of the standard deviation of the noise. The growing neighborhood will hopefully exclude high-contrast edges, lines, and points, since these will differ too greatly from the surrounding average; it may thus have a highly irregular shape. The growth process can be stopped after a specified number of steps, if desired. This method is applicable only in homogeneous regions of the nonnoisy picture; it would be much more complicated to generalize it to textured regions.

Averaging that avoids edges (or lines) will be less effective in reducing the noise level near edges. Noise is, however, generally less conspicuous when it is located near edges; hence this limitation may not be serious.

A selective averaging technique can also be used to remove quantization noise from a picture. If a picture has been quantized too coarsely, i.e., to a set of levels that are spaced too far apart (say Δ apart) and are thus easily discriminable, the picture will contain conspicuous "false contours" separating regions that have different levels (see Section 4.3). We can remove the false contours by averaging the picture and quantizing the averages to a larger set of levels, closer together than Δ. To avoid blurring real edges in the picture, we can suppress the averaging at edges where the gray level changes by a multiple of Δ; but we allow averaging at edges where the change is only Δ, since such edges are likely to be false contours.

Another way to remove false contours is to add noise (of magnitude less than Δ) to the picture, again using a finer set of quantization levels; this will tend to break up the false contours. This last approach is essentially the same as the Roberts compression technique mentioned in Section 5.3. In effect, it trades quantization noise for salt-and-pepper noise.

6.6 BIBLIOGRAPHICAL NOTES

For general revies of image quality and enhancement see Linfoot [18], Huang [12], Levi [17], McCamy [20], Prewitt [27], and Biberman [4]. Two recent discussions of the use of visual system models in the design of image-processing systems, particularly as regards the nonlinear brightness response of the eye and its dependence on spatial and temporal frequency, are Budrikis [5] and Stockham [29]. The space-domain effects of frequency-domain errors are treated by Anderson and Huang [27]. On histogram modification see Hall et al. [10], Troy et al. [31], and Haralick et al. [11]; see also the references on tapered and optimum quantization in Section 4.4.

Much work on image enhancement, including both gray scale and geometric correction techniques, as well as filtering techniques for sharpening and smoothing, has been done at NASA's Jet Propulsion Laboratory (see, e.g., O'Handley and Green [23]). For further discussion of geometrical operations on digital pictures see Johnston and Rosenfeld [13].

The use of derivative operations for sharpening a picture is very old (e.g., Goldmark and Hollywood [7]; Kovasznay and Joseph [15, 16]). A more recent review by Prewitt [27] also discusses Laplacians and their modifications (those based on the second derivative in the gradient direction are due to Gabor). Transfer function modification by high-emphasis filtering is also a classical technique; for a recent discussion see Arguello et al. [3]. Regarding filtering based on a logarithmic transformation, see Oppenheim et al. [24] and Stockham [29]. On the use of derivative operations for edge and line detection see Sections 8.2 and 8.3.

A global approach to smoothing, using mathematical programming, is discussed by Martelli and Montanari [19]. Techniques for removing "salt-and-pepper" noise (in the two-valued case) are described by Dinneen [6]. On smoothing by averaging multiple copies of an image see Kohler and Howell [14]. The application to symmetric objects is given by Agrawal et al. [1] and Norman [22]. For early work on smoothing of film-grain noise see Thiry [30]. On noise removal by shrinking and reexpanding see reference [35] in Section 9.5. Regarding averaging in the absence of, or along, edges and lines, see Graham [9]; for a generalization to texture edges see Newman and Dirilten [21]. Variable-neighborhood averaging is discussed by Pizer and Vetter [25, 26] for the case of quantum-limited pictures; they also describe methods of deblurring such pictures by shifting dots in the direction for which local dot density is increasing fastest. Averaging based on region growing is treated by Roetling [28].

A major topic not treated here is that of reconstructing pictures from their one-dimensional projections, and three-dimensional objects from their two-dimensional projections. A good review of methods for solving the latter problem, for the case of nonopaque objects, is given by Gordon and Herman [8]. The related topic of estimating point spread functions from line spread functions is treated in Section 7.1.2, where additional references are given.

REFERENCES

1. H. O. Agrawal, J. W. Kent, and D. M. MacKay, Rotation technique in electron microscopy of viruses, *Science* **148**, 1965, 638–640.
2. G. B. Anderson and T. S. Huang, Frequency-domain image errors, *Pattern Recognit.* **3**, 1961, 185–196.
3. R. J. Arguello, H. R. Sellner, and J. Z. Stuller, Transfer function compensation of sampled imagery, *IEEE Trans. Comput.* **C-21**, 1972, 812–818.
4. L. M. Biberman (ed.), "Perception of Displayed Information." Plenum Press, New York, 1973.
5. Z. L. Budrikis, Visual fidelity criterion and modeling, *Proc. IEEE* **60**, 1972, 771–779.
6. G. P. Dinneen, Programming pattern recognition, *Proc. Western Joint Comput. Conf.* 1955, 94–100.
7. P. C. Goldmark and J. M. Hollywood, A new technique for improving the sharpness of television pictures, *Proc. IRE* **39**, 1951, 1314–1322.
8. R. Gordon and G. T. Herman, Three-dimensional reconstruction from projections: a review of algorithms, *Int. Rev. Cytol.* **38**, 1974, 111–151.
9. R. E. Graham, Snow removal—a noise-stripping process for picture signals, *IRE Trans. Informat. Theory* **IT-8**, 1962, 129–144.
10. E. L. Hall, R. P. Kruger, S. J. Dwyer III, D. L. Hall, R. W. McLaren, and G. S. Lodwick, A survey of preprocessing and feature extraction techniques for radiographic images, *IEEE Trans. Comput.* **C-20**, 1971, 1032–1044.
11. R. M. Haralick, K. Shanmugam, and I. Dinstein, Textural features for image classification, *IEEE Trans. Syst. Man Cybernet.* **SMC-3**, 1973, 610–621.
12. T. S. Huang, Image enhancement: a review, *Opto-Electron.* **1**, 1969, 49–59.
13. E. G. Johnston and A. Rosenfeld, Geometrical operations on digitized pictures, *in* "Picture Processing and Psychopictorics" (B. S. Lipkin and A. Rosenfeld, eds.), pp. 217–240. Academic Press, New York, 1970.
14. R. Kohler and H. Howell, Photographic image enhancement by superimposition of multiple images, *Photogr. Sci. Eng.* **7**, 1963, 241–245.
15. L. S. G. Kovasznay and H. M. Joseph, Processing of two-dimensional patterns by scanning techniques, *Science* **118**, 1953, 475–477.
16. L. S. G. Kovasznay and H. M. Joseph, Image processing ,*Proc. IRE* **43**, 1955, 560–570.
17. L. Levi, On image evaluation and enhancement, *Opt. Acta* **17**, 1970, 59–76.
18. E. H. Linfoot, "Fourier Methods in Optical Image Evaluation." Focal Press, New York, 1964.
19. A. Martelli and U. Montanari, Optimal smoothing in picture processing: an application to fingerprints, *Proc. IFIP Congr.* **71** Booklet TA-2, 86–90.

20. C. S. McCamy, The evaluation and manipulation of photographic images, *in* "Picture Processing and Psychopictorics" (B. S. Lipkin and A. Rosenfeld, eds.), pp. 57–74. Academic Press, New York, 1970.

21. T. G. Newman and H. Dirilten, A nonlinear transformation for digital picture processing, *IEEE Trans. Comput.* **C-22**, 1973, 869–873.

22. R. S. Norman, Rotation technique in radially symmetric electron micrographs: mathematical analysis, *Science* **152**, 1966, 1238–1239.

23. D. A. O'Handley and W. B. Green, Recent developments in digital image processing at the Image Processing Laboratory at the Jet Propulsion Laboratory, *Proc. IEEE* **60**, 1972, 821–828.

24. A. V. Oppenheim, R. W. Schafer, and T. G. Stockham, Jr., Nonlinear filtering of multiplied and convolved signals, *Proc. IEEE* **56**, 1968, 1264–1291.

25. S. M. Pizer and H. G. Vetter, Perception and processing of medical radio-isotope scans, *in* "Pictorial Pattern Recognition" (G. C. Cheng et al., eds.), pp. 147–156. Thompson, Washington, D. C., 1968.

26. S. M. Pizer and H. G. Vetter, Processing quantum-limited images, *in* "Picture Processing and Psychopictorics" (B. S. Lipkin and A. Rosenfeld, eds.), pp. 165–176. Academic Press, New York, 1970.

27. J. M. S. Prewitt, Object enhancemant and extraction, *in* "Picture Processing and Psychopictorics" (B. S. Lipkin and A. Rosenfeld, eds.), pp. 75–149. Academic Press, New York, 1970.

28. P. G. Roetling, Image enhancement by noise suppression, *J. Opt. Soc. Amer.* **60**, 1970, 867–869.

29. T. G. Stockham, Jr., Image processing in the context of a visual model, *Proc. IEEE* **60**, 1972, 828–842.

30. H. Thiry, Some qualitative and quantitative results on spatial filtering of granularity, *Appl. Opt.* **3**, 1964, 39–43.

31. E. B. Troy, E. S. Deutsch, and A. Rosenfeld, Gray-level manipulation experiments for texture analysis, *IEEE Trans. Syst. Man Cybernet.* **SMC-3**, 1973, 91–98.

Chapter 7

Restoration

Picture restoration deals with images that have been recorded in the presence of one or more sources of degradation. There are many sources of degradation in imaging systems. Some types of degradation affect only the gray levels of the individual picture points, without introducing spatial blur; these are some-times called *point degradations*. Other types which do involve blur are called *spatial degradations*. Still other types involve chromatic or temporal effects. In this chapter, we will only be concerned with point and spatial degradations. Such degradations occur in a variety of applications. In aerial reconnaissance, astronomy, and remote sensing the pictures one obtains are degraded by atmospheric turbulence, aberrations of the optical system, and relative motion between the camera and the object. Electron micrographs are often degraded by the spherical aberration of the electron lens. Medical radiographic images are of low resolution and contrast due to the nature of x-ray imaging systems.

Given an ideal picture $f(x, y)$ and the corresponding degraded picture $g(x, y)$, we will assume that g and f are related by

$$g(x, y) = \int \int h(x, y, x', y') f(x', y') \, dx' \, dy' + v(x, y) \qquad (1)$$

where $h(x, y, x', y')$ is the degradation function and $v(x, y)$ the random noise that may be present in the output picture. Evidently, in the absence of noise the degraded image of a point source described by $f(x', y') = \delta(x' - \alpha, y' - \beta)$

would be given by $h(x, y, \alpha, \beta)$; this can be seen by substitution in (1). There-fore, $h(x, y, \alpha, \beta)$ is a point spread function which, in general, is dependent on the position (α, β) of the point in the ideal picture.

The assumption that the observed picture $g(x, y)$ is a linear function of the ideal picture $f(x, y)$ as in (1) is approximately correct only over a small dynamic range of gray levels. For example, in a photographic system what is recorded is usually a nonlinear function of $f(x, y)$. If, however, the nonlinear characteristic (H–D curve for photographic emulsions [22]) of the film is known, it can be used to recover $f(x, y)$ from what is recorded on the film over a large dynamic range of gray levels (see Section 6.2.2).

The assumption that noise is additive is also subject to criticism (see Section 6.1.4). Many noise sources may be individually modeled as additive. When, however, additive noise is followed by a nonlinear transformation, its effect on the function $g(x, y)$ can be assumed additive only over a small dynamic range. Nevertheless, because the assumption of the additivity of noise makes the problem mathematically tractable, it is common to most work on picture restoration. In some cases multiplicative noise can be converted into an additive form by applying a logarithmic transformation to $g(x, y)$ (see refer-ence [24] in Section 6.6).

If, except for translation, the degraded image of a point is independent of the position of the point, then the point spread function (PSF) takes the form $h(x - x', y - y')$ and (1) becomes

$$g(x, y) = \int \int h(x - x', y - y') f(x', y') \, dx' \, dy' + v(x, y) \qquad (2)$$

In this case the degradation is termed *shift invariant* (Section 2.1.2). In this chapter, we will further restrict ourselves to pictures that have suffered this type of degradation.

In the absence of noise, Eq. (2) becomes

$$g(x, y) = \int \int h(x - x', y - y') f(x', y') \, dx' \, dy' \qquad (3)$$

Fourier transforming both sides, and using the convolution theorem [Eq. (23a) of Chapter 2], we obtain

$$G(u, v) = H(u, v) F(u, v) \qquad (4)$$

where $G(u, v)$, $F(u, v)$ and $H(u, v)$ are, respectively, the Fourier transforms of $g(x, y), f(x, y)$ and $h(x, y)$. The function $H(u, v)$ is the transfer function of the system that transforms the ideal picture $f(x, y)$ into the degraded picture $g(x, y)$.

7.1   THE *A PRIORI* KNOWLEDGE REQUIRED IN RESTORATION

From a mathematical standpoint, given the model described by (2) and the degraded picture $g(x, y)$, the aim of picture restoration is to make as good an estimate as possible of the original picture or scene $f(x, y)$. Evidently, any such estimation procedure must require some form of knowledge concerning the degradation function $h(x, y)$ in (2). In some cases, the physical phenomenon underlying the degradation can be used to determine $h(x, y)$. Examples of this approach are given in Section 7.1.1. In other situations, it may be possible to determine $h(x, y)$ from the degraded picture itself, e.g., when it is known *a priori* that a certain portion of the degraded picture is the image of a point, line, or edge in the original picture, as described in Section 7.1.2. Section 7.1.3 briefly discusses the *a priori* information about noise that is needed for restoration.

Recently some restoration techniques have been suggested in which some form of *a priori* knowledge is used to place constraints on the solution of the restoration algorithm. For example, the constraint that the solution be a nonnegative function is directly or indirectly implied in the various schemes that have been suggested to date [18, 19, 55, 64].

7.1.1   PSF's of Some Specific Degradations

We first consider the degradation due to diffraction in spatially incoherent optical imaging systems that are diffraction limited. If the exit pupil is denoted by a function as follows

$$P_e(x, y) = \begin{cases} 1 & \text{for} \quad (x, y) \text{ in the pupil} \\ 0 & \text{otherwise} \end{cases}$$

then the transfer function of the system is given by

$$H(u, v) = \int_{-\infty}^{\infty} \int_{-\infty}^{\infty} P_e(\xi, \eta) \, P_e(\xi - \lambda D_i u, \eta - \lambda D_i v) \, d\xi \, d\eta$$

where λ is the wavelength of the light used and D_i the distance from the exit pupil to the image plane. For a derivation of this result see [22].

Next let us consider the degradation caused in photography by relative motion between the camera and the scene. Let us further assume that the image is invariant in time except for the motion. If the relative motion is approximately the same as would be produced by motion of the recording film in its plane, then, as we will show, the degradation can be modeled by (3). The total exposure at any point of the film can be obtained by integrating the instantaneous exposures over the time interval during which the shutter is open. We will assume that the shutter requires negligible time to change from closed to open and vice versa. If $\alpha(t)$ and $\beta(t)$ are, respectively, the x- and y-components of the displacement, we have

$$G(x, y) = \int_{-T/2}^{T/2} f(x - \alpha(t), y - \beta(t))\, dt \tag{5}$$

where T is the duration of the exposure, for convenience assumed to be from $-T/2$ to $T/2$. Fourier transforming both sides of (5), we obtain

$$G(u, v) = \int dx \int dy \exp[-j2\pi(ux + vy)] \int_{-T/2}^{T/2} dt\, f(x - \alpha(t), y - \beta(t))$$

$$= \int_{-T/2}^{T/2} dt \int dx \int dy\, f(x - \alpha(t), y - \beta(t)) \exp[-j2\pi(ux + vy)]$$

By using the transformation

$$x - \alpha(t) = \xi, \qquad y - \beta(t) = \eta$$

the preceding equation can be expressed as

$$G(u, v) = \int_{-T/2}^{T/2} dt \int d\xi\, d\eta\, f(\xi, \eta)$$

$$\times \exp[-j2\pi(u\xi + v\eta)] \exp[-j2\pi(\alpha(t)u + \beta(t)v)]$$

$$= F(u, v) \int_{-T/2}^{T/2} \exp[-j2\pi(u\alpha(t) + v\beta(t))]dt$$

$$\equiv F(u, v)\, H(u, v) \tag{6}$$

which proves that the degradation can be modeled by (4) or equivalently (3). We see that the transfer function $H(u, v)$ of this degradation is given by

$$H(u, v) = \int_{-T/2}^{T/2} \exp[-j2\pi(u\alpha(t) + v\beta(t))]\, dt \tag{7}$$

For example, if the motion is uniform in the x-direction with velocity V, then

$$\alpha(t) = Vt, \qquad \beta(t) = 0 \tag{8}$$

Substituting (8) in (7), we obtain for this case

$$H(u, v) = (\sin \pi uVT)/\pi uV = T \operatorname{sinc}(uVT) \tag{9}$$

The PSF, $h(x, y)$, may be obtained by inverse Fourier transforming (9); it is $(1/V^2T) \operatorname{rect}(x/VT)$ (see Exercise 1 in Section 2.1.3).

Another case in which the PSF can be inferred from the underlying physical process is that of atmospheric turbulence. Hufnagel and Stanley [39] have shown that for long exposures the transfer function $H(u, v)$ may be approximated by $\exp[-c(u^2+v^2)^{5/6}]$, where c is a constant depending on the nature of the turbulence. For short exposures, however, the problem becomes more difficult and the transfer function, which can still be assumed to be position invariant provided the angle subtended by the object on the film is small, becomes probabilistic [15, 16, 39, 40, 54].

7.1.2   *A Posteriori* Determination of the PSF

If the degradation is of an unknown nature or if the phenomenon underlying the degradation is too complex for the analytical determination of $h(x, y)$, the only possible alternative is to estimate it from the degraded picture itself. For example, if there is any reason to believe that the original scene contains a sharp point, then the image of that point in the degraded picture is the PSF. This would be the case in an astronomical picture, where the image of a faint star could be used as an estimate of the PSF.

If the original scene contains sharp lines, then it is sometimes possible to determine $h(x, y)$ from the images of these lines. To show how it can be done, let us assume an ideal line source parallel to the x-axis in the original scene. We saw in Section 6.1.1 that the image of such a line source, denoted by $h_l(y)$, is related to the PSF by

$$h_l(y) = \int_{-\infty}^{\infty} h(x, y)\, dx \tag{10}$$

In other words the image of a line source is constant in the direction along the line, and its dependence on the perpendicualr direction is given by integrating the PSF along the line. It is clear that if the PSF is not circularly symmetric, then the image of a line source depends on the orientation of the line.

Let the Fourier transform of $h_l(y)$ be $H_l(v)$; then

$$H_l(v) = \int_{-\infty}^{\infty} h_l(y)\, e^{-j2\pi vy}\, dy \tag{11}$$

Now

$$H(u, v) = \int\!\!\!\int_{-\infty}^{\infty} h(x, y)\, \exp[-j2\pi(ux+vy)]\, dx\, dy$$

If we substitute $u = 0$ in this equation and use (10) and (11), we obtain

$$H(0, v) = \int_{-\infty}^{\infty} \left[\int_{-\infty}^{\infty} h(x, y)\, dx\right] e^{-j2\pi vy}\, dy = H_l(v) \tag{12}$$

This shows that if the image of a line parallel to the x-axis is Fourier transformed, the result gives the values of the transfer function $H(u, v)$ along the line $u = 0$ in the uv-plane. Similarly, it can be shown that the Fourier transform

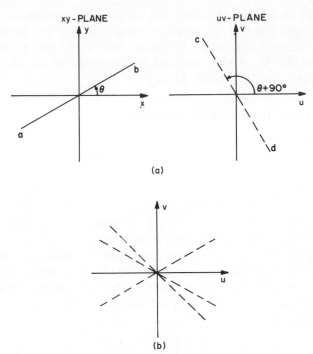

xy - PLANE

uv - PLANE

(a)

(b)

Fig. 1 (a) Fourier transform of the image of a line source parallel to *ab* in the *xy*-plane yields the values of the transfer function $H(u, v)$ along the line *cd* in the *uv*-plane. (b) If the original scene contains lines at various orientations, then from the images of these lines the values of $H(u, v)$ can be determined along radial lines such as those shown here.

of the image of a line oriented at an angle θ to the *x*-axis furnishes in the *uv*-plane the values of $H(u, v)$ along a line of slope $\theta + 90°$. This result is illustrated in Fig. 1. Therefore, if a scene that has been photographed contains lines at various orientations $\theta_1, \theta_2, \ldots, \theta_n$, then from the photograph of such a scene we can derive the values of $H(u, v)$ along radial lines through the origin at slopes $\theta_1 + 90°$, $\theta_2 + 90°$, ..., $\theta_n + 90°$.

If there is reason to believe that the PSF is circularly symmetric, then $H(u, v)$ is also circularly symmetric (Section 2.1.3, Exercise 2), so that it needs only to be known along one radial line for it to be known everywhere. If no such *a priori* knowledge is available, $H(u, v)$ would, in general, have to be known along many closely spaced radial lines. If the *uv*-plane can be covered sufficiently densely with such lines, we can construct a close approximation to $H(u, v)$, and $h(x, y)$ can then be obtained by simple Fourier inversion. This computation usually requires $H(u, v)$ to be interpolated on a rectangular lattice of points from the values on the radial lines, which can be a source of error.

The problem of determining the PSF from the images of lines at various orientations is identical to the problem of reconstructing objects from their projections. The method that was previously discussed is only one of many that can be used. Two other methods which have proved popular are the algebraic reconstruction technique and the convolution method, the latter being the fastest and most accurate from the computational standpoint. More on these methods can be found in [5, 8, 12, 21, 24, 51, 63, 70, 77].

Usually the original scene will not contain sharp points or lines. It is quite likely, however, that it will contain sharp edges. We will now show that the derivative of the image of an edge is equal to the image of a line source parallel to the edge.

Let us consider an ideal edge along the x-axis. Such an edge can be mathematically represented by $S(y)$, where $S(y)$ is a unit step function which is equal to 0 for $y < 0$ and to 1 for $y \geqslant 0$. Let $h_e(x, y)$ be the image of this edge; then

$$h_e(x, y) = \int\!\!\!\int_{-\infty}^{\infty} h(x-x', y-y') S(y') \, dx' \, dy'$$

$$= \int\!\!\!\int_{-\infty}^{\infty} h(x', y') S(y-y') \, dx' \, dy'$$

Since the image of the edge parallel to the x-axis is independent of x, we will write $h_e(y)$ instead of $h_e(x, y)$; thus

$$h_e(y) = \int\!\!\!\int_{-\infty}^{\infty} h(x', y') S(y-y') \, dx' \, dy'$$

Taking the partial derivative of both sides with respect to y and interchanging the order of the integral and derivative operators on the right-hand side, we obtain[†]

$$\frac{\partial h_e(y)}{\partial y} = \int\!\!\!\int_{-\infty}^{\infty} h(x', y') \frac{\partial}{\partial y} S(y-y') \, dx' \, dy'$$

$$= \int\!\!\!\int_{-\infty}^{\infty} h(x', y') \delta(y-y') \, dx' \, dy' = \int_{-\infty}^{\infty} h(x', y) \, dx' \qquad (13)$$

[†] We use here the fact that the derivative of a step function is a δ function. This can be justified by noting that the step is a limit of an increasingly steep ramp function of the form

and the derivative of such a ramp function is a rect function; the limit of such rect functions is a delta function.

By comparing (10) and (13), we see that

$$h_l(y) = \partial h_e(y)/\partial y \tag{14}$$

In other words, the image of a line is the derivative of the image of an edge parallel to the line. Therefore, if a picture contains edges in various orientations, then the methods previously discussed can be used to determine the PSF from the derivatives of the images of these edges. A major obstacle to determining the PSF in this way from a photograph is film grain noise which strongly affects the values of the derivative. A number of investigations have been devoted to the effects of noise and methods for overcoming it [4, 6, 7, 27, 48–50, 57, 66, 81].

Exercise 1. The operator $\partial/\partial y$ is linear and shift invariant. Prove that its transfer function is $2\pi jv$. *Hint:*

$$\frac{\partial}{\partial y} f(x, y) = \frac{\partial}{\partial y} \left[\int_{-\infty}^{\infty} F(u, v) \exp[j2\pi(ux + vy)] \, du \, dv \right] \quad \blacksquare$$

There is yet another method [38] for estimating the transfer function $H(u, v)$ from the degraded picture itself. Let the degraded picture be divided into n regions, all identical in size. Let the gray level in each region be given by $g_i(x, y), i = 1, 2, \ldots, n$. Let $f_i(x, y)$ be the gray level in the corresponding region in the undegraded picture. Let us assume that the portion of the plane in which the PSF of the degradation has values significantly different from zero is small compared to the size of the regions. Then, ignoring edge effects, we have

$$g_i(x, y) = \int\!\!\!\int_{-\infty}^{\infty} h(x - x', y - y') f_i(x', y') \, dx' \, dy', \qquad i = 1, 2, \ldots, n$$

Taking the Fourier transform of both sides, we obtain

$$G_i(u, v) = H(u, v) F_i(u, v), \qquad i = 1, 2, \ldots, n$$

Taking the product over i gives

$$\prod_{i=1}^{n} G_i(u, v) = \prod_{i=1}^{n} F_i(u, v) H^n(u, v)$$

or

$$H(u, v) = \left[\prod_{i=1}^{n} G_i(u, v) \right]^{1/n} \bigg/ \left[\prod_{i=1}^{n} F_i(u, v) \right]^{1/n} \tag{15}$$

The logarithm of the denominator on the right-hand side of (15) is

$$(1/n) \sum_{i=1}^{n} \ln F_i(u, v)$$

which is the average of the logarithms of the Fourier transforms of the f_i's. If we assume that an average of a large number of $\ln F_i(u, v)$'s tends to be approximately constant, then the denominator in (15) is essentially constant, so that (15) determines $H(u, v)$ up to a multiplicative constant. More realistically, if we write

$$\ln F_i(u, v) = \ln |F_i(u, v)| + \varphi_i(u, v)$$

where $\varphi_i(u, v)$ is the phase of $F_i(u, v)$, then it may be reasonable to assume that the average of the φ_i's is approximately constant, but we still need to know something about the average of $\ln |F_i(u, v)|$ in order to obtain $H(u, v)$ from (15). More work needs to be done to determine the classes of pictures for which this technique gives reasonable results.

7.1.3 Noise in Restoration

To restore a picture in the presence of noise, in addition to a knowledge of the PSF, one needs to know (at least in theory) both the statistical properties of the noise and how it is correlated with the picture. In practice the most common assumptions about noise are that it is white, i.e., its spectral density is constant, and that it is uncorrelated with the picture. Both these assumptions can be seriously questioned. The concept of white noise is a mathematical abstraction, but it is a convenient, though somewhat inaccurate, model provided the noise bandwidth (i.e., the region of the spatial frequency plane where the noise spectrum has values significantly different from zero) is much larger than the picture bandwidth. As far as the picture and noise being uncorrelated is concerned, there are many examples where this is not true, e.g., in the case of film grain noise [14, 36, 59] or in the case of quantum-limited images, such as x-ray and nuclear scan pictures in medicine. Other examples of signal-dependent noise are given in Section 6.1.4. On methods of restoring quantum-limited images see references [25, 26] in Chapter 6.

Different restoration techniques require different amounts of *a priori* information about the noise. For example, the restoration filter derived in Section 7.3, based on Wiener theory, requires the characterization of the noise process in terms of its spectral density. On the other hand, for the constrained deconvolution procedure discussed in Section 7.4, only the variance of the noise need be known.

7.2 INVERSE FILTERING

Assume that the degraded picture $g(x, y)$ and original $f(x, y)$ obey model (3); then, in the absence of noise, the Fourier transforms of $g(x, y), f(x, y),$

and the PSF, $h(x, y)$, satisfy Eq. (4), repeated here for convenience:

$$G(u,v) = H(u,v)\,F(u,v)$$

or equivalently,

$$F(u,v) = G(u,v)/H(u,v) \tag{16}$$

This implies that if $H(u,v)$ is known, we can restore $f(x, y)$ by multiplying the Fourier transform $G(u,v)$ of the degraded picture by $1/H(u,v)$ and then inverse Fourier transforming. In other words, the filter transfer function is $1/H(u,v)$.

A number of problems arise when one attempts to make practical use of (16). There may be points or regions in the uv-plane where $H(u,v) = 0$. In the absence of noise, the transform $G(u,v)$ of the degraded image would also be zero at these frequencies, leading to indeterminate ratios. So we see that even in the absence of noise, it is, in general, impossible to reconstruct $f(x, y)$ exactly if $H(u,v)$ has zeros in the uv-plane. (One notes, however, that if $H(u,v)$ has at most a countably infinite number of zeros, then $f(x, y)$ is, in principle, perfectly recoverable. That is because the inverse Fourier transform of $G(u,v)/H(u,v)$ is merely an integration process, to which any single frequency component contributes no area.) In the presence of noise, the zeros of $G(u,v)$ and $H(u,v)$ will not coincide. Therefore, in the neighborhoods of zeros of $H(u,v)$, the division in (16) would result in very large values. In fact, when noise is present we have

$$G(u,v) = H(u,v)\,F(u,v) + N(u,v)$$

where $N(u,v)$ is the Fourier transform of $v(x, y)$, so that applying the restoring filter gives

$$\frac{G(u,v)}{H(u,v)} = F(u,v) + \frac{N(u,v)}{H(u,v)} \tag{17}$$

In the neighborhood of zeros of $H(u,v)$, it may take on values much smaller than those of $N(u,v)$. Thus the term $N(u,v)/H(u,v)$ may have much larger magnitude than $F(u,v)$ in such a neighborhood. The inverse transform of $G(u,v)/H(u,v)$ will then be strongly influenced by these large terms, and will no longer resemble $f(x, y)$). Instead, it will contain many rapid noiselike variations, and will not be a meaningful restoration of $f(x, y)$.

Since the right-hand side of (16) cannot normally be evaluated numerically because of the zeros of $H(u,v)$, a compromise must be made. Often the most that one can do is to restore only those frequencies in the uv-plane where the signal-to-noise ratio is high. This means that in practice the inverse filter is not $1/H(u,v)$ but some other function of u and v; let us call it $M(u,v)$. Of course, if $H(u,v)$ has no zeros and noise is not present we have

$$M(u,v) = 1/H(u,v)$$

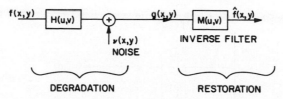

Fig. 2 Schematic representation of the degradation of a picture $f(x, y)$ by transfer function $H(u, v)$ and noise $v(x, y)$, and the restoration of the degraded picture by transfer function $M(u, v)$.

The degradation (including noise) and the restoration operation can be represented in Fig. 2. The overall transfer function of both the degradation and the restoration is given by the product $H(u, v) M(u, v)$. If $\hat{f}(x, y)$ represents the restored picture and $\hat{F}(u, v)$ its Fourier transform, then

$$\hat{F}(u, v) = (H(u, v) M(u, v)) F(u, v) \qquad (18)$$

Sometimes $H(u, v)$ is referred to as the "input transfer function," $M(u, v)$ as the "processing transfer function," and $H(u, v) M(u, v)$ as the "output transfer function."

There is considerable arbitrariness in the selection of $M(u, v)$. For example, for the case of motion blur discussed in Section 7.1.1 the function $H(u, v)$ is equal to $(\sin \pi V T u)/\pi V u$. This function, which only depends on u, is shown in Fig. 3a. Two choices of $M(u, v)$ discussed by Harris [29] are shown in Figs. 3b and 3d. The restoration corresponding to Fig. 3b consists of multiplying each frequency component by a constant times the frequency and also performing a 90° phase shift at all frequencies ($\Delta\theta = \pi/2$ in Fig. 3b). This restoring filter is mathematically described by $M(u, v) = \pi V u e^{j\pi/2}$. It can be shown that applying this filter is equivalent to differentiation in the x-direction (in the xy-plane); see Exercise 1. The output transfer function corresponding to $M(u, v)$ in Fig. 3b is shown in Fig. 3c and is of the form $\sin \pi V T u$. This is a rather extreme departure from the flat spectrum associated with ideal restoration, and very poor restoration might be expected. An examination of the PSF associated with the output transfer function, however, reveals that this method can give satisfactory results if the displacement caused by the motion is greater than the dimensions of the objects in the picture. The other choice of the restoration filter shown is in Fig. 3d; it is mathematically described by $M(u, v) = V u \pi \sin \pi V T u$. The output transfer function for this filter is shown in Fig. 3e.

Note that in both of the preceding cases the restoration filter places very little weight on those parts of the spectrum of the degraded image that are in the neighborhood of zeros of $H(u, v)$. In the vicinity of these points the signal-to-noise ratio is likely to be very low. It must be mentioned at this point that

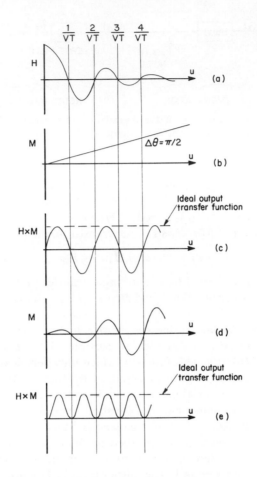

Fig. 3 (a) $H(u,v)$ for the degradation caused by uniform motion at velocity V in the x-direction. T is the exposure time. Note that $H(u,v)$ is independent of v. (b) A choice of $M(u,v)$ for the restoration filter. $M(u,v)$ also depends on u only. (c) The resulting "output transfer function" $H(u,v) M(u,v)$. (d) Another choice for the restoration filter $M(u,v)$. (e) The output transfer function $H(u,v) M(u,v)$ for the restoration filter in (d). (From [29].)

for the case of motion degradation previously discussed, it is possible under certain conditions (which include the absence of noise) to recover $f(x, y)$ perfectly from $g(x, y)$, in spite of the fact that $H(u,v)$ is zero at some points. A method for doing this is described by Slepian [73]. A variation of this technique in the presence of noise has been given by Cutrona and Hall [13]; see also Sondhi [75].

In many cases the magnitude of $H(u,v)$ drops rapidly with distance from the origin in the uv-plane. The transform $N(u,v)$ of the noise, on the other hand, while it will not in practice have constant magnitude (see Section 7.1.3), is likely to drop off much less rapidly. Thus as we get far from the origin, the quotient $N(u,v)/H(u,v)$ in (17) will become large, while (for most pictures) $F(u,v)$ will become small. This suggests that if we do not want the noise to dominate when we apply an inverse restoring filter $1/H(u,v)$, we should apply this filter only in a neighborhood of the origin; in other words, we should use a restoring filter of the form

$$M(u,v) = \begin{cases} 1/H(u,v), & u^2 + v^2 \leqslant w_0{}^2 \\ 1, & u^2 + v^2 > w_0{}^2 \end{cases}$$

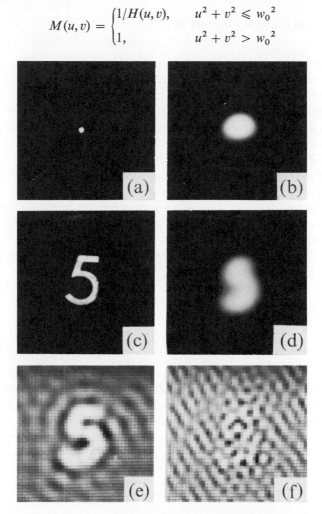

Fig. 4 (a) Picture of a point without degradation. (b) Picture of a point with degradation. This picture is approximately the PSF of the degradation. (c) Picture of the numeral 5 undegraded. (d) "5" degraded. (e) "5" restored by using Eq. (16) with division being done only for spatial frequencies below 2 cycles/mm. (f) Result of doing the division for spatial frequencies up to 3 cycles/mm. (From B. L. McGlamery, Restoration of turbulence degraded images, *J. Opt. Amer.* **57**, 1967, 295.)

for some w_0. Of course, w_0 should be chosen so as to exclude the zeros of $H(u, v)$ from the neighborhood. An example of restoration done in this way is shown in Figs. 4a–4e. The neighborhood used in Fig. 4e is small enough to avoid the zeros of $H(u, v)$, so that the restoration is fairly good. When a larger neighborhood is used, the results become very poor, as shown in Fig. 4f.

7.3 LEAST SQUARES FILTERING

One way to avoid arbitrariness in inverse filtering is to find a restoration $\hat{f}(x, y)$ of the picture $f(x, y)$ that minimizes some measure of the difference between $\hat{f}(x, y)$ and $f(x, y)$. If the restored pictures are meant for a human observer, this measure should in some manner incorporate the properties of the human visual system as discussed in Chapter 3. Such properties, however, are very hard to describe mathematically and therefore cannot be incorporated into any simple optimization procedure. One measure that has been extensively used on account of its mathematical simplicity is the mean squared error. In this section we shall derive a restoration filter that minimizes (in a statistical sense) the mean squared error between the original picture $f(x, y)$ and its restoration $\hat{f}(x, y)$. This filter is called the *least squares filter* or the *Wiener filter*.

Let the undegraded picture, the corresponding degraded picture, and the noise belong to the random fields $\mathbf{f}(\vec{r})$, $\mathbf{g}(\vec{r})$, and $\mathbf{v}(\vec{r})$, respectively, where $\vec{r}$ is the position vector in the xy-plane. Then in line with the model for degradation described by (2), the following relationship holds:

$$\mathbf{g}(\vec{r}) = \int\int h(\vec{r} - \vec{r}')\,\mathbf{f}(\vec{r}')\,d\vec{r}' + \mathbf{v}(\vec{r}) \qquad (19)$$

where $d\vec{r}'$ represents the area element $dx'\,dy'$ in the xy-plane and $h(\vec{r})$ is the point spread function of the degradation. As with the equations involving random fields in Section 2.4, (19) is a family of equations, one for each picture in the ensemble represented by the random field $\mathbf{f}(\vec{r})$.

In (19), $\mathbf{v}(\vec{r})$ is not known exactly, although its statistical properties are assumed to be known. Hence, given $\mathbf{g}(\vec{r})$, (19) cannot be exactly solved for $\mathbf{f}(\vec{r})$. What we will do here is find an $\hat{\mathbf{f}}(\vec{r})$ such that the mean square error

$$e^2 = E\{[\mathbf{f}(\vec{r}) - \hat{\mathbf{f}}(\vec{r})]^2\} \qquad (20)$$

is minimized. $\hat{\mathbf{f}}(\vec{r})$ will be called the *least squares estimate* of $\mathbf{f}(\vec{r})$ given $\mathbf{g}(\vec{r})$.

It is well known that if no restrictions are placed on the solution to the preceding problem, the least squares estimate $\hat{\mathbf{f}}(\vec{r})$ turns out to be the conditional expectation of $\mathbf{f}(\vec{r})$ given $\mathbf{g}(\vec{r})$ (Ref. [9] of Chapter 2, p. 388). This is in general a nonlinear function of the gray levels $\mathbf{g}(\vec{r})$. Moreover, for its evaluation one needs the joint probability density over the random fields $\mathbf{f}(\vec{r})$ and $\mathbf{g}(\vec{r})$, making the method complicated from an analytical as well as practical standpoint.

The problem becomes mathematically tractable if one minimizes (20) subject to the constraint that the estimate $\hat{\mathbf{f}}(r)$ be a linear function of the gray levels in $\mathbf{g}(\vec{r})$. Such an estimate is called the *linear least squares estimate*. Evidently, in general, such an estimate does not absolutely minimize (20), but of all the linear estimates it yields the smallest value for e^2. There is one case in which the optimum nonlinear estimate is the same as the linear least squares estimate, namely when the random fields $\mathbf{f}(\vec{r})$, $\mathbf{g}(\vec{r})$, and $\mathbf{v}(\vec{r})$ in addition to being homogeneous are jointly Gaussian, an assumption that is not generally valid for pictorial data.

If the estimate $\hat{\mathbf{f}}(\vec{r})$ is to be a linear function of the $\mathbf{g}(\vec{r})$, it can be expressed as

$$\hat{\mathbf{f}}(\vec{r}) = \int\int m(\vec{r}, \vec{r}')\,\mathbf{g}(\vec{r}')\,d\vec{r}' \tag{21}$$

where the yet to be determined function $m(\vec{r}, \vec{r}')$ is the weight to be given to the gray level in the degraded picture at the point $\vec{r}'$ for the computation of $\hat{\mathbf{f}}$ at $\vec{r}$. It is not hard to see that if all the random fields involved are homogeneous, this weighting function should only depend on $\vec{r} - \vec{r}'$ (Ref. [9] of Chapter 2, p. 403). Therefore, in such a case (21) can be written as

$$\hat{\mathbf{f}}(\vec{r}) = \int\int m(\vec{r} - \vec{r}')\,\mathbf{g}(\vec{r}')\,d\vec{r}' \tag{22}$$

Substituting (22) in (20), we obtain

$$e^2 = E\left\{\left[\mathbf{f}(\vec{r}) - \int\int m(\vec{r} - \vec{r}')\,\mathbf{g}(\vec{r}')\,d\vec{r}'\right]^2\right\} \tag{23}$$

Our aim is to find a function $m(\vec{r})$, which is the point spread function of the restoration filter, such that (23) is minimized. We will now show that a function satisfying

$$E\left\{\left[\mathbf{f}(\vec{r}) - \int\int m(\vec{r} - \vec{r}')\,\mathbf{g}(\vec{r}')\,d\vec{r}'\right]\mathbf{g}(\vec{s})\right\} = 0 \tag{24}$$

for all position vectors $\vec{r}$ and $\vec{s}$ in the xy-plane will minimize (23). In order to prove this result, we will show that any other choice for the restoration filter would result in a mean square error larger than that given by (23) with $m(\vec{r})$ satisfying (24). To show this, let $m(\vec{r})$ be a function that does satisfy (24). Then for any other choice $m'(\vec{r})$ for the PSF of the restoration filter the mean square error is given by

$$e'^2 = E\left\{\left[\mathbf{f}(\vec{r}) - \int\int m'(\vec{r} - \vec{r}')\,\mathbf{g}(\vec{r}')\,d\vec{r}'\right]^2\right\} \tag{25}$$

We will now show that (25) is minimized when $m'(\vec{r})$ is equal to $m(\vec{r})$. Equation (25) can be written as

$$E\left\{\left[\mathbf{f}(\vec{r}) - \int\int m(\vec{r}-\vec{r}')\mathbf{g}(\vec{r}')\,d\vec{r}'\right] + \int\int [m(\vec{r}-\vec{r}') - m'(\vec{r}-\vec{r}')]\mathbf{g}(\vec{r}')\,d\vec{r}'\right\}^2$$

which can be expressed as

$$e^2 + E\left\{\int\int [m(\vec{r}-\vec{r}') - m'(\vec{r}-\vec{r}')]\mathbf{g}(\vec{r}')\,d\vec{r}'\right\}^2$$

$$+ 2E\left\{\left[\mathbf{f}(\vec{r}) - \int\int m(\vec{r}-\vec{r}')\mathbf{g}(\vec{r}')\,d\vec{r}'\right]\right.$$

$$\left. \times \left[\int\int (m(\vec{r}-\vec{r}') - m'(\vec{r}-\vec{r}'))\mathbf{g}(\vec{r}')\,d\vec{r}'\right]\right\} \tag{26}$$

In this expression, the middle term is always nonnegative. The last term in (26) is a product of two expressions, each involving integration with respect to $\vec{r}'$. Changing the variable of integration in the second of these expressions from $\vec{r}'$ to $\vec{s}$, and interchanging the order of integration and expectation, (26) can be written as

$$e^2 + \text{a nonnegative number} + \int\int E\left\{\left[\mathbf{f}(\vec{r}) - \int\int m(\vec{r}-\vec{r}')\right.\right.$$

$$\left.\left. \times \mathbf{g}(\vec{r}')\,d\vec{r}'\right]\mathbf{g}(\vec{s})\right\}[m(\vec{r}-\vec{s}) - m'(\vec{r}-\vec{s})]\,d\vec{s} \tag{27}$$

Since $m(\vec{r})$ satisfies (24), the third term in (27) is zero. Therefore, (27) becomes

$$e^2 + \text{a nonnegative number} \tag{28}$$

so that $e'^2 \geqslant e^2$. In other words, the mean square error for an arbitrary m is always at least as great as that for an m that satisfies (24). Thus an m that satisfies (24) will give (23) its minimum possible value. It can be shown that the converse is also true, that is, an m that minimizes (23) must also satisfy (24).

Equation (24) can be written as

$$\int\int m(\vec{r}-\vec{r}')\,E\{\mathbf{g}(\vec{r}')\mathbf{g}(\vec{s})\}\,d\vec{r}' = E\{\mathbf{f}(\vec{r})\mathbf{g}(\vec{s})\} \tag{29}$$

for every $\vec{r}$ and $\vec{s}$ in the xy-plane. Making use of the definitions of the auto-correlation and cross-correlation of a random field (see Eqs. (89) and (91) of Chapter 2), (29) can be written as

$$\int\int m(\vec{r}-\vec{r}')\,R_{gg}(\vec{r}',\vec{s})\,d\vec{r}' = R_{fg}(\vec{r},\vec{s}) \tag{30}$$

for all position vectors $\vec{r}$ and $\vec{s}$ in the xy-plane. Since the random fields have been assumed to be homogeneous, the autocorrelation function $R_{gg}(\vec{r}', \vec{s})$ and the cross-correlation function $R_{fg}(\vec{r}, \vec{s})$ can be expressed as $R_{gg}(\vec{r}' - \vec{s})$ and $R_{fg}(\vec{r} - \vec{s})$, respectively (see Section 2.4.3). Thus (30) becomes

$$\int \int m(\vec{r} - \vec{r}') \, R_{gg}(\vec{r}' - \vec{s}) \, d\vec{r}' = R_{fg}(\vec{r} - \vec{s}) \tag{31}$$

for every $\vec{r}$ and $\vec{s}$ in the xy-plane. Letting $\vec{r}' - \vec{s} = \vec{i}$ and $\vec{r} - \vec{s} = \vec{\tau}$, we obtain

$$\int \int m(\vec{\tau} - \vec{i}) \, R_{gg}(\vec{i}) \, d\vec{i} = R_{fg}(\vec{\tau}) \tag{32}$$

for all position vectors $\vec{\tau}$ in the xy-plane, where $d\vec{i}$ represents an elemental area $dx\,dy$. If we denote the coordinates of $\vec{i}$ by (x, y) and of $\vec{\tau}$ by (α, β), we can write (32) as

$$\int_{-\infty}^{\infty} \int_{-\infty}^{\infty} m(\alpha - x, \beta - y) \, R_{gg}(x, y) \, dx\,dy = R_{fg}(\alpha, \beta),$$
$$-\infty < \alpha < \infty, \quad -\infty < \beta < \infty \tag{33}$$

As mentioned before, $m(x, y)$ is the point spread function of the restoration filter, and its Fourier transform $M(u, v)$ is the transfer function. If we take the Fourier transform of both sides of (33), then by the convolution theorem (Eq. (23) of Chapter 2), we have, using Eqs. (102a) and (102b) of Chapter 2,

$$M(u, v) \, S_{gg}(u, v) = S_{fg}(u, v)$$

or

$$M(u, v) = S_{fg}(u, v) / S_{gg}(u, v) \tag{34}$$

where $S_{gg}(u, v)$ is the spectral density of the degraded picture $\mathbf{g}(x, y)$, and $S_{fg}(u, v)$ is the cross spectral density of the degraded and undegraded pictures.

It is clear from (34) that, in general, the derivation of the least squares restoration filter $M(u, v)$ requires a knowledge of the cross-correlation statistics between the undegraded and degraded pictures. The filter takes a simpler form for the case where the pictures $\mathbf{f}(\vec{r})$ and noise $\mathbf{v}(\vec{r})$ are uncorrelated and where either $\mathbf{f}(\vec{r})$ or $\mathbf{v}(\vec{r})$ has zero mean, so that

$$E\{\mathbf{f}(\vec{r})\mathbf{v}(\vec{r})\} = E\{\mathbf{f}(\vec{r})\} E\{\mathbf{v}(\vec{r})\} = 0 \tag{35}$$

For such a case

$$R_{fg}(\vec{r}, \vec{s}) = E\{\mathbf{f}(\vec{r})\mathbf{g}(\vec{s})\} = \int \int h(\vec{s} - \vec{r}') \, E\{\mathbf{f}(\vec{r})\mathbf{f}(\vec{r}')\} \, d\vec{r}' \tag{36}$$

where we have made use of (19) and (35). Making use of the homogeneity of the random fields and the definition of the autocorrelation function, we obtain

$$R_{fg}(\vec{r} - \vec{s}) = \int \int h(\vec{s} - \vec{r}') \, R_{ff}(\vec{r} - \vec{r}') \, d\vec{r}'. \tag{37}$$

By using transformations similar to those employed in deriving (32) from (31), (37) can be written as

$$R_{fg}(\hat{\imath}) = \int \int h(\tilde{\tau} - \hat{\imath}) R_{ff}(\tilde{\tau}) \, d\tilde{\tau} \tag{38}$$

where $\hat{\imath}$ and $\tilde{\tau}$ are position vectors in the xy-plane and $d\tilde{\tau}$ is the elemental area $dx \, dy$. If (x, y) and (α, β) are the coordinates of $\hat{\imath}$ and $\tilde{\tau}$, respectively, we can write (38) as

$$R_{fg}(x, y) = \int_{-\infty}^{\infty} \int_{-\infty}^{\infty} h(\alpha - x, \beta - y) R_{ff}(\alpha, \beta) \, d\alpha \, d\beta \tag{39}$$

The right-hand side in (39) is the cross correlation of the two deterministic (and real) functions $h(\alpha, \beta)$ and $R_{ff}(\alpha, \beta)$. The Fourier transform of this cross correlation is given by Eq. (27) of Chapter 2. Therefore, Fourier transforming both sides of (39), we obtain

$$S_{fg}(u, v) = H^*(u, v) S_{ff}(u, v) \tag{40}$$

Also, when (35) is true, $S_{gg}(u, v)$ is given by Eq. (112) of Chapter 2:

$$S_{gg}(u, v) = S_{ff}(u, v) |H(u, v)|^2 + S_{vv}(u, v)$$

where $S_{vv}(u, v)$ is the spectral density of the noise.

Substituting Eq. (112) of Chapter 2 and (40) in (34), we obtain for the restoration filter

$$M(u, v) = \frac{H^*(u, v) S_{ff}(u, v)}{S_{ff}(u, v) |H(u, v)|^2 + S_{vv}(u, v)} \tag{41a}$$

$$= \frac{1}{H(u, v)} \frac{|H(u, v)|^2}{|H(u, v)|^2 + [S_{vv}(u, v)/S_{ff}(u, v)]} \tag{41b}$$

Note that in the absence of noise $S_{vv} = 0$, so that (41b) reduces to the ideal inverse filter $1/H(u, v)$. We thus see that the term in square brackets in (41b) can be regarded as a "modification" function which smooths $1/H(u, v)$ in order to provide optimum restoration (in the mean square sense) in the presence of noise.

Often the noise can be assumed to be spectrally white, i.e., $S_{vv}(u, v) = a$ constant. This assumption is approximately correct if $S_{ff}(u, v)$ falls off much faster in the uv-plane than $S_{vv}(u, v)$. The constant value of $S_{vv}(u, v)$ can be estimated by considering it to be equal to $S_{vv}(0, 0)$, which is equal to

$$S_{vv}(0, 0) = \int_{-\infty}^{\infty} \int_{-\infty}^{\infty} R_{vv}(x, y) \, dx \, dy \tag{42}$$

where we have used Eq. (102a) from Chapter 2. If the noise process is ergodic

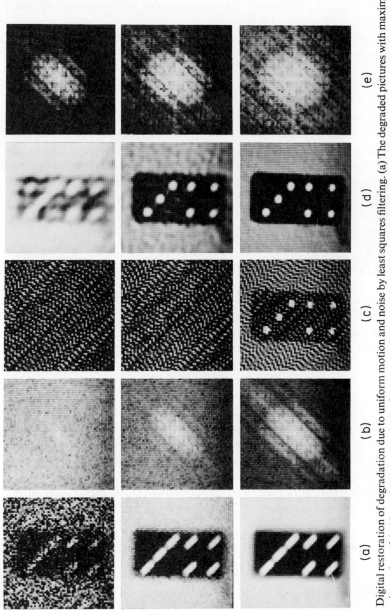

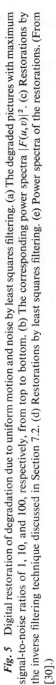

Fig. 5   Digital restoration of degradation due to uniform motion and noise by least squares filtering. (a) The degraded pictures with maximum signal-to-noise ratios of 1, 10, and 100, respectively, from top to bottom. (b) The corresponding power spectra $|F(u,v)|^2$. (c) Restorations by the inverse filtering technique discussed in Section 7.2. (d) Restorations by least squares filtering. (e) Power spectra of the restorations. (From [30].)

(Section 2.4.6), $R_{vv}(x, y)$ can be determined by applying Eq. (28) of Chapter 2 to a picture containing noise only, i.e., to an output picture $g(x, y)$ obtained when the input picture $f(x, y)$ is zero.

If no statistical properties of the random processes involved are known, it is not uncommon to approximate (41) by

$$M(u, v) = \frac{1}{H(u, v)} \frac{|H(u, v)|^2}{|H(u, v)|^2 + \Gamma}$$

where Γ, the noise-to-signal power density ratio, is approximated by a suitable constant. The value of this constant evidently reflects some *a priori* knowledge about the relative magnitudes of signal and noise power in the picture. The optimality of this form of the restoration filter is open to question and superior restoration may be achieved by other techniques [10].

Figure 5, due to Harris [30], illustrates least squares restoration of motion degraded pictures. The results are compared with those obtained by simple inverse filtering as discussed in Section 7.2.

7.4   CONSTRAINED DECONVOLUTION

In the Wiener or the least squares restoration filter derived in the preceding section, the following basic assumptions were made: that the undegraded picture and the noise belonged to homogeneous random fields and that their power spectra were known. In many situations one may not have *a priori* knowledge to this extent.

If all that is known (aside from the nature of the degradation) is the variance of the noise, then the method to be described in this section is applicable. This method was first formulated by Phillips [62] and later refined by Twomey [78, 79] in the one-dimensional case. The two-dimensional case of pictures can, in principle, be reduced to a one-dimensional problem and then solved by the method of Phillips and Twomey. The actual implementation, however, is made difficult by the large size of the matrices involved. Hunt [41–43], recognizing the special structure of degradation operators, successfully solved this problem of implementation.

7.4.1   The One-Dimensional Case

We first present the method in the one-dimensional case. For this case the equation corresponding to (2) is

$$g(x) = \int_0^x h(x - x') f(x') \, dx' + v(x) \tag{43}$$

where $g(x)$ is the degraded signal, $h(x)$ the point spread function (known as the impulse response in the one-dimensional case), $f(x)$ the ideal signal, and $v(x)$ the uncertainty in the measurement of $g(x)$. Note that $v(x)$ is a deterministic function like the other functions in (43), the difference being that it is only known in terms of its averages; this will be explained in greater detail later. In (43) we are also assuming that the signals are zero outside finite intervals, and that the origin of the coordinates is positioned such that

$$f(x) = 0 \qquad \text{except for } 0 \leqslant x \leqslant A$$

$$h(x) = 0 \qquad \text{except for } 0 \leqslant x \leqslant B$$

where these definitions imply that

$$g(x) \quad \text{and} \quad v(x) = 0 \qquad \text{except for } 0 \leqslant x \leqslant A + B$$

Of course, within the finite intervals previously defined there may be points where any of the four functions takes zero value.

The discrete approximation to (43) can be expressed as

$$g(p) = \sum_{i=0}^{p} h(p-i)\, f(i) + v(p) \qquad \text{for} \quad p = 0, 1, 2, ..., M + J - 2$$

(44)

where we have assumed that there are M numbers in the sequence $f(i)$, and J in the sequence $h(i)$, so that there are $M+J-1$ numbers in the sequence $g(p)$. Evidently, the number of terms in each of the sequences depends on the sampling interval Δx. It is known that if the noise $v(p)$ is ignored in (44) and straightforward matrix inversion techniques are used to solve for $f(i)$ given $g(p)$, the solutions become more accurate as the sampling interval Δx becomes smaller, but eventually they begin to get worse. The solutions begin to get worse sooner, the larger the magnitude of the noise $v(x)$, since this is the uncertainty in the knowledge of the true value of $g(x)$.

Equation (44) can be written in the matrix form

$$\vec{g} = [\mathfrak{H}]\vec{f} + \vec{v}$$

(45)

where $\vec{g}$, $\vec{f}$, and $\vec{v}$ are vectors composed, respectively, of the sequences $g(p)$, $f(i)$, and $v(j)$; and $[\mathfrak{H}]$ is a matrix whose (p, i)th element is

$$\mathfrak{H}(p, i) = \begin{cases} h(p-i) & \text{if } 0 \leqslant p - i \leqslant J - 1 \\ 0 & \text{otherwise} \end{cases}$$

(46)

for $p = 0, 1, 2, ..., M+J-2$ and $i = 0, 1, 2, ..., M-1$. For example, if $M = 3$

and $J = 2$, the $[\mathfrak{H}]$ matrix looks like

$$\begin{bmatrix} h(0) & 0 & 0 \\ h(1) & h(0) & 0 \\ 0 & h(1) & h(0) \\ 0 & 0 & h(1) \end{bmatrix}$$

We shall not assume that the function $v(p)$ is known but only that some of its statistical or average properties are known. In particular we shall assume that

$$\vec{v}^T \vec{v} = \sum_{p=0}^{M+J-2} v^2(p) = \varepsilon \tag{47}$$

a known constant, where the superscript T indicates the transpose. Note that if $v(p)$ has zero mean, i.e., $\sum_{p=0}^{M+J-2} v(p) \simeq 0$, then

$$\varepsilon \simeq (M+J-1)\sigma_v^2$$

where σ_v^2 is the variance of v. (In picture restoration the variance of noise can be estimated either from those regions of the picture that are almost constant in brightness or from the physical processes responsible for the noise in the measurement of $\vec{g}$.)

We can now formulate the restoration problem as follows: Given $\vec{g}$, $[\mathfrak{H}]$, and ε we want to find an $\vec{f}$ such that the "residual" $\vec{g} - [\mathfrak{H}]\vec{f}$ has the same average properties as $\vec{v}$. In particular, the residual must satisfy

$$(\vec{g} - [\mathfrak{H}]\vec{f})^T (\vec{g} - [\mathfrak{H}]\vec{f}) = \varepsilon \tag{48}$$

There may exist many functions $\vec{f}$ such that (48) is satisfied. Of all such $\vec{f}$ some other constraint must be used to select the "optimum" solution. Such a constraint must have some *a priori* plausibility. For example, if the original undegraded signals $\vec{f}$ are known to be smooth functions, then the additional constraint could be the minimization of a smoothness measure such as the second derivative. The numerical approximation to the second-order derivative at a point i is

$$f(i+1) - 2f(i) + f(i-1)$$

Therefore, the criterion for the selection of the optimum solution can be expressed as

$$\text{minimize} \left\{ \sum_{i=-1}^{M} (f(i+1) - 2f(i) + f(i-1))^2 \right\} \tag{49}$$

It can easily be verified that in matrix notation (49) can be expressed as

$$\text{minimize} \{ \vec{f}^T [C]^T [C] \vec{f} \} \tag{50}$$

where

$$[C] = \begin{bmatrix} 1 & & & & & & & & & \\ -2 & 1 & & & & & & & & \\ 1 & -2 & 1 & & & & & & & \\ & 1 & -2 & 1 & & & & & & \\ & & 1 & -2 & & & & & & \\ & & & 1 & & & & & & \\ & & & & \ddots & & & & & \\ & & & & & 1 & & & & \\ & & & & & -2 & 1 & & & \\ & & & & & 1 & -2 & 1 & & \\ & & & & & & 1 & -2 & & \\ & & & & & & & 1 & & \end{bmatrix} \qquad (51)$$

is an $(M+2) \times M$ matrix.

The restoration problem then is to find the $\vec{f}$ that minimizes (50) subject to constraint (48).[†] The required solution can be obtained by the method of Lagrange multipliers. If λ is the Lagrange multiplier, the required solution must satisfy the system of linear equations

$$\frac{\partial}{\partial f(i)} \{\lambda (\vec{g} - [\mathfrak{H}] \vec{f})^{\mathrm{T}} (\vec{g} - [\mathfrak{H}] \vec{f}) + \vec{f}^{\mathrm{T}} [C]^{\mathrm{T}} [C] \vec{f}\} = 0, \qquad i = 0, 1, ..., M - 1$$

(52)

This system, together with (48), yields $M + 1$ equations for the $M + 1$ unknowns $f(0), ..., f(M-1)$ and λ. The reader may verify that if we perform the differentiations indicated in (52), we obtain a set of equations that can be combined into the single matrix equation[‡]

$$\lambda ([\mathfrak{H}]^{\mathrm{T}} [\mathfrak{H}] \vec{f} - [\mathfrak{H}]^{\mathrm{T}} \vec{g}) + [C]^{\mathrm{T}} [C] \vec{f} = 0 \qquad (53)$$

which immediately gives the required solution

$$\vec{f} = ([\mathfrak{H}]^{\mathrm{T}} [\mathfrak{H}] + \gamma [C]^{\mathrm{T}} [C])^{-1} [\mathfrak{H}]^{\mathrm{T}} \vec{g} \qquad (54)$$

where $\gamma = 1/\lambda$.

The value of the parameter γ in (54) can be determined iteratively as follows. One chooses a value for γ, computes $\vec{f}$ using (54), and computes the residual

$$([\mathfrak{H}] \vec{f} - \vec{g})^{\mathrm{T}} ([\mathfrak{H}] \vec{f} - \vec{g}) \qquad (55)$$

[†] The corresponding problem in the continuous case would be to find f that minimizes $\int |\nabla^2 f|^2 \, dx$ subject to the constraint $\int (g - h * f)^2 \, dx = \varepsilon$, where $\varepsilon = \int v^2(x) \, dx$ is a known constant.

[‡] On taking derivatives of matrix equations, see, for example, Section 3.1 of "Estimation Theory" by R. Deutsch, Prentice-Hall, Englewood Cliffs, New Jersey, 1965.

If the value of γ is correct, then by (48) the residual should equal the known constant ε. If the computed value of (55) exceeds ε, γ is decreased; if the value is less than ε, it is increased. This is based on the fact that by substituting (54) in (55) ε can be shown to be a monotonically increasing function of γ [43].

7.4.2 Formulation in Terms of Circulant Matrices

We will show in Section 7.4.3 that the two-dimensional problem of picture restoration can be reduced to the one-dimensional case and then solved by the method in Section 7.4.1. In this case, however, M may take very large values, on the order of 40,000. Since the size of the matrix $\mathfrak{H}$ is greater than $M \times M$, it is clear that the matrix inversion in (54) can, in practice, prove to be very difficult and time consuming. In what follows, we will show that this difficulty in implementation can be circumvented by making use of the properties of circulant matrices. This was first pointed out by Hunt [42].

We first need to express (44) or (45) in a slightly different form. For an integer $P \geqslant M + J - 1$, we form new extended sequences $f_e(i)$, $g_e(i)$, $h_e(i)$, and $v_e(i)$ by padding the original sequences with zeros as follows:

$$f_e(i) = \begin{cases} f(i) & \text{for} \quad 0 \leqslant i \leqslant M - 1 \\ 0 & \text{for} \quad M \leqslant i \leqslant P - 1 \end{cases}$$

$$h_e(i) = \begin{cases} h(i) & \text{for} \quad 0 \leqslant i \leqslant J - 1 \\ 0 & \text{for} \quad J \leqslant i \leqslant P - 1 \end{cases}$$

$$g_e(i) = \begin{cases} g(i) & \text{for} \quad 0 \leqslant i \leqslant M + J - 2 \\ 0 & \text{for} \quad M + J - 1 \leqslant i \leqslant P - 1 \end{cases} \tag{56}$$

$$v_e(i) = \begin{cases} v(i) & \text{for} \quad 0 \leqslant i \leqslant M + J - 2 \\ 0 & \text{for} \quad M + J - 1 \leqslant i \leqslant P - 1 \end{cases}$$

Note that all the extended sequences are of the same length P.

Now let us construct a matrix $[H_e]$ as follows:

$$[H_e] = \begin{bmatrix} h_e(0) & h_e(P-1) & h_e(P-2) & \cdots & h_e(1) \\ h_e(1) & h_e(0) & h_e(P-1) & \cdots & h_e(2) \\ \vdots & \vdots & \vdots & & \vdots \\ h_e(P-1) & h_e(P-2) & & \cdots & h_e(0) \end{bmatrix} \tag{57}$$

It can easily be seen by substitution from (56) that with $[H_e]$ defined as above, the expression

$$\vec{g}_e = [H_e]\vec{f}_e + \vec{v}_e \tag{58}$$

is identical to (45) if $P \geqslant M + J - 1$.

The structure of the matrix in (57) is of particular importance. Note that each row can be obtained by cyclically shifting the row immediately above one position to the right. The rightmost element of the row immediately above now appears in the leftmost position. Matrices such as this are called *circulant matrices.*

Expression (50) which we want to minimize can be expressed in an equivalent form involving only circulant matrices. For example, if we want to minimize the second derivative, $[C]$ is given by (51). Let the sequence $c(i)$ be $1, -2, 1$. The extended sequence $c_e(i)$ is then given by

$$c_e(i) = \begin{cases} c(i) & \text{for } 0 \leqslant i \leqslant 2 \\ 0 & \text{for } 3 \leqslant i \leqslant P-1 \end{cases}$$

A matrix $[C_e]$ is formed from the extended sequence $c_e(i)$ in exactly the same manner as $[H_e]$ is formed from $h_e(i)$, and expression (50) is rewritten as

$$\text{minimize}\, \vec{f}^{\mathrm{T}}\,[C_e]^{\mathrm{T}}[C_e]\,\vec{f}_e \tag{59}$$

It can be verified that the expression in the brackets in (59) is equal to the corresponding expression in (50) for $P \geqslant M+J-2$.

Writing the constraint (47) as

$$v_e^{\mathrm{T}} v_e = \text{known constant } \varepsilon \tag{60}$$

it is clear from the preceding subsection that the solution of (58) that would minimize (59) subject to this constraint is given by

$$\vec{f}_e = ([H_e]^{\mathrm{T}}[H_e] + \gamma [C_e]^{\mathrm{T}}[C_e])^{-1}([H_e]^{\mathrm{T}}\vec{g}_e) \tag{61}$$

One final word about the value of P. We mentioned previously that for the convolution in (58) to be identical to that in (45), P can be any integer provided $P \geqslant M+J-1$. In other words, sufficient numbers of zeros must be inserted in the extended sequences to eliminate the "wrap-around" effect caused by $[H_e]$ being circulant. Now if we look at (61), there is a double convolution involved in it. This becomes apparent if we write it in the form

$$[H_e]^{\mathrm{T}}[H_e]\,\vec{f}_e + \gamma [C_e]^{\mathrm{T}}[C_e]\,\vec{f}_e = [H_e]^{\mathrm{T}}\vec{g}_e$$

since $[H_e]$ operating on $\vec{f}_e$ is a convolution, and $[H_e]^{\mathrm{T}}$ operating on $[H_e]\vec{f}_e$ is another convolution. To prevent the "wrap-around" effect in the first convolution, a sufficient number of zeros must be inserted in the extended sequence $\vec{f}_e$ so that the minimum value of P is $M+J-1$. To prevent the "wrap-around" caused by the second convolution, the number of zeros must be further increased by $J-1$. (Note that J is the length of the sequence $h(j)$.) Therefore, for solution (61) to be identical to (54), P must have the lower bound

$$P \geqslant M + 2J - 2 \tag{62}$$

This can also be verified by direct substitution.

Note let us define $\lambda_h(k)$, $\lambda_c(k)$, and a vector $\vec{w}(k)$ as follows:

$$\lambda_h(k) = h_e(0) + h_e(P-1)\exp\left(j\frac{2\pi}{P}k\right) + h_e(P-2)\exp\left(j\frac{2\pi}{P}2k\right)$$

$$+ \cdots + h_e(1)\exp\left(j\frac{2\pi}{P}(P-1)k\right) \tag{63a}$$

$$\lambda_c(k) = c_e(0) + c_e(P-1)\exp\left(j\frac{2\pi}{P}k\right) + c_e(P-2)\exp\left(j\frac{2\pi}{P}2k\right)$$

$$+ \cdots + c_e(1)\exp\left(j\frac{2\pi}{P}(P-1)k\right) \tag{63b}$$

$$\vec{w}_k = \begin{bmatrix} 1 \\ \exp\left(j\dfrac{2\pi}{P}k\right) \\ \exp\left(j\dfrac{2\pi}{P}2k\right) \\ \vdots \\ \exp\left(j\dfrac{2\pi}{P}(P-1)k\right) \end{bmatrix}, \qquad k = 0, 1, 2, \ldots, P-1 \tag{63c}$$

It can be directly verified that

$$[H_e]\vec{w}_k = \lambda_h(k)\vec{w}_k, \qquad k = 0, 1, 2, \ldots, P-1 \tag{64}$$

In other words, the vectors $\vec{w}_k$ are the eigenvectors of the circulant matrix $[H_e]$, and the $\lambda_h(k)$s are the corresponding eigenvalues. Similarly, $\vec{w}(k)$ and $\lambda_c(k)$ are the eigenvectors and the corresponding eigenvalues of the circulant matrix $[C_e]$.

Let $[W]$ denote the matrix formed by the column vectors $\vec{w}_k$:

$$[W] = [\vec{w}_0, \vec{w}_1, \vec{w}_2, \ldots, \vec{w}_{P-1}]$$

The (i, k) element of $[W]$ is obviously

$$W(i,k) = \exp\left(j\frac{2\pi}{P}ik\right) \tag{65}$$

It is easily seen that the inverse matrix $[W]^{-1}$ has the form

$$W^{-1}(i,k) = \frac{1}{P}\exp\left(-j\frac{2\pi}{P}ik\right) \tag{66}$$

Let $[D_h]$ and $[D_c]$ be $P \times P$ diagonal matrices whose elements are given by

$$D_h(k,k) = \lambda_h(k) \qquad (67)$$

and

$$D_c(k,k) = \lambda_c(k) \qquad (68)$$

Equations (64) can be written in a combined matrix form as

$$[H_e][W] = [W][D_h]$$

so that multiplying both sides on the right by $[W]^{-1}$, we have

$$[H_e] = [W][D_h][W]^{-1} \qquad (69)$$

Similarly, we have

$$[C_e] = [W][D_c][W]^{-1} \qquad (70)$$

By taking the transpose of both sides of (69) and (70) and making use of the fact that $[H_e]$ and $[C_e]$ are real matrices, we obtain[†]

$$[H_e]^T = [W][D_h]^*[W]^{-1} \qquad (71)$$

$$[C_e]^T = [W][D_c]^*[W]^{-1} \qquad (72)$$

Substituting (67), (68), (71), and (72) in (61), we can express the result as

$$[[D_h]^*[D_h] + \gamma [D_c]^*[D_c]][[W]^{-1}\vec{f}_e] = [D_h]^*([W]^{-1}\vec{g}_e) \qquad (73)$$

Note that $\vec{f}_e$ is a vector of length P, and from (65),

$$([W]^{-1}\vec{f}_e)_u = \frac{1}{P}\sum_{i=0}^{P-1} f_e(i) \exp\left(-j\frac{2\pi}{P}ui\right), \qquad u = 0, 1, 2, ..., P - 1 \qquad (74)$$

where $([W]^{-1}\vec{f}_e)_u$ is the uth element in the vector $([W]^{-1}\vec{f}_e)$. The right-hand side of (74) is identical to the discrete Fourier transform (DFT) of the sequence $f_e(j)$ (see Eq. (37) of Chapter 2 for the two-dimensional definition). If we

---

[†] For example, by taking the transpose of both sides of (69) we obtain

$$[H_e]^T = ([W]^{-1})^T[D_h][W]^T$$

where we have made use of the fact that $[D_h]$ is a diagonal matrix. Since $[W]^{-1}$ and $[W]$ are symmetrical matrices, we have

$$[H_e]^T = [W]^{-1}[D_h][W]$$

Taking the complex conjugate of both sides, we obtain

$$[H_e]^T = ([W]^{-1})^*[D_h]^*([W])^*$$

where we have used the fact that $[H_e]$ is a real matrix. From (66) it is clear that

$$([W]^{-1})^* = (1/P)[W] \qquad \text{and} \qquad [W]^* = P[W]^{-1}$$

from which (71) follows.

denote the DFT of the sequence $f_e(i)$ by $F(u)$, we can write

$$([W]^{-1}\vec{f_e})_u = F(u) \tag{75}$$

Similarly,

$$([W]^{-1}\vec{g_e})_u = G(u), \qquad u = 0, 1, 2, ..., P-1 \tag{76}$$

where $G(u)$ is the DFT of the sequence $g_e(i)$. By using the definitions of $\lambda_h(k)$ and $\lambda_c(k)$ in (63), it is equally easy to show that

$$D_h(u, u) = PH(u) \tag{77}$$

and

$$D_c(u, u) = PC(u), \qquad u = 0, 1, 2, ..., P-1 \tag{78}$$

where $H(u)$ and $C(u)$ are the DFT's of the sequences $h_e(i)$ and $c_e(i)$, respectively.

Substituting (75)–(78) in (73), it can be verified that we obtain the following result in the frequency domain:

$$P^2(|H(u)|^2 + \gamma|C(u)|^2)F(u) = PH^*(u)G(u)$$

which immediately gives

$$F(u) = \frac{H^*(u)G(u)}{P(|H(u)|^2 + \gamma|C(u)|^2)}, \qquad u = 0, 1, 2, ..., P-1 \tag{79}$$

If we let $M(u)$ denote the restoration filter, (79) gives us the result

$$M(u) = \frac{H^*(u)}{P(|H(u)|^2 + \gamma|C(u)|^2)} \tag{80}$$

Since fast discrete Fourier transform algorithms are available on most computers, (79) should be much easier and less time consuming to implement than (54) or (61a).

7.4.3   Application to Pictures

The equation that corresponds to (44) for the two-dimensional case is

$$g(p, q) = \sum_{i=0}^{p} \sum_{k=0}^{q} h(p-i, q-k) f(i, k) + v(p, q) \tag{81}$$

We will assume that the size of the ideal picture matrix $[f]$ is $M \times N$, and the size of the point spread function matrix $[h]$ is $J \times K$. Thus the degraded picture matrix $[g]$ and the noise matrix $[v]$ are each of size $(M+J-1) \times (N+K-1)$.

The constraint equation that corresponds to (48) for the two-dimensional case is

$$\sum_{p=0}^{M+J-2} \sum_{q=0}^{N+K-2} v^2(p,q) = \varepsilon \qquad \text{(a known constant)} \qquad (82)$$

A smoothness criterion that corresponds to (49) is

$$\text{minimize} \sum \sum [f(i-1,k)+f(i,k-1)+f(i+1,k)+f(i,k+1)-4f(i,k)]^2 \qquad (83)$$

which is obtained by convolving the discrete Laplacian operator

$$[l] \equiv \begin{bmatrix} 0 & 1 & 0 \\ 1 & -4 & 1 \\ 0 & 1 & 0 \end{bmatrix} \qquad (84)$$

with the matrix $[f]$ and summing the squares of the results at each point (i,k).

For the two-dimensional case, the restoration problem can now be stated as follows: Find a solution $[f]$ of (81) that minimizes (83) subject to the constraint (82). Evidently the smoothness criterion in (83) is one of many possible criteria.

Hunt [43] solved this problem by first expressing (81)–(84) in vector–matrix form and then using the solution derived in Section 7.4.1. We will follow Hunt's approach here. The steps that are involved in expressing these equations in vector–matrix form are as follows:

Step 1: Choose $P \geqslant M+J-1$ and $Q \geqslant N+K-1$. Form new extended matrices $[f_a]$, $[h_a]$, $[g_a]$, $[v_a]$ and $[l_a]$ as follows:

$$f_a(i,k) = \begin{cases} f(i,k) & \text{for} \quad 0 \leqslant i \leqslant M-1 \text{ and } 0 \leqslant k \leqslant N-1 \\ 0 & \text{for} \quad M \leqslant i \leqslant P-1 \text{ and } N \leqslant k \leqslant Q-1 \end{cases}$$

$$h_a(i,k) = \begin{cases} h(i,k) & \text{for} \quad 0 \leqslant i \leqslant J-1 \text{ and } 0 \leqslant k \leqslant K-1 \\ 0 & \text{for} \quad J \leqslant i \leqslant P-1 \text{ and } K \leqslant k \leqslant Q-1 \end{cases}$$

$$g_a(i,k) = \begin{cases} g(i,k) & \text{for} \quad 0 \leqslant i \leqslant M+J-2 \text{ and } 0 \leqslant k \leqslant N+K-2 \\ 0 & \text{for} \quad M+J-1 \leqslant i \leqslant P-1 \text{ and } N+K-1 \leqslant k \leqslant Q-1 \end{cases} \qquad (85)$$

$$v_a(i,k) = \begin{cases} v(i,k) & \text{for} \quad 0 \leqslant i \leqslant M+J-2 \text{ and } 0 \leqslant k \leqslant N+K-2 \\ 0 & \text{for} \quad M+J-1 \leqslant i \leqslant P-1 \text{ and } N+K-1 \leqslant k \leqslant Q-1 \end{cases}$$

$$l_a(i,k) = \begin{cases} l(i,k) & \text{for} \quad 0 \leqslant i \leqslant 2 \text{ and } 0 \leqslant k \leqslant 2 \\ 0 & \text{for} \quad 3 \leqslant i \leqslant P-1 \text{ and } 3 \leqslant k \leqslant Q-1 \end{cases}$$

Step 2:   Create column vectors $\vec{f}_a$, $\vec{g}_a$, $\vec{v}_a$, each of length PQ, by lexico-graphically ordering the matrices $[f_a]$, $[g_a]$, and $[v_a]$, respectively. In this construction, the first row of a matrix becomes the first segment of the corresponding vector, the second row the second segment, etc. Thus, for example,

$$\vec{f}_a = \begin{bmatrix} f_{a0} \\ \cdots \\ f_{a1} \\ \cdots \\ f_{a2} \\ \cdots \\ \vdots \\ \cdots \\ f_{a,P-1} \end{bmatrix} \qquad (86)$$

where the segment f_{aj} is formed by transposing the jth row of the matrix $[f_a]$. The vectors $\vec{g}_a$ and $\vec{v}_a$ are constructed in a similar manner.

Step 3:   Construct $PQ \times PQ$ matrices $[\mathscr{H}]$ as follows: $[\mathscr{H}]$ consists of P^2 blocks, each block being of size $Q \times Q$. Thus

$$[\mathscr{H}] = \begin{bmatrix} [\mathscr{H}_0] & [\mathscr{H}_{P-1}] & [\mathscr{H}_{P-2}] & \cdots & [\mathscr{H}_1] \\ [\mathscr{H}_1] & [\mathscr{H}_0] & [\mathscr{H}_{P-1}] & \cdots & [\mathscr{H}_2] \\ [\mathscr{H}_2] & [\mathscr{H}_1] & [\mathscr{H}_0] & \cdots & [\mathscr{H}_3] \\ \vdots & \vdots & \vdots & \vdots & \vdots \\ [\mathscr{H}_{P-1}] & & & \cdots & [\mathscr{H}_0] \end{bmatrix} \qquad (87)$$

where each block matrix $[\mathscr{H}_i]$ is constructed from the ith row of $[h_a]$ according to

$$[\mathscr{H}_i] = \begin{bmatrix} h_a(i,0) & h_a(i,Q-1) & h_a(i,Q-2) & \cdots & h_a(i,1) \\ h_a(i,1) & h_a(i,0) & h_a(i,Q-1) & \cdots & h_a(i,2) \\ h_a(i,2) & h_a(i,1) & h_a(i,0) & \cdots & h_a(i,3) \\ \vdots & \vdots & \vdots & \vdots & \vdots \\ h_a(i,Q-1) & h_a(i,Q-2) & h_a(i,Q-3) & \cdots & h_a(i,0) \end{bmatrix} \qquad (88)$$

Similarly, construct a $PQ \times PQ$ matrix $[\mathscr{L}]$ from $[l_a]$.

With the vectors $\vec{f}_a$, $\vec{g}_a$, $\vec{v}_a$, and matrices $[\mathscr{H}]$ and $[\mathscr{L}]$ as previously defined, it can be verified directly that Eqs. (81)–(83) can be represented by the following vector–matrix expressions:

$$\vec{g}_a = [\mathscr{H}]\,\vec{f}_a + \vec{v}_a \qquad (89)$$

$$\vec{v}_a^{\mathrm{T}}\vec{v}_a = (\vec{g}_a - [\mathscr{H}]\,\vec{f}_a)^{\mathrm{T}}(\vec{g}_a - [\mathscr{H}]\,\vec{f}_a) = \varepsilon \qquad (90)$$

$$\text{minimize } \vec{f}_a^{\mathrm{T}}[\mathscr{L}]^{\mathrm{T}}[\mathscr{L}]\,\vec{f}_a \qquad (91)$$

provided $P \geqslant M+J-1$ and $Q \geqslant N+K-1$. The restoration problem then consists of finding a solution $\vec{f}_a$ of (89) that minimizes (91) and at the same time satisfies the constraint (90). This problem is identical to the one-dimensional problem solved in Section 7.4.1. The solution is, therefore, given by (54), that is,

$$\vec{f}_a = ([\mathscr{H}]^{\mathrm{T}}[\mathscr{H}]+\gamma[\mathscr{L}]^{\mathrm{T}}[\mathscr{L}])^{-1}[\mathscr{H}]^{\mathrm{T}}\vec{g}_a \qquad (92)$$

Because of the fact that there are double convolutions involved in (92), and by the same reasoning that was used to arrive at (62), the lower bounds on P and Q must be increased to

$$P \geqslant M + 2J - 2, \qquad Q \geqslant N + 2K - 2 \qquad (93)$$

Considering that there is a matrix inversion involved in (92), if the picture matrix is large, it may not always be possible to obtain the solution by (92). For example, if the picture $f(x, y)$ is sampled on a 200×200 grid, the matrices in (92) are larger than $40,000 \times 40,000$, making direct inversion impractical.

It was because of the difficulty of implementing (92) directly that we reformulated the one-dimensional restoration problem in Section 7.4.2 in terms of circulant matrices and showed how discrete Fourier transforms could be used to implement (54) in the transform domain. Equation (92) can also be expressed in the transform domain by a method very similar to that in Section 7.4.2. Note the special structure of matrices $[\mathscr{H}]$ and $[\mathscr{L}]$ as defined in Step 3. They are called *block circulant*. Just as a circulant matrix is diagonalized by the one-dimensional discrete Fourier transform, a block circulant matrix is diagonalized by the two-dimensional discrete Fourier transform.

Let $[\mathscr{W}]$ be a $PQ \times PQ$ matrix that consists of P^2 blocks, each of size $Q \times Q$. The (m, n)th block of $[\mathscr{W}]$ is denoted by $[\mathscr{W}]_{mn}$ and is defined as

$$[\mathscr{W}]_{mn} = \exp\left(j\frac{2\pi}{P}mn\right)[W], \qquad m, n = 0, 1, 2, ..., P - 1 \qquad (94)$$

where $[W]$ is a $Q \times Q$ matrix, the (j, k)th element of which is given by

$$W(i, k) = \exp\left(j\frac{2\pi}{Q}ik\right), \qquad i, k = 0, 1, 2, ..., Q - 1 \qquad (95)$$

It can be demonstrated by direct substitution that the inverse matrix $[\mathscr{W}]^{-1}$ consists of P^2 blocks each of size $Q \times Q$. If $[\mathscr{W}]_{mn}^{-1}$ denotes the (m, n)th block of the matrix $[\mathscr{W}]^{-1}$, we have

$$[\mathscr{W}]_{mn}^{-1} = \frac{1}{P}\exp\left(-j\frac{2\pi}{P}mn\right)[W]^{-1}, \qquad m, n = 0, 1, 2, ..., P - 1 \qquad (96)$$

where $[W]^{-1}$ is a $Q \times Q$ matrix, the (i, k)th element of which is given by

$$W^{-1}(i, k) = \frac{1}{Q}\exp\left(-j\frac{2\pi}{Q}ik\right), \qquad i, k = 0, 1, ..., Q - 1 \qquad (97)$$

Let $[F_a]$, $[G_a]$, $[H_a]$, and $[L_a]$ denote the two-dimensional discrete Fourier transforms of $[f_a]$, $[g_a]$, $[h_e]$, and $[l_e]$, respectively, defined as in Eq. (37) of Chapter 2. Evidently each of the transforms is of size $P \times Q$. Now let the vector $\vec{F}_a$ of length PQ be obtained by lexicographically ordering the matrix $[F_a]$, just as $\vec{f}_a$ was obtained from $[f_a]$ in (86). Let the vector $\vec{G}_a$ be similarly obtained from $[G_a]$.

The following can now be proved (see Hunt [43] for details):

$$\vec{F}_a = [\mathscr{W}]^{-1}\vec{f}_a, \qquad \vec{G}_a = [\mathscr{W}]^{-1}\vec{g}_a$$

$$[\mathscr{H}] = [\mathscr{W}][D_h][\mathscr{W}]^{-1}, \qquad [\mathscr{H}]^{\mathrm{T}} = [\mathscr{W}][D_h]^*[\mathscr{W}]^{-1} \qquad (98)$$

$$[\mathscr{L}] = [\mathscr{W}][D_l][\mathscr{W}]^{-1}, \qquad [\mathscr{L}]^{\mathrm{T}} = [\mathscr{W}][D_l]^*[\mathscr{W}]^{-1}$$

where $[D_h]$ and $[D_l]$ are $PQ \times PQ$ diagonal matrices. The elements of these diagonal matrices are given by

$$D_h(k,k) = PQH_a([k/P], k \bmod Q),$$
$$\qquad\qquad\qquad\qquad\qquad\qquad k = 0, 1, 2, ..., PQ - 1 \quad (99)$$
$$D_l(k,k) = PQL_a([k/P], k \bmod Q),$$

where $[x]$ denotes the greatest integer less than x and $k \bmod Q$ is the remainder obtained by dividing k by Q. The matrices $[H_a]$ and $[L_a]$ were defined earlier; $H_a(u,v)$ and $L_a(u,v)$, $u = 0, 1, ..., P-1$; $v = 0, 1, ..., Q-1$ are the two-dimensional discrete Fourier transforms of $[h_a]$ and $[l_a]$, respectively. Note that by (99) the Fourier transform $H_a(u,v)$ is ordered lexicographically along the diagonal of the matrix $[D_h]$.

Substituting the last four equations of (98) in (92), making use of the first two equations in (98), and the relationship between the vectors $\vec{F}_a$ and $\vec{G}_a$ and the two-dimensional discrete Fourier transforms $[F_a]$, $[G_a]$, we obtain

$$F_a(u,v) = \frac{1}{PQ} \frac{H_a^*(u,v)\, G_a(u,v)}{|H_a(u,v)|^2 + \gamma |L_a(u,v)|^2} \qquad (100)$$

for $u = 0, 1, 2, ..., P-1$, $v = 0, 1, 2, ..., Q-1$. This equation in the frequency domain corresponds to (92) in the space domain and can be used at great savings in computation time relative to (92).

The correct value of γ in (100) can be obtained by the iterative procedure discussed in Section 7.4.1. Hunt [43] has discussed a Newton–Raphson-like procedure for incrementing γ in the iterative approach discussed earlier. Hunt points out that, depending on the initial value of γ, 3–12 iterations may be required to converge to the proper value of γ.

The transfer function $M(u,v)$ of the restoration filter for the restoration formula in (100) is given by

$$M(u,v) = \frac{1}{H_a(u,v)} \frac{|H_a(u,v)|^2}{PQ(|H(u,v)|^2 + \gamma |L_a(u,v)|^2)} \qquad (101)$$

for $u = 0, 1, 2, ..., P-1$, $v = 0, 1, 2, ..., Q-1$.

Even though this formula looks somewhat similar to the Wiener filter in (41b) (which can also be implemented in the discrete case by taking a sufficiently large number of samples in the uv-plane), there are important differences between the two. While the Wiener filter gives the best restoration in an average sense for a family of pictures, the formula here gives an optimum restoration for the *one* degraded picture that is to be restored. Also, the derivation of the Wiener filter required the basic assumption that the random fields be homogeneous and their spectral densities be known. Strictly speaking, finite pictures do not even obey the assumption of homogeneity. (When a filter is to be designed for an ensemble of pictures, however, this usually would not seriously degrade the optimality of the filter.) The filter here, on the other hand, makes no such assumptions. It does, however, require an intelligent choice for the optimality criterion, one particular form of which was given in (83) and (84).

One might think that in the formulation presented in this section, since the smoothest solution is being sought, the result should be a restored picture with blurred edges. An examination of (90), however, which the restored picture $\hat{f}_a$ must satisfy, leads to the conclusion that the degree to which the edges get blurred should to a large extent depend on the magnitude of ε, which is a measure of the uncertainty in the knowledge of the gray levels in the degraded picture.

Figure 6, due to Hunt [43], shows an example of restoration by constrained deconvolution. Figure 6a shows a picture of a resolution chart. The chart was digitized on a 450×450 grid and was blurred by a radially symmetric Gaussian shaped point spread function with a standard deviation of approximately 24 samples. To this was added random noise drawn from a uniform distribution on the interval $[0, 0.5]$. The noisy blurred picture is shown in Fig. 6b. Figure 6c shows the restoration with $\gamma = 0$ while Fig. 6d shows the restoration with the constraint on the residual satisfied. Note the improvement in quality over Fig. 6c.

7.5   RECURSIVE FILTERING

We will now introduce a technique of picture restoration in which a picture is represented by a Markov process corrupted by white noise. In this technique an attempt is made to recover the original uncorrupted picture from its degraded version by the method of least squares estimation. The optimum estimate at each point can be expressed in terms of the optimum estimate at the neighboring points and the data at that point. This permits a rapid "on-line" implementation.

A word of caution to the reader is in order here. The theory of statistical recursive filtering in two dimensions is still in its infancy and some of the

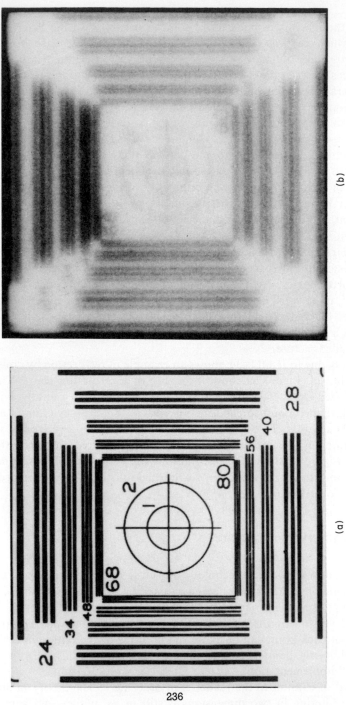

(b)

(a)

236

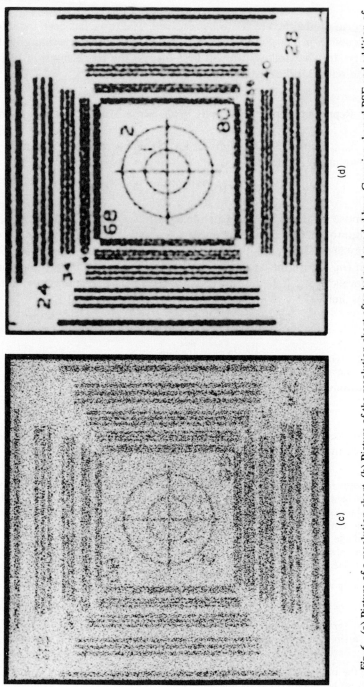

(c)

(d)

Fig. 6 (a) Picture of a resolution chart. (b) Picture of the resolution chart after being degraded by a Gaussian shaped PSF and addition of random noise. (c) Restoration with $\gamma = 0$. (d) Restoration with the constraint on the residual satisfied. (From [43].)

237

theoretical difficulties are yet to be resolved. The topic has been included here for discussion primarily to illustrate some of the basic ideas that go into such a theory. In what follows we will first discuss the representation of pictures by wide-sense Markov random fields.

7.5.1 Representation of Pictures by Wide-Sense Markov Random Fields[†]

Let $X_{m,n}$ denote all the points in an L-shaped region of a picture matrix as shown in Fig. 7, that is,

$$X_{m,n} = ((i,j) \,|\, i < m \quad \text{or} \quad j < n) \tag{102}$$

Suppose we are given an uncorrupted discrete random field $\mathbf{f}(m,n)$, assumed to be homogeneous, defined on an $M \times N$ array of points. Given that we know the gray levels $\mathbf{f}(i,j)$ at all points within $X_{m,n}$, we want to estimate the gray level at (m,n) with the constraint that this estimate be a linear function of the gray levels in $X_{m,n}$. In other words, if $\hat{\mathbf{f}}(m,n)$ is the optimum estimate for $\mathbf{f}(m,n)$, then

$$\hat{\mathbf{f}}(m,n) = \sum_{\substack{\text{for all } (i,j) \\ \text{such that} \\ (m-i,\,n-j)\,\in\,X_{m,n}}} c_{i,j}\,\mathbf{f}(m-i, n-j), \tag{103}$$

where the coefficients $c_{i,j}$ must be determined such that the mean square estimation error

$$e_{m,n} = E\{[\mathbf{f}(m,n) - \hat{\mathbf{f}}(m,n)]^2\} \tag{104}$$

is minimized; $\hat{\mathbf{f}}(m,n)$ is called the "linear least squares estimate of $\mathbf{f}(m,n)$."

Substituting (103) in (104), differentiating with respect to each $c_{i,j}$, and

[†] Markov random fields may be defined as either strict sense or wide sense. The strict-sense Markov fields are defined in terms of conditional distribution functions; and the wide-sense Markov fields are defined directly in terms of least squares estimates. For the case of Gaussian random fields the two definitions are equivalent.

The shape of the region $X_{m,n}$ is the same as that used by Abend et al. [1] for defining strict-sense Markov random fields. They have shown that if a strict-sense Markov random field is defined as one which satisfies

$$p(\mathbf{f}(m,n)\,|\,\text{gray levels at points in } X_{m,n}) = p(\mathbf{f}(m,n)\,|\,\mathbf{f}(m-1,n),\,\mathbf{f}(m-1,n-1),\,\mathbf{f}(m,n-1))$$

where $p(\mathbf{f}(m,n)\,|\,A)$ is the conditional probability density of $\mathbf{f}(m,n)$ given A, then

$$p\left(\mathbf{f}(m,n)\,\middle|\, \begin{matrix} \text{gray levels at all points} \\ \text{in the picture matrix} \\ \text{except the point } (m,n) \end{matrix} \right) = p\left(\mathbf{f}(m,n)\,\middle|\, \begin{matrix} \mathbf{f}(m-1,n-1) & \mathbf{f}(m,n-1) & \mathbf{f}(m+1,n-1) \\ \mathbf{f}(m-1,n) & & \mathbf{f}(m+1,n) \\ \mathbf{f}(m-1,n+1) & \mathbf{f}(m,n+1) & \mathbf{f}(m+1,n+1) \end{matrix} \right)$$

that is, the probability of the gray level at the point (m,n), given the rest of the picture, depends explicitly only on its eight neighbors.

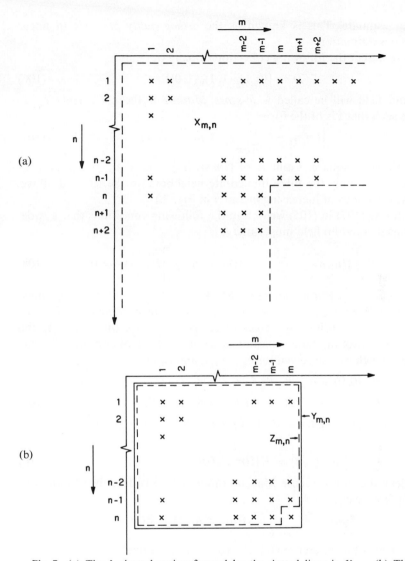

Fig. 7 (a) The L-shaped region formed by the dotted lines is $X_{m,n}$. (b) The regions $Y_{m,n}$ and $Z_{m,n}$.

setting each derivative equal to zero, we obtain the following set of simultaneous equations for the unknowns $c_{i,j}$:

$$E\{[\mathbf{f}(m,n)-\hat{\mathbf{f}}(m,n)]\,\mathbf{f}(i,j)\} = 0 \qquad \text{for} \quad \text{all } (i,j) \in X_{m,n} \qquad (105)$$

which says that the coefficients $c_{i,j}$ must be such that the estimation error $\mathbf{f}(m,n)-\hat{\mathbf{f}}(m,n)$ is statistically orthogonal to each $\mathbf{f}(i,j)$ that is used to form the linear estimate. This is known as the *orthogonality principle* in linear least squares estimation.

Let D represent the following collection of pairs (i, j):

$$D = \{(0, 1), (1, 1), (1, 0)\} \tag{106}$$

A random field will be called *wide-sense Markov* if the coefficients $c_{i,j}$ in (103) are such that $\hat{\mathbf{f}}$ is of the form

$$\hat{\mathbf{f}}(m, n) = \sum_{(i, j) \in D} c_{i, j} \mathbf{f}(m - i, n - j) \tag{107}$$

That is, the least squares estimate of $\mathbf{f}(m, n)$ in terms of $X_{m, n}$ is the same as that in terms of only the three immediate neighbors on the left and above. (Note the directions of increasing m and n in Fig. 7.)

Substituting (107) in (105) we obtain the following conditions that a wide-sense Markov random field must satisfy:

$$E\left\{\left[\mathbf{f}(m, n) - \sum_{(i, j) \in D} c_{i, j} \mathbf{f}(m - i, n - j)\right] \mathbf{f}(p, q)\right\} = 0 \tag{108}$$

for all $(p, q) \in X_{m, n}$. For a wide-sense Markov field, the coefficients $c_{i, j}$ must be such that (108) is satisfied for all $(p, q) \in X_{m, n}$. In particular, (108) must be satisfied for the following values of (p, q): $(m - 1, n)$, $(m - 1, n - 1)$, and $(m, n - 1)$. Substituting these values of (p, q) in (108), we obtain the following equations, which can be solved for $c_{1, 0}$, $c_{1, 1}$, and $c_{0, 1}$:

$$c_{1, 0} R_{ff}(0, 0) + c_{1, 1} R_{ff}(0, 1) + c_{0, 1} R_{ff}(-1, 1) = R_{ff}(-1, 0)$$

$$c_{1, 0} R_{ff}(0, 1) + c_{1, 1} R_{ff}(0, 0) + c_{0, 1} R_{ff}(-1, 0) = R_{ff}(-1, 1) \tag{109}$$

$$c_{1, 0} R_{ff}(1, -1) + c_{1, 1} R_{ff}(1, 0) + c_{0, 1} R_{ff}(0, 0) = R_{ff}(0, -1)$$

where

$$R_{ff}(\alpha, \beta) = E\{\mathbf{f}(m, n)\mathbf{f}(m + \alpha, n + \beta)\}. \tag{110}$$

It is clear that a wide-sense Markov random field is described by the following difference equation:

$$\mathbf{f}(m, n) - \sum_{(i, j) \in D} c_{i, j} \mathbf{f}(m - i, n - j) = \xi(m, n)$$

where $\xi(m, n)$ is the difference $\mathbf{f}(m, n) - \hat{\mathbf{f}}(m, n)$ at each point.

The difference equation is driven by white noise. By this we mean that $\xi(m, n)$ is an array of uncorrelated random variables. The following theorem addresses itself to this point.

Theorem. A discrete random field $\mathbf{f}(m, n)$ is wide-sense Markov if and only if it satisfies the following difference equation for all points (m, n) with $m > 1$ and $n > 1$:

$$\mathbf{f}(m, n) - \sum_{(i, j) \in D} c_{i, j} \mathbf{f}(m - i, n - j) = \xi(m, n) \tag{111}$$

and, for the points on the topmost row and the leftmost columns,

$$\mathbf{f}(1,1) = \xi(1,1) \tag{112a}$$

$$\mathbf{f}(m,1) - \mathscr{A}\mathbf{f}(m-1,1) = \xi(m,1), \qquad m > 1 \tag{112b}$$

$$\mathbf{f}(1,n) - \mathscr{B}\mathbf{f}(1,n-1) = \xi(1,n), \qquad n > 1 \tag{112c}$$

where the $c_{i,j}$ are the solutions of (109), and

$$\mathscr{A} = R_{ff}(1,0)/R_{ff}(0,0), \qquad \mathscr{B} = R_{ff}(0,1)/R_{ff}(0,0) \tag{113}$$

and where $\xi(m,n)$ is a discrete random field of orthogonal random variables, that is,

$$E\{\xi(m,n)\xi(p,q)\} = 0, \qquad m \neq p \quad \text{or} \quad n \neq q \tag{114}$$

For a zero-mean random field the random variables $\xi(m,n)$ are uncorrelated.

Proof: First a few words about Eqs. (112). The difference equation (112b) for points on the topmost row is consistent with (111). In keeping with our definition of a wide-sense Markov random field, we insist that the least squares estimate $\hat{\mathbf{f}}(m,1)$ of $\mathbf{f}(m,1)$ in terms of all the points in $X_{m,1}$ be the same as that in terms of only $\mathbf{f}(m-1,1)$, that is,

$$\hat{\mathbf{f}}(m,1) = \mathscr{A}\mathbf{f}(m-1,1) \tag{115}$$

where, by arguments similar to those leading to (108), the coefficient $\mathscr{A}$ must be such that

$$E\{[\mathbf{f}(m,1) - \mathscr{A}\mathbf{f}(m-1,1)]\mathbf{f}(p,1)\} = 0, \qquad \text{for all } (p,1) \in X_{m,1} \tag{116}$$

With $p = m-1$, this relationship yields

$$\mathscr{A} = R_{ff}(-1,0)/R_{ff}(0,0) = R_{ff}(1,0)/R_{ff}(0,0)$$

Clearly, then,

$$\mathbf{f}(m,1) = \mathscr{A}\mathbf{f}(m-1,1) + \xi(m,1)$$

where $\xi(m,1)$ is the error $\mathbf{f}(m,1) - \hat{\mathbf{f}}(m,1)$.

It can, similarly, be shown that the points on the leftmost column should satisfy the following orthogonality condition:

$$E\{[\mathbf{f}(1,n) - \mathscr{B}\mathbf{f}(1,n-1)]\mathbf{f}(1,q)\} = 0 \qquad \text{for all } (1,q) \in X_{1,q} \tag{117}$$

As for the case of topmost row points, (117) implies that points on the leftmost column should obey (112c).

Coming to the proof of the theorem, let us assume that the field is wide-sense Markov; we then want to show that (114) is true. Note that from (108)

[(116) and (117) for points on the top and left boundary] each $\xi(m,n)$ is orthogonal to $\mathbf{f}(p,q)$ for all $(p,q) \in X_{m,n}$. Now $\xi(k,l)$ for any $(k,l) \in X_{m,n}$ is a linear combination of $\mathbf{f}(p,q)$'s in $X_{m,n}$. Therefore, $\xi(m,n)$ must be orthogonal to $\xi(k,l)$ for all $(k,l) \in X_{m,n}$. Since this must be true for every (m,n) in the picture, it follows that

$$E\{\xi(p,q)\xi(m,n)\} = 0, \qquad p \neq m \quad \text{or} \quad q \neq n$$

which is (114).

Conversely, if (114) is true for $\xi(m,n)$ as defined by (111) and (112), then we must show that the field $\mathbf{f}(m,n)$ is wide-sense Markov. We proceed as follows. From (112b) and (112a), $\mathbf{f}(2,1) = A\xi(1,1)+\xi(2,1)$. Similarly, $\mathbf{f}(1,2) = \mathscr{B}\xi(1,1)+\xi(1,2)$. Substituting these two expressions in (111), we obtain

$$\mathbf{f}(2,2) = (c_{1,1}+\mathscr{B}c_{1,0}+\mathscr{A}c_{0,1})\xi(1,1) + c_{1,0}\xi(1,2) + c_{0,1}\xi(2,1) + \xi(2,2)$$

Similarly, any $\mathbf{f}(m,n)$ can be expressed in terms of $\xi(p,q)$'s with $(p,q) \in Y_{m,n}$ defined in Fig. 7b. By (114) this implies that $\mathbf{f}(m,n)$ is orthogonal to any $\xi(k,l)$ for which $k > m$ or $l > n$. From this it follows that the estimation error $\mathbf{f}(k,l)-\hat{\mathbf{f}}(k,l)$ $[=\xi(k,l)]$ is orthogonal to all $\mathbf{f}(p,q)$ with $(p,q) \in X_{k,l}$; this, with $\hat{\mathbf{f}}(k,l)$ defined by (107), is exactly (108). Therefore, the random field $\mathbf{f}(m,n)$ is wide-sense Markov.

Note that for a zero-mean random field $\mathbf{f}(m,n)$, (111) and (112) imply that the random variables $\xi(m,n)$ also have zero mean. In this case, therefore, (114) implies that the random variables $\xi(m,n)$ are uncorrelated. ∎

A wide-sense Markov random field obeys the difference equation (111). Now suppose we are given a homogeneous random field, not necessarily wide-sense Markov, whose autocorrelation function is known. We would like to represent this random field by (111) to a good approximation. This can be done by using the known autocorrelation function for $R_{ff}(\alpha,\beta)$ in Eqs. (109) to solve for the $c_{i,j}$'s. Note that if the given random field is not wide-sense Markov, the random variables $\xi(m,n)$ may now not be completely uncorrelated. It is obvious that the degree of correlation among the $\xi(m,n)$ should serve as a measure of the goodness of the representation of the given random field by (111).

Under certain conditions this procedure of modeling a non-wide-sense Markov field by (111) may not be mathematically valid. The modeling is valid only if the autocorrelation function of $\xi(m,n)$ is a positive-definite function; otherwise one has to place further restrictions on the solutions $c_{i,j}$ of (109). This has been discussed in detail by Wood [80]. In this section we will not be concerned with the validity of the modeling. We will assume that the $c_{i,j}$'s calculated from (109) result in a valid solution and, if not, that these $c_{i,j}$'s are good approximations to those necessary for a valid solution.

Let us assume that the autocorrelation function of a given random field can be approximated by Eq. (47) of Chapter 3. We shall further assume that the random field is normalized in that the mean value of the random field has been subtracted from every picture and each picture has been divided by $\sqrt{R_{ff}(0,0)}$. The autocorrelation function can then be written as

$$R_{ff}(\alpha, \beta) = \exp(-c_1 |\alpha| - c_2 |\beta|) \tag{118}$$

which can also be expressed as

$$R_{ff}(\alpha, \beta) = \rho_h^{|\alpha|} \rho_v^{|\beta|} \tag{119}$$

where $\rho_h = e^{-c_1}$ and $\rho_v = e^{-c_2}$ are measures of the horizontal and vertical correlation, respectively. Substituting this $R_{ff}(\alpha, \beta)$ in (109) we obtain the following solution:

$$c_{1,0} = \rho_h, \qquad c_{0,1} = \rho_v, \qquad c_{1,1} = -\rho_h \rho_v \tag{120}$$

It can easily be verified that with the $c_{i,j}$s as given and for the autocorrelation function (118), the conditions in (108) are exactly satisfied. Therefore, (114) is also exactly satisfied.

7.5.2 Restoration in the Presence of Additive Noise

Consider a random field whose autocorrelation function can be approximated by (119) and which can be modeled by the difference equation (111) with the $c_{i,j}$'s given by (120). Let the random field be corrupted by additive noise $\mathbf{v}(m, n)$ so that the observed random field is given by

$$\mathbf{g}(m,n) = \mathbf{f}(m,n) + \mathbf{v}(m,n) \tag{121}$$

where we assume the noise to be white, which implies

$$E\{\mathbf{v}(m,n)\,\mathbf{v}(i,j)\} = \begin{cases} 0 & \text{if } m \neq i \text{ or } n \neq j \\ \sigma^2 & \text{if } m = i \text{ and } n = j \end{cases} \tag{122}$$

and that it is uncorrelated with the picture:

$$E\{\mathbf{f}(m,n)\,\mathbf{v}(i,j)\} = 0 \qquad \text{for all } m, n, i, j \tag{123}$$

Given $\mathbf{g}(m,n)$, the problem is to find a linear least squares estimate $\hat{\mathbf{f}}(m,n)$ for the original random field $\mathbf{f}(m,n)$ at every point (m,n). For the purpose of implementation in real time we would like this estimate to have the following two properties:

(i) If $\mathbf{g}(p,q)$ is scanned top to bottom, one row at a time, then the estimate $\hat{\mathbf{f}}(m,n)$ at a point (m,n) should be a linear function of only those $\mathbf{g}(p,q)$ that have already been scanned.

(ii) For rapid implementation it would be desirable to express $\hat{\mathbf{f}}(m,n)$

in terms of neighboring, and already determined, estimates $\hat{\mathbf{f}}(m-1,n)$, $\hat{\mathbf{f}}(m-1,n-1)$, $\hat{\mathbf{f}}(m,n-1)$, and $\mathbf{g}(m,n)$ at the "current" point.

Let us denote by $Y_{m,n}$ the collection of all the points shown in Fig. 7b. The estimate will have property (i) if it is of the general form

$$\hat{\mathbf{f}}(m,n) = \sum_{(p,q) \in Y_{m,n}} d_{m,n}(p,q)\mathbf{g}(p,q) \tag{124}$$

where for each (m,n) the coefficients $d_{m,n}(p,q)$ must be such that

$$E\{[\mathbf{f}(m,n)-\hat{\mathbf{f}}(m,n)]^2\} \tag{125}$$

is minimized. Substituting (124) in (125) and differentiating with respect to each $d_{m,n}(p,q)$, $p = 1,2,...,m$; $q = 1,2,...,n$; we obtain the following set of orthogonality conditions that must be satisfied by the estimate:

$$E\{[\mathbf{f}(m,n)-\hat{\mathbf{f}}(m,n)]\mathbf{g}(p,q)\} = 0 \qquad \text{for all} \quad (p,q) \in Y_{m,n} \tag{126}$$

The estimate would have property (ii) if (124) could, for example, be expressed in the following form:

$$\hat{\mathbf{f}}(m,n) = \sum_{(i,j) \in D} d_{m,n}(i,j)\hat{\mathbf{f}}(m-i,n-j) + \eta_{m,n}\mathbf{g}(m,n) \tag{127}$$

where for each (m,n), $d_{m,n}(i,j)$, $(i,j) \in D$, and $\eta_{m,n}$ are the four coefficients of the filter. We will now show, in a heuristic manner, that under certain conditions the linear estimate (124) may be approximated by the form in (127). Note that if for each (m,n) we could find four coefficients $d_{m,n}(i,j)$, $(i,j) \in D$, and $\eta_{m,n}$ so that (126) is exactly satisfied, then (127) would be the required linear least squares estimate. This follows from the easily shown fact that $\hat{\mathbf{f}}(m,n)$ in (127) is indeed a linear function of $\mathbf{g}(p,q) \in Y_{m,n}$, and from a uniqueness property of the least squares estimate. The uniqueness property, which has not been proved here, implies that a linear function of $\mathbf{g}(p,q)$, $(p,q) \in Y_{m,n}$, is a least squares estimate if and only if it satisfies (126). (The proof of this can be found in, for example, [11] of Chapter 2, p. 116.)

Substituting (127) in (126), we have

$$E\left\{\left[\mathbf{f}(m,n) - \sum_{(i,j) \in D} d_{m,n}(i,j)\hat{\mathbf{f}}(m-i,n-j) - \eta_{m,n}\mathbf{g}(m,n)\right]\mathbf{g}(p,q)\right\} = 0 \tag{128}$$

for all $(p,q) \in Z_{m,n}$ and

$$E\left\{\left[\mathbf{f}(m,n) - \sum_{(i,j) \in D} d_{m,n}(i,j)\hat{\mathbf{f}}(m-i,n-j) - \eta_{m,n}\mathbf{g}(m,n)\right]\mathbf{g}(m,n)\right\} = 0 \tag{129}$$

where the region $Z_{m,n}$ contains all the points in $Y_{m,n}$ except the point (m,n). Equation (128) expresses the orthogonality conditions in (126) at all points

(p, q) in $Z_{m,n}$, and (129) express the condition at the remaining point (m, n). We can express (128) in the form

$$E\left\{\left[(1-\eta_{m,n})\mathbf{f}(m,n) - \sum_{(i,j)\in D} d_{m,n}(i,j)\mathbf{f}(m-i,n-j)\right.\right.$$
$$\left.\left. + \sum_{(i,j)\in D} d_{m,n}(i,j)(\mathbf{f}(m-i,n-j) - \hat{\mathbf{f}}(m-i,n-j))\right]\mathbf{g}(p,q)\right\} = 0$$

(130)

for $(p, q) \in Z_{m,n}$, where we have used (121) and (123).

Equation (126) implies that

$$E\{[\mathbf{f}(m-i,n-j) - \hat{\mathbf{f}}(m-i,n-j)]\mathbf{g}(p,q)\} = 0 \qquad (131)$$

for $(i, j) \in D$ and $(p, q) \in Y_{m-1, n-1}$. Therefore, for all $(p, q) \in Y_{m-1, n-1}$ (130) becomes

$$E\left\{\left[(1-\eta_{m,n})\mathbf{f}(m,n) - \sum_{(i,j)\in D} d_{m,n}(i,j)\mathbf{f}(m-i,n-j)\right]\mathbf{g}(p,q)\right\} = 0$$

(132)

Now, the random field $\mathbf{f}(m, n)$ satisfies (108). Because of (123), (108) can be written as

$$E\left\{\left[\mathbf{f}(m,n) - \sum_{(i,j)\in D} c_{i,j}\mathbf{f}(m-i,n-j)\right]\mathbf{g}(p,q)\right\} = 0, \qquad (p,q) \in X_{m,n}$$

(133)

Since (133) is true for all $(p, q) \in X_{m,n}$, it is also true for all $(p, q) \in Y_{m-1, n-1}$. A comparison of (132) and (133) reveals that the former is true provided that

$$d_{m,n}(i,j)/(1-\eta_{m,n}) = c_{i,j}, \qquad (i,j) \in D \qquad (134)$$

which at each (m, n) supplies us with three equations for the four unknowns $d_{m,n}(i,j)$, $(i,j) \in D$, and $\eta_{m,n}$. The fourth equation can be obtained from (129), which yields

$$R_{ff}(0,0) = \sum_{(i,j)\in D} d_{m,n}(i,j) E\{\hat{\mathbf{f}}(m-i,n-j)\mathbf{f}(m,n)\}$$
$$+ \eta_{m,n}(R_{ff}(0,0)+\sigma^2) \qquad (135)$$

where we have used (121)–(123). Substituting (120) and (134) in (135), and using the notation

$$C_{p,q}(m,n) = E\{\hat{\mathbf{f}}(p,q)\mathbf{f}(m,n)\}, \qquad (p,q) \in Z_{m,n} \qquad (136)$$

we obtain

$$R_{ff}(0,0) = (1-\eta_{m,n})[\rho_h C_{m-1,n}(m,n) - \rho_h \rho_v C_{m-1,n-1}(m,n)$$
$$+ \rho_v C_{m,n-1}(m,n)] + \eta_{m,n}[R_{ff}(0,0)+\sigma^2] \qquad (137)$$

Let

$$\mathfrak{C}_p(m,n) = \rho_h\, C_{m-1,n}(m,n) - \rho_h\,\rho_v\, C_{m-1,n-1}(m,n)$$

$$+ \rho_v\, C_{m,n-1}(m,n) \tag{138}$$

Then from (137), we have

$$\eta_{m,n} = \frac{R_{ff}(0,0) - \mathfrak{C}_p(m,n)}{R_{ff}(0,0) + \sigma^2 - \mathfrak{C}_p(m,n)} \tag{139}$$

The form of (139) is intuitively appealing. Note that $\sigma^2/R_{ff}(0,0)$ is the noise-to-signal power ratio in the picture. As this ratio approaches zero, then regardless of the values taken by $C_p(m,n)$ the coefficient $\eta_{m,n}$ approaches unity, while at the same time by (134) the other three coefficients $d_{m,n}(i,j)$ approach zero. This means that in estimate (127) most of the contribution comes from the "current" data point. This is as it should be, because as the noise in a picture diminishes one can place greater confidence in the gray levels $g(m,n)$ in the observed picture. Conversely, as the noise-to-signal power ratio becomes large, (139) indicates that $\eta_{m,n}$ should be a small fraction of unity, leading to relatively larger values for $d_{m,n}(i,j)$. This in (127) means little confidence in the current data point and a greater smoothing effect in restoring the noisy picture.

Calculation of $\eta_{m,n}$ requires that $\mathfrak{C}_p(m,n)$ be known, which in turn can be determined from (138) provided the $C_{p,q}(m,n)$ are known. These can be obtained recursively from the following relationship obtained by substituting (127) and (134) in (136):

$$C_{p,q}(m,n) = (1 - \eta_{p,q})\,[\rho_h\, C_{p-1,q}(m,n) - \rho_v\,\rho_h\, C_{p-1,q-1}(m,n)$$

$$+ \rho_v\, C_{p,q-1}(m,n)] + \eta_{p,q}\, R_{ff}(m-p,n-q), \qquad (p,q) \in Z_{m,n} \tag{140}$$

Note that in an "on-line" implementation, when estimate (127) at the point (m,n) is being calculated, the coefficients $\eta_{p,q}$ for $(p,q) \in Z_{m,n}$ have already been determined. Equation (140) is recursive in the indices (p,q) and its use to calculate $C_{m-p,n-q}(m,n)$ for $(p,q) \in D$ requires that the following boundary values be known:

$$C_{1,1}(m,n) \tag{141a}$$

$$C_{p,1}(m,n) \qquad \text{for all } p \tag{141b}$$

and

$$C_{1,q}(m,n) \qquad \text{for all } q \tag{141c}$$

Since the point $(1,1)$ does not have any neighbors to the left or above, it is clear from (127) that

$$\hat{\mathbf{f}}(1,1) = \eta_{1,1}\, \mathbf{g}(1,1) \tag{142}$$

where, by using (126) with $(m, n) = (1, 1)$, it immediately follows that

$$\eta_{1,1} = R_{ff}(0,0)/[R_{ff}(0,0)+\sigma^2] \qquad (143)$$

From (136) and (142),

$$C_{1,1}(m,n) = E\{\eta_{1,1}\,\mathbf{g}(1,1)\,\mathbf{f}(m,n)\} = \eta_{1,1}\,R_{ff}(m-1,n-1) \qquad (144)$$

where we have used (121) and (123). This determines (141a).

At the points on the top row estimate (127) takes the form

$$\hat{\mathbf{f}}(p,1) = (1-\eta_{p,1})\,\rho_h\,\hat{\mathbf{f}}(p-1,1) + \eta_{p,1}\,\mathbf{g}(p,1) \qquad (145)$$

Substituting (145) in (136), we have

$$C_{p,1}(m,n) = (1-\eta_{p,1})\,\rho_h\,C_{p-1,1}(m,n) + \eta_{p,1}\,R_{ff}(m-p,n-1) \qquad (146)$$

for $p = 2, 3, \ldots$. Clearly, once $\eta_{1,1}$ and $C_{1,1}(m,n)$ are known from (143) and (144), (138) can be used to determine $\mathfrak{C}_p(2,1)$, since $\mathfrak{C}_p(2,1) = \rho_h\,C_{1,1}(m,n)$, and then (139) to determine $\eta_{2,1}$. This substituted in (146) yields $C_{2,1}(m,n)$. One can similarly find $C_{p,1}(m,n)$ for all values of p. This determines (141b). Similarly, one can write a recursive relationship for $C_{1,q}(m,n)$, which determines (141c).

Once the boundary elements in (141a)–(141c) are known, (140) can then be used to express $C_{2,2}(m,n)$ in terms of the now known $C_{1,2}(m,n)$, $C_{1,1}(m,n)$ and $C_{2,1}(m,n)$; and one continues in this way until the three quantitites $C_{m-p,n-q}(m,n)$ for $(p,q) \in D$ are determined.

Note the excessive amount of computation required to determine $\eta_{m,n}$.

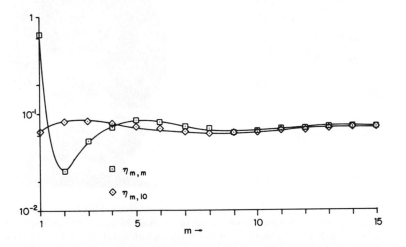

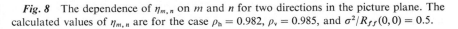

Fig. 8 The dependence of $\eta_{m,n}$ on m and n for two directions in the picture plane. The calculated values of $\eta_{m,n}$ are for the case $\rho_h = 0.982$, $\rho_v = 0.985$, and $\sigma^2/R_{ff}(0,0) = 0.5$.

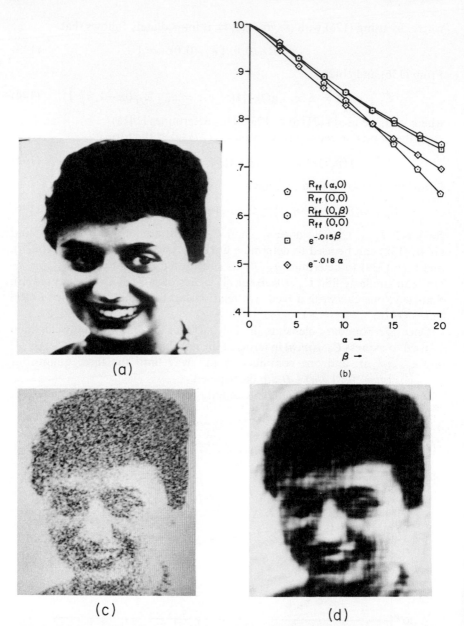

(a)

(b)

(c)   (d)

Fig. 9 (a) The undegraded picture which can be modeled by $\rho_h = 0.982$ and $\rho_v = 0.985$. (b) $R_{ff}(\alpha, 0)/R_{ff}(0,0)$ and $R_{ff}(0,\beta)/R_{ff}(0,0)$ for the picture in (a) after the mean value of the gray levels is subtracted from it. (c) Result of adding zero-mean Gaussian white noise to the picture in (a). The variance of noise is such that $\sigma^2/R_{ff}(0,0) = 0.5$. (d) Restoration achieved by the technique discussed in Section 7.5.2.

For every point (m, n), one has to start at the top left corner of the picture and go through the previously mentioned recursion. Fortunately, it turns out that in practice this is not necessary. Only a few points away from the edges of the picture $\eta_{m,n}$ attains practically a constant value. Hence, one needs to compute $\eta_{m,n}$ for only a few rows and columns near the edges of a picture. To illustrate this fact we have in Fig. 8 shown the variation of $\eta_{m,n}$ along two directions in the picture plane: one along a diagonal and the other along a line parallel to the x-axis. The calculated $\eta_{m,n}$ are for the recursive filtering example discussed later.

Equation (127) is the filter for the recursive restoration of noisy pictures, the coefficients of the filter being given by (134) and (139). In the derivation of the filter we were supposed to satisfy the orthogonality conditions in (128) at all points $(p, q) \in Y_{m,n}$. We ended up, however satisfying these conditions at only the subset $Y_{m-1, n-1}$ and the point (m, n). This is clear from the qualifying remark preceding (132). Therefore, we have really not satisfied the orthogonality conditions (128) for the points in the mth column from $(m, 1)$ through $(m, n-1)$; and for the points in the nth row for $(1, n)$ through $(m-1, n)$. It can be shown, however, that in practice as the sampling density increases in a continuous picture of a given size, the error caused by this becomes insignificant. The theoretical difficulty mentioned here also exists in an earlier contribution [26] on the recursive filtering of pictures.

In Fig. 9a we have shown an undegraded picture. This picture was sampled on a 256×256 array of points and the gray level of each sample was represented by an 8-bit word. Taken as a single picture its autocorrelation function $R_{ff}(\alpha, \beta)$ was calculated by using Eq. (28) of Chapter 2. In Fig. 9b we have plotted $R_{ff}(\alpha, 0)/R_{ff}(0, 0)$ and $R_{ff}(0, \beta)/R_{ff}(0, 0)$. Before these quantities were calculated the mean value of the gray levels in the picture was subtracted from it. It is clear from Fig. 9a that $R_{ff}(\alpha, 0)/R_{ff}(0, 0)$ may be approximated by $e^{-(0.018)\alpha}$, while $R_{ff}(0, \beta)/R_{ff}(0, 0)$ may be approximated by $e^{-(0.015)\beta}$. Therefore, the picture in Fig. 9a may be "assumed" to belong to an ergodic random field with $c_1 = 0.018$ and $c_2 = 0.015$. The constants ρ_h and ρ_v for these c_1 and c_2 are given by 0.982 and 0.985, respectively. In Fig. 9c zero-mean Gaussian white noise has been added to the picture in Fig. 9a. The variance of the noise is such that $\sigma^2/R_{ff}(0, 0) = 0.5$. Figure 9d shows the restored picture after it was filtered by the technique discussed in this section.

7.6 BIBLIOGRAPHICAL NOTES

In addition to the techniques that have been presented in this chapter for solving Eq. (2), several other approaches are worth noting. Jansson [45–47] has proposed an iterative method for solving (2), which can best be implemented digitally. In a method due to MacAdam [55], Eq. (2) is first converted into a vector–matrix form as in Eq. (89), and the solution is then obtained by dividing the z-transform of the left hand side by the z-transform of the equivalent one-dimensional impulse response function implied in Eq. (89). The solution is obtained with the constraint that it lies within a certain range (for example, the constraint could be that all samples of the solution be positive). This method suffers from the disadvantage that when the picture matrix is of large size, the computation time may become unreasonably long. It, however, does have the attractive feature that the restored picture can be guaranteed to be positive. The positivity constraint is also implied in the methods proposed by Frieden [18, 19], which use maximum likelihood and maximum entropy algorithms for restoration, as well as those proposed by Richardson [64], which assume that the picture is a two-dimensional probability density function and then use Bayes' theorem for restoration. More recently Frieden [20] has proposed a restoration technique in which the degraded image is restored by convolving it with a weighted sum of the shifted versions of the point spread function of the degradation. The number of terms in this weighted sum is small, and the weights are calculated by minimizing the sidelobes of the output point spread function. The advantage of this method lies in the small number of mathematical steps required for its implementation. While Frieden's method for the selection of weights tends to minimize the distortion between the original ideal picture and its restored version, it also suffers from increased susceptibility to noise. By using the mathematical methods developed by Backus and Gilbert [2], Saleh [67] has studied possible choices of the superposition weights that effect a compromise between distortion and noise.

The reader has probably noticed that in the restoration solutions derived in this chapter the restored picture does not contain those frequency components at which the transfer function $H(u, v)$ of the degradation has zero values. One would think that if $H(u, v)$ is zero outside a certain bounded region in the frequency plane (as is the case with all optical imaging), then we cannot hope to recover the spectral frequencies in the original $f(x, y)$ that are outside of this bounded region. Actually this is not the case, at least in theory. For all bounded functions $f(x, y)$ the Fourier transform $F(u, v)$ is an analytic function. This means that if $F(u, v)$ is known for a certain range of (u, v), then the techniques of analytic continuation can be used to extend the solution to all (u, v). Harris [28] has discussed several practical methods for carrying out the analytic continuation. From a theoretical standpoint, the method proposed by Slepian and Pollak [71] (see also [38]), which makes use of prolate spheroidal wave functions, is most satisfying. The reason why these techniques for increasing the resolution in the restored picture have not proved popular is because of their extreme susceptibility to noise [11, 17]. In one example [31] it was found that in order to succeed in analytic continuation a noise-to-signal amplitude ratio of 1:1000 was required.

In Section 7.1.1 we gave some examples of cases in which the point spread function of the degradation can be determined from an understanding of the underlying physical process. To add to those examples, the point spread function of a defocused lens can be calculated, taking into account both geometrical optics and the effects of diffraction [22, 23]. It can be shown that for low spatial frequencies the transfer function calculated on the basis of geometrical optics agrees well with that calculated when diffraction effects are involved [34, 35, 76]. The point spread function due to lens aberration can also be calculated theoretically [3, 19, 56]. Only spherical aberration can be considered to be shift invariant, the others being definitely shift variant.

In Section 7.2.1 we also calculated the point spread function of linear motion degradation. Many investigators have worked on this problem [52, 68, 69, 73, 74] In our derivation we assumed that the image that is incident on the film is invariant in time except for a displacement of the origin. This is not necessarily true even when the camera moves perpendicularly to the optical axis. For example, the images of objects at different distances from the film move by different amounts, thereby producing a spatially varying blur. Several investigators have also considered the effect of camera shutter operation on the point spread function of motion degradation [33, 61, 69].

In Section 7.2.2 we discussed the problem of determining the point spread function $h(x, y)$ *a posteriori* from the degraded image $h(x, y)$. The method for estimating the transfer function $H(u, v)$ given by Eq. (15) is a result of Stockham's work on the restoration of old Caruso recordings.

In this chapter we have assumed that the point spread function $h(x, y)$, whether it is known *a priori* or estimated *a posteriori*, is a deterministic function. The problem of least squares filtering of pictures with a deterministic PSF was first solved by Helstrom [32]. Slepian [72] has solved the least mean square estimation problem for the case where $h(x, y)$ is stochastic. The optimum filter was found to again be given by Eq. (41), except that H^* and $|H|^2$ are replaced by $E\{H^*\}$ and $E\{|H|^2\}$, where $E\{\ \}$ denotes the ensemble average.

In Section 7.5 we discussed on-line recursive estimation of pictures from noisy observations. A basic assumption was made there that the random field is wide-sense Markov. Specification of the Markov property requires that the "past," the "present," and the "future" be defined. It is, of course, trivial to do so for one-dimensional signals. For pictorial data, however, there is no unique way to specify these. In our definitions of the Markov property the region $X_{m,n}$ is the past, the point (m, n) the present, and the rest of the picture the future. For other ways to specify these regions, see, for example, Jain and Angel [44] and Woods [80].

Our discussion of two-dimensional recursive estimation can be found in greater detail in Panda and Kak [60], where a discussion of the application of the theory to pictures with convolutional degradation is also included. The filter presented here is similar to that derived by Habibi [26]. A picture may also be represented as a one-dimensional stochastic process, the time signal being the output of a scanner. One can then use the one-dimensional theories of recursive estimation [58]. For another approach to statistical estimation of pictures, see Jain and Angel [44].

In this chapter we considered only degradations that are shift invariant. Many important types of degradations do not, however, fall into this category. The reader is referred to the literature for techniques that can be used to handle shift-variant degradations [25, 37, 53, 65, 68].

REFERENCES

1. K. Abend, T. J. Harley, and L. N. Kanal, Classification of binary random patterns, *IEEE Trans. Informat. Theory* **IT-11**, 1965, 538–544.
2. G. Backus and F. Gilbert, Uniqueness in the inversion of inaccurate gross earth data, *Phil. Trans. Roy. Soc. London* 266, Ser. A, 1970, 123–192.
3. R. Barakat, Numerical results concerning the transfer functions and the total illuminance for optimum balanced fifth-order spherical aberration, *J. Opt. Soc. Amer.* **54**, 1964, 38–44.
4. M. A. Berkovitz, Edge gradient analysis OTF accuracy study, *Proc. SPIE Seminar Modulat. Transfer Function*, Boston, Massachusetts, 1968.
5. M. V. Berry and D. F. Gibbs, The interpretation of optical projections, *Proc. Roy. Soc. Ser. A* **314**, 1970, 143–152.
6. E. S. Blackman, Effects of noise on the determination of photographic system modulation transfer function, *Photogr. Sci. Eng.* 12, 1968, 244–250.
7. E. S. Blackman, Recovery of system transfer functions from noisy photographic records, *Proc. SPIE Seminar Image Informat. Recovery*, 1969.
8. R. N. Bracewell and A. C. Riddle, Inversion of fan-bean scans in radio astronomy, *Astrophys. J.* **150**, 1967, 427–434.
9. C. P. C. Bray, Comparative analysis of geometric vs. diffraction heterochromatic lens evaluations using optical transfer function theory, *J. Opt. Soc. Amer.* **55**, 1965, 1136–1138,
10. E. O. Brigham, H. W. Smith, F. X. Bostick, and W. C. Dusterhoeft, An iterative technique for dertmining inverse filters, *IEEE Trans. Geosci. Electron.* **GE-6**, 1968, 86–96.
11. G. J. Buck and J. J. Gustincic, Resolution limitations of a finite aperture, *IEEE Trans. Antennas Propagat.* **AP-15**, 1967, 376–381.
12. R. A. Crowther, D. J. Rosier, and A. Klug, The reconstruction of a three-dimensional structure from its projections and its applications to electron microscopy, *Proc. Roy. Soc. London A* **317**, 1970, 319–340.
13. L. J. Cutrona and W. D. Hall, Some considerations in post-facto blur removal, *in Evaluation of Motion-Degraded Images*, NASA Publ. SP-193, pp. 139–148, December 1968.
14. D. G. Falconer, Image enhancement and film-grain noise, *Opt. Acta* 17, 1970, 693–705.
15. D. L. Fried, Optical resolution through a randomly inhomogeneous medium for very long and very short exposures, *J. Opt. Soc. Amer.* **56**, 1966, 1372–1379.
16. D. L. Fried, Limiting resolution looking down through the atmosphere, *J. Opt. Soc. Amer.* **56**, 1966, 1380–1384.
17. B. R. Frieden, Bandlimited reconstruction of optical objects and spectra, *J. Opt. Soc. Amer.* **57**, 1967, 1013–1019.

18. B. R. Frieden, Restoring with maximum likelihood and maximum entropy, *J. Opt. Soc. Amer.* **62**, 1972, 511–518.

19. B. R. Frieden and J. J. Burke, Restoring with maximum entropy II: Superresolution of photographs of diffraction-blurred images, *J. Opt. Soc. Amer.* **62**, 1972, 1207–1210.

20. B. R. Frieden, Image restoration by discrete deconvolution of minimal length, *J. Opt. Soc. Amer.* **64**, 1974, 682–686.

21. P. Gilbert, Iterative methods for reconstruction of three-dimensional objects from projections, *J. Theor. Biol.* **36**, 1972, 105–117.

22. J. W. Goodman, "Introduction to Fourier Optics," McGraw-Hill, New York, 1968.

23. J. W. Goodman, Use of a large-aperture optical system as a triple interferometer for removal of atmospheric image degradations, *in* Evaluation of Motion-Degraded Images, NASA Publ. SP-193, pp. 89–93, December 1968.

24. R. Gordon, R. Bender, and G. T. Herman, Algebraic reconstruction techniques for three-dimensional electron microscopy and X-ray photography, *J. Theor. Biol.* **29**, 1970, 471–481.

25. E. M. Granger, Restoration of images degraded by spatially varying smear, *in* Evaluation of Motion Degraded Images, NASA Publ. SP-193, pp. 161–174. December 1968.

26. A. Habibi, Two-dimensional Bayesian estimate of images, *Proc. IEEE* **60**, 1972, 878–883.

27. H. B. Hammill and C. Holladay, The effect of certain approximations in image quality evaluation from edge traces, *SPIE J.* **8**, 1970, 223–228.

28. J. L. Harris, Diffraction and resolving power, *J. Opt. Soc. Amer.* **54**, 1964, 931–936.

29. J. L. Harris, Sr., Image evaluation and restoration, *J. Opt. Soc. Amer.* **56**, 1966, 569–574.

30. J. L. Harris, Sr., Potential and limitations of techniques for processing linear motion-degraded imagery, *in* Evaluation of Motion-Degraded Images, NASA Publ. SP-193, pp. 131–138, December 1968.

31. J. L. Harris, Information extraction from diffraction limited imagery, *Pattern Recognit.* **2**, 1970, 69–77.

32. C. W. Helstrom, Image restoration by the method of least squares, *J. Opt. Soc. Amer.* **57**, 1967, 297–303.

33. L. O. Hendeberg and W. E. Welander, Experimental transfer characteristics of image motion and air conditions in aerial photography, *Appl. Opt.* **2**, 1963, 379–386.

34. H. H. Hopkins, The frequency response of a defocused optical system, *Proc. Roy. Soc. Ser. A* **231**, 1955, 91–103.

35. H. H. Hopkins, Interferometric methods for the study of diffraction images, *Opt. Acta* **2**, 1955, 23–29.

36. T. S. Huang, Some notes on film grain noise, Restoration of Atmospherically-Degraded Images, NSF Summer Study Rep., Woods Hole, Massachusetts, 1966.

37. T. S. Huang, Digital computer analysis of linear shift-variant systems, *in* Evaluation of Motion-Degraded Images, NASA Publ. SP-193, pp. 83–87, December 1968.

38. T. S. Huang, W. F. Schreiber, and O. J. Tretiak, Image processing, *Proc. IEEE* **59**, 1971, 1586–1609.

39. R. E. Hufnagel and N. R. Stanley, Modulation transfer function associated with image transmission through turbulent media, *J. Opt. Soc. Amer.* **54**, 1964, 52–61.

40. R. E. Hufnagel, An improved model for turbulent atmosphere, Restoration of Atmospherically Degraded Images, NSF Summer Study Rep., Woods Hole, Massachusetts, 1966.

41. B. R. Hunt, A matrix theory proof of the discrete convolution theorem, *IEEE Trans. Audio Electroacoust.* **AU-19**, 1971, 285–288.

42. B. R. Hunt, Deconvolution of linear systems by constrained regression and its relationship to the Wiener theory, *IEEE Trans. Autom. Contr.* 1972, 703–705.

43. B. R. Hunt, The application of constrained least squares estimation to image restoration by digital computer, *IEEE Trans. Comput.* **C-22**, 1973, 805–812.

44. A. K. Jain and E. Angel, Image restoration, modeling, and reduction of dimensionality, *IEEE Trans. Comput.* **C-23**, 1974, 470–476.

45. P. A. Jansson, R. H. Hunt, and E. K. Plyler, Response function for spectral resolution enhancement, *J. Opt. Soc. Amer.* **58**, 1968, 1665–1666.

46. P. A. Jansson, Method for determining the response function of a high-resolution infrared spectrometer, *J. Opt. Soc. Amer.* **60**, 1970, 184–191.

47. P. A. Jansson, Resolution enhancement of spectra, *J. Opt. Soc. Amer.* **60**, 1970, 596–599.

48. R. A. Jones, An automated technique for deriving MTF's from edge traces, *Photogr. Sci. Eng.* **11**, 1967, 102–106.

49. R. A. Jones, Accuracy test procedure for image evaluation techniques, *Appl. Opt.* **7**, 1968, 133–136.

50. R. A. Jones and E. C. Yeadon, Determination of the spread function from noisy edge scans, *Photogr. Sci. Engr.* **13**, 1969, 200–204.

51. A. C. Kak, C. V. Jakowatz, N. A. Baily, and R. Keller, Computerized tomography using video recorded fluoroscopic images, *IEEE Trans. Biomed. Engrg.*, January 1977.

52. A. W. Lohmann and D. P. Paris, Influence of longitudinal vibrations on image quality, *Appl. Opt.* **4**, 1965, 393–397.

53. A. W. Lohmann and D. P. Paris, Space-variant image formation, *J. Opt. Soc. Amer.* **55**, 1965, 1007–1013.

54. R. F. Lutomirski and H. T. Yura, Modulation-transfer function and phase-structure function of an optical wave in a turbulent medium—1, *J. Opt. Soc. Amer.* **59**, 1969, 999–1000.

55. D. P. MacAdam, Digital image restoration by constrained deconvolution, *J. Opt. Soc. Amer.* **60**, 1970, 1617–1627.

56. L. Miyamoto, Wave optics and geometrical optics in optical design, *Progr. Opt.* **1**, 1961.

57. J. C. Moldon, High resolution image estimation in a turbulent enviroment, *Pattern Recognit.* **2**, 1970, 79–90.

58. N. E. Nahi, Role of recursive estimation in statistical image enhancement, *Proc. IEEE* **60**, 1972, 872–877.

59. E. L. O'Neill, "Introduction to Statistical Optics." Addison-Wesley, Reading, Massachusetts, 1963.

60. D. P. Panda and A. C. Kak, Recursive Filtering of Pictures, Tech. Rep. No. TR-EE-76, School of Elec. Eng., Purdue Univ., Lafayette, Indiana, 1976.

61. D. P. Paris, Influence of image motion on the resolution of a photographic system—II, *Photogr. Sci. Eng.* **7**, 1963, 233–236.

62. D. L. Phillips, A technique for the numerical solution of certain integral equations of the first kind, *J. ACM* **9**, 1962, 84–97.

63. G. N. Ramachandran and V. A. Lakshminarayanan, Three-dimensional reconstruction for radiographs and electron micrographs: Application of convolutions instead of Fourier transforms, *Proc. Nat. Acad. Sci. U.S.* **68**, 1971, 2236–2240.

64. W. H. Richardson, Bayesian-based iterative method of image restoration, *J. Opt. Soc. Amer.* **62**, 1972, 55–59.

65. G. M. Robbins and T. S. Huang, Inverse filtering for linear shift-variant imaging systems, *Proc. IEEE* **60**, 1972, 862–872.

66. P. G. Roetling, R. C. Haas, and R. E. Kinzly, Some practical aspects of measurement and restoration of motion-degraded images, *Proc. NASA/ERC Seminar*, Cambridge, Massachusetts, December 1968.

67. E. A. Saleh, Trade-off between resolution and noise in restoration by superposition of images, *Appl. Opt.* **13**, 1974, 1833–1838.

68. A. A. Sawchuk, Space-variant image motion degradation and restoration, *Proc. IEEE* **60**, 1972, 854–861.

69. R. V. Shack, The influence of image motion and shutter operation on the photographic transfer function, *Appl. Opt.* **3**, 1964, 1171–1181.

70. L. A. Shepp and B. F. Logan, Reconstructing interior head tissue from X-ray transmissions, *IEEE Trans. Nucl. Sci.* **NS-21**, 1974, 228–236.

71. D. Slepian and H. O. Pollak, Prolate spheroidal wave functions, Fourier analysis and uncertainty—I, *BSTJ* **40**, 1961, 43–63.

72. D. Slepian, Linear least squares filtering of distorted images, *J. Opt. Soc. Amer.* **57**, 1967, 918–922.

73. D. Slepian, Restoration of photographs blurred by image motion, *BSTJ* **46**, 1967, 2353–2362.

74. S. C. Som, Analysis of the effects of linear smear, *J. Opt. Soc. Amer.* **61**, 1971, 859–864.

75. M. M. Sondhi, Image restoration: The removal of spatially invariant degradations, *Proc. IEEE* **60**, 1972, 842–853.

76. P. A. Stokseth, Properties of a defocused optical system, *J. Opt. Soc. Amer.* **59**, 1969, 1314–1321.

77. K. Tanabe, Projection method of solving a singular system of linear equations and its applications, *Numer. Math.* **17**, 1971, 203–214.

78. S. Twomey, On the numerical solution of Fredholm integral equations of the first kind by the inversion of the linear system produced by quadrature, *J. ACM* **10**, 1963, 97–101.

79. S. Twomey, The application of numerical filtering to the solution of integral equations encountered in indirect sensing measurements, *J. Franklin Inst.* **297**, 1965, 95–109.

80. J. W. Woods, Two-dimensional discrete Markovian fields, *IEEE Trans. Informat. Theory* **IT-18**, 1972, 232–240.

81. E. C. Yeadon, R. A. Jones, and J. T. Kelley, Confidence limits for individual modulation transfer function measurements based upon the phase transfer function, *Photogr. Sci. Eng.* **12**, 1968, 244–250.

Chapter 8

Segmentation

In image compression or enhancement, the desired output is a picture—an approximation to, or an improved version of, the input picture. Another major branch of picture processing deals with *image analysis* or *scene analysis*; here the input is still pictorial, but the desired output is a *description* of the given picture or scene. The following are examples of image-analysis problems which have been extensively studied:

(1) The input is text (machine printed or handwritten), and it is desired to read the text; here the desired description of the input consists of a sequence of characters.

(2) The input is a nuclear bubble chamber picture, and it is desired to detect and locate certain types of "events" (e.g., particle collisions); here the description consists of a set of coordinates.

(3) The input is a photomicrograph of a mitotic cell; the desired output is a karyotype, or "map" showing the chromosomes arranged in a standard order. Note that here the output is pictorial, but its construction requires location and identification of the chromosomes.

(4) The input is an aerial photograph of terrain; the desired output is a map showing specific types of terrain features (forests, urban areas, bodies of water, roads, etc.). Here again the output is pictorial, and is even in registration with the input; but construction of the output requires location and identification of the desired terrain types.

(5)   The input is a television image of a pile of blocks; the desired output is a plan of action that can be used by a robot to build a tower out of the blocks. This evidently requires identification and location of individual blocks in the scene.

In all of these examples, the description refers to specific *parts* (regions or objects) in the picture or scene; to generate the description, it is necessary to *segment* the picture into these parts. Thus to identify the individual characters in text, they must first be singled out; to locate bubble chamber events, the bubble tracks and their ends or branches must be found; to make a karyotype, the individual chromosomes must be "scissored out"; and so on. This chapter discusses picture and scene segmentation techniques. The remaining chapters of the book discuss how picture segments—regions or objects—are used in picture descriptions, which involve properties of objects and relationships among objects.

Some segmentation operations can be applied directly to any picture; others can only be applied to a picture that has already been partially segmented, since they depend on the geometry of the parts that have already been extracted from the picture. For example, a chromosome picture can be (crudely) segmented by thresholding its gray level—dark points probably belong to chromosomes, while light points are probably background. Once this has been done, further segmentation into individual chromosomes can be attempted, based on connectedness, size, and shape criteria. This chapter will deal primarily with the initial segmentation of pictures; methods for further segmenting an already segmented picture will be discussed in Chapter 9.

It should be emphasized that there is no single standard approach to segmentation. Many different types of picture or scene parts can serve as the segments on which descriptions are based, and there are many different ways in which one can attempt to extract these parts from the picture. The perceptual processes involved in segmentation of a scene by the human visual system, e.g., the Gestalt laws of organization, are not yet well understood (Section 3.4). For this reason, no attempt will be made here to define criteria for successful segmentation; success must be judged by the utility of the description that is obtained using the resulting objects.

In order to "extract" an object from a picture explicitly, we must somehow mark the points that belong to the object in a special way. This marking process can be thought of as creating a "mask" or "overlay," congruent with the picture, in which there are marks at positions corresponding to object points. We can regard this overlay as a two-valued picture (e.g., 1's at object points, 0's elsewhere); the overlay thus represents the "characteristic function" of the object, i.e., the function which has the value 1 for points belonging to the object, and value 0 elsewhere. (There are many other ways

of representing objects or regions, e.g., using their boundaries or their
"skeletons"; see Sections 1.2.3 and 9.2.3.)

In practice, the objects that we try to extract from pictures are not always
clearly defined, and it is not always wise to attempt to map these objects into
clear-cut, two-valued overlays. Very often, our objects are defined only in a
"fuzzy" sense, and it would be more appropriate to represent them by con-
tinuous-valued "membership functions." These are functions that can take
on any value between 0 and 1; a point that has value 1 definitely belongs to
the object, a point with value 0 definitely does not belong, while points with
intermediate values have intermediate "degrees of membership" in the object.
For an introduction to the theory of fuzzy sets see Zadeh [48]. It would be of
interest to extend the concepts in this and the following chapters from clear-
cut objects to fuzzy objects.

8.1 THRESHOLDING

The most common way to extract objects from a picture is to *threshold* the
picture. If the given picture f has gray level range $[z_1, z_K]$, and t is any number
between z_1 and z_K, the result of thresholding f at t is the two-valued picture
f_t defined by

$$f_t(x, y) = \begin{cases} 1 & \text{if } f(x, y) \geq t \\ 0 & \text{if } f(x, y) < t \end{cases}$$

We can also consider thresholding operations that map other specified
ranges of gray levels into 1, and levels outside these ranges into 0. For example,
the two-valued pictures $f^{(u)}$ and $f_{[u, v]}$ defined by

$$f^{(u)}(x, y) = \begin{cases} 1 & \text{if } f(x, y) \leq u \\ 0 & \text{if } f(x, y) > u, \end{cases}$$

$$f_{[u, v]}(x, y) = \begin{cases} 1 & \text{if } u \leq f(x, y) \leq v \\ 0 & \text{otherwise} \end{cases}$$

can also be regarded as "thresholded" versions of f. More generally, if Z is
any *set* of gray levels, $Z \subseteq [z_1, z_K]$, we can define a generalized "thresholding" operation as mapping the gray levels in Z into 1 and those not in Z into 0,
i.e., as defining the two-valued picture

$$f_Z(x, y) = \begin{cases} 1 & \text{if } f(x, y) \in Z \\ 0 & \text{otherwise} \end{cases}$$

In this section we discuss the use of thresholding operations, and various

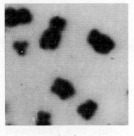

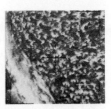

(a)   (b)   (c)

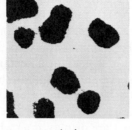

(d)   (e)   (f)

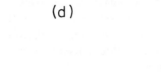

(g)   (h)   (i)

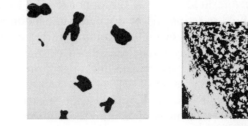

(j)   (k)   (l)

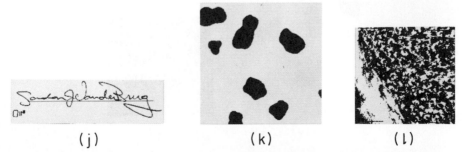

Fig. 1 Thresholding. (a)–(c) Original pictures. (d)–(f) Results of thresholding too low (at 11, 25, and 10, respectively, on a scale of 0–63). (g)–(i) Results of thresholding too high (at 45, 52, and 35). (j)–(l) Results of using good threshold levels (28, 37, and 29).

259

generalizations of such operations, in extracting objects from pictures. Special attention is given to the problem of automatically selecting appropriate thresholds, and to the treatment of cases in which no single threshold works well throughout the given picture.

8.1.1   Thresholding a Picture or a Processed Picture

If the given picture f consists of objects that are, say, darker than their background, then simple thresholding of f is a natural way to extract the objects. For example, if f is a picture of some printing or writing, the characters are generally darker than the page, and one could attempt to extract them by thresholding. Similar remarks apply to the tracks in a bubble chamber picture and to the chromosomes in a metaphase spread. Analogously, the clouds in an aerial or space photograph are brighter than most terrain types, so that they too should be extractable by thresholding. Some examples of object extraction by simple thresholding are shown in Fig. 1. In this figure, the output values 0 and 1 are displayed as white and black, respectively.

It is seen from Fig. 1 that proper selection of the threshold is critical. If we threshold too high, too many object points will be classified as background; if we threshold too low, the opposite will happen. Methods of automatic threshold selection will be discussed in Section 8.1.2.

A useful variation on thresholding is what might be called *semithresholding*. Here certain gray levels, e.g., those below a threshold, become 0, but the remaining gray levels remain unchanged, rather than becoming 1. Applied to a picture that consists of objects on a background, this technique can be used to "zero out" the background while preserving the gray levels within the objects. Two examples of semithresholding are shown in Fig. 2; in one example the background becomes white, in the other it becomes black.

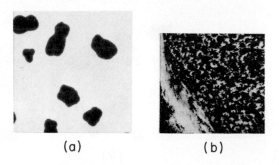

(a)   (b)

Fig. 2   Semithresholding. (a) Upper semithresholding: In Fig. 1b, gray levels below 37 have been mapped into white, while the remaining gray levels are preserved intact. (b) Lower semithresholding: In Fig. 1c, gray levels 29 and above have been mapped into black.

Thresholding can sometimes be used to extract the edges of objects, rather than the objects themselves. (On other methods of edge detection see Section 8.2.) For example, suppose that the objects are dark and the background light, and that intermediate gray levels occur only on the borders between objects and background. If we map these intermediate levels into 1, and all other levels into 0, we obtain outlines of the objects. This technique is illustrated in Fig. 3. More generally, mapping narrow ranges of gray levels into 1, and the

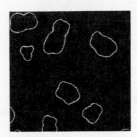

Fig. 3 Outlining by thresholding. In Fig. 1b, gray levels 35–39 have been mapped into white, all other levels into black.

levels outside these ranges into 0, produces interesting gray level "contour maps" of the given picture.

Simple gray level thresholding is effective in extracting objects when the objects have a characteristic range or set of gray levels. More generally, we can use thresholding techniques to extract objects which have characteristic patterns of gray levels, or characteristic textures. We can do this by first converting the given picture f into a new picture f' in which the desired objects now have characteristic gray level ranges, so that the objects can now be extracted by simple thresholding of f'. Several examples of this approach will be given in the following paragraphs.

Suppose first that the desired objects have a known *shape*, or a known *pattern* of gray levels. Using matching techniques (Section 8.3), we can produce a new picture f' whose value at a given point represents the degree of match between the original f and the given pattern at that point. Thus f' has high values at points where there is a close match, so that thresholding f' can be used to detect occurrences of the desired pattern in f. Note that in this case, the thresholding does not produce 1's at all points of the pattern, but only at isolated points marking the position(s) of the pattern in f. This approach is appropriate when we are looking for objects whose shapes, sizes, and orientations are known, e.g., characters from a specific type font.

Edges, lines, and spots are important examples of simple patterns that we may want to detect in a picture. A wide variety of techniques for detecting these types of features will be described in Sections 8.2 and 8.3. For the present we note only that they can be detected by performing operations that yield high values at points belonging to such features, and low values elsewhere (e.g., derivative operations), and then thresholding the results.

Another type of example in which it is useful to threshold a processed picture f', rather than the original picture f, is illustrated in Fig. 4. Here the

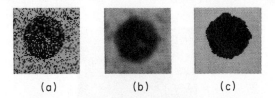

(a)   (b)   (c)

Fig. 4 Extraction of a textured region by smoothing and thresholding. (a) Original picture: densely dotted region on a sparsely dotted background. Since all dots have the same gray level, the densely dotted region cannot be extracted by thresholding. (b) Results of smoothing (a); the gray level at point (i, j) of (b) is the average of the gray levels in a 7×7 square centered at (i, j) in (a). (c) Results of thresholding (b) at 32. The densely dotted region has now been extracted.

only gray levels in the picture are black and white, but the probability of black in the object is higher than in the background, so that the object has darker average gray level than the background. Thus if we apply local averaging to f, we obtain an f' in which the object now consists of darker gray levels than the background, so that it can be extracted by simple thresholding. (See Section 6.5 for other smoothing operations that we could have applied, rather than local averaging.)

In the example just given, an object that had different average gray level than its background was extracted by locally averaging and then thresholding. More generally, suppose that the object differs from its background with respect to the average value of some other local property; for example, suppose that the gradient or Laplacian (Section 6.4) has a high average value at points of the object, and a low average value at background points. Two cases where this is true are shown in Fig. 5. (In both cases, the object consists of black-and-white salt-and-pepper noise; while the background in the first case is solid gray, and in the second case it consists of randomly chosen gray levels.) To extract the object in these cases, we need only compute the gradient (say) at every point, obtaining a new picture f' in which the object has higher average value than the background. The method of the previous paragraph can now be applied, i.e., we first locally average f to obtain a smoothed picture f'' in which the object is darker than the background, and we can then extract the object from f'' by simple thresholding.

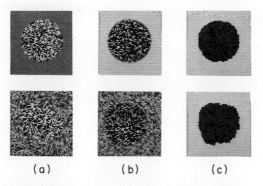

(a)   (b)   (c)

Fig. 5 Extraction of textured regions by preprocessing, smoothing, and thresholding. (a) Original pictures: salt-and-pepper noise (0's and 63's) on a background of solid gray or of uniformly distributed random gray levels. The average gray levels of the region and background are the same, so that the region cannot be extracted by smoothing and thresholding. (b) Results of computing a digital gradient (see Section 8.2) at each point of (a). (c) Results of smoothing (b), using averaging over 11×11 squares, and thresholding at 37.

The general approach just described—applying a local operation at every point, then locally averaging, and finally thresholding—can be used to extract a wide variety of objects that differ in *texture* from their backgrounds. The description of texture in terms of average values of local properties will be discussed further in Section 10.2. It should be noted that if a picture contains three regions R_1, R_2, R_3, where the average gray level (or local property value) of R_2 is intermediate between those of R_1 and R_3, and the latter two are adjacent to one another, then averaging and thresholding cannot be used to extract R_2, since a strip along the border between R_1 and R_3 will have the same average as R_2.

8.1.2 Threshold Selection

We now consider methods of automatically selecting a threshold at which to segment a given picture. As before, we denote the given picture by f, and the result of thresholding f at t by f_t.

If the desired f_t has some known property, we can try various f's until we obtain an f_t that satisfies this property. For example, if f is a picture of a printed page, we may know that the characters should occupy a certain fraction θ of the space on the page. Knowing this, we should threshold at that value of t such that just fraction θ of the picture points have gray levels $\geqslant t$; this fraction is called the $(1-\theta)$-tile of the gray level histogram. This *p-tile method* is rarely applicable in practice, since we hardly ever know the areas of the desired objects in advance. Other types of information about f_t may,

```
                                             ++
                                           ++++
                                          +++++
                                         ++++++
                                        +++++++
                                       ++++++++
                                      +++++++++
                                     ++++++++++
                                    +++++++++++
                 +                 +++++++++++++
                 +                +++++++++++++
                 +               ++++++++++++++
                 +              +++++++++++++++
                 +             ++++++++++++++++
                 +            +++++++++++++++++
                 +           ++++++++++++++++++
                 +          +++++++++++++++++++
                 +         ++++++++++++++++++++
                 +        +++++++++++++++++++++
                 +       ++++++++++++++++++++++
                 +      +++++++++++++++++++++++
                 +     ++++++++++++++++++++++++
                 +    +++++++++++++++++++++++++
                 +   ++++++++++++++++++++++++++
                 +  +++++++++++++++++++++++++++
                 + +++++++++++++++++++++++++++++
                 ++++++++++++++++++++++++++++++
                 +++++++++++++++++++++++++++++
                 +++++++++++++++++++++++++++
                 +++++++++++++++++++++++ +
                 +++++++++++++++ ++
                 +++++++++ ++        +              +
                 ++ +       +                       ++
                 +                                 +++
                                                    ++
                                                     +
```

(a)

bin	count	cum%
0	3818	12
2	5104	28
4	5900	47
6	6020	67
8	4312	80
10	1827	86
12	763	89
14	288	90
16	173	90
18	123	91
20	85	91
22	83	91
24	82	91
26	74	92
28	69	92
30	106	92
32	85	93
34	68	93
36	93	93
38	121	93
40	134	94
42	154	94
44	212	95
46	252	96
48	295	97
50	313	98
52	249	99
54	161	99
56	97	99
58	36	99
60	71	100

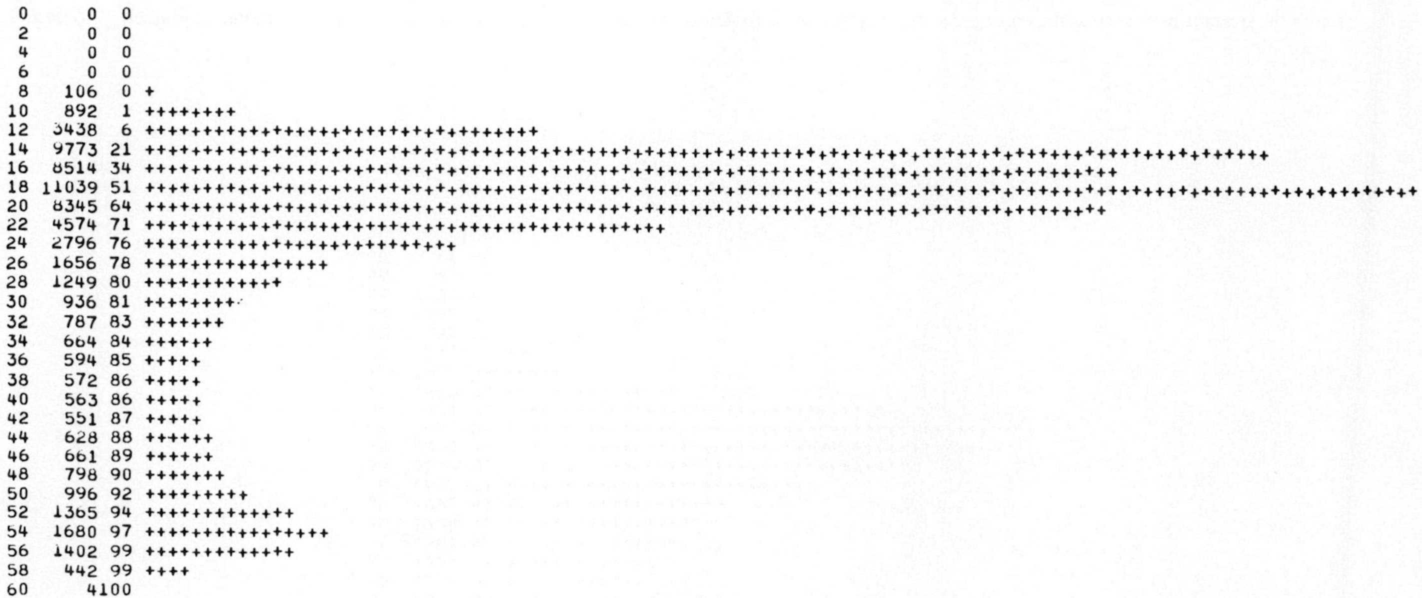

0	0	
2	0	
4	0	
6	0	
8	106	0
10	892	1
12	3438	6
14	9773	21
16	8514	34
18	11039	51
20	8345	64
22	4574	71
24	2796	76
26	1656	78
28	1249	80
30	936	81
32	787	83
34	664	84
36	594	85
38	572	86
40	563	86
42	551	87
44	628	88
46	661	89
48	798	90
50	996	92
52	1365	94
54	1680	97
56	1402	99
58	442	99
60	4100	

(b)

Fig. 6 Threshold selection by the mode method. (a), (b) Gray level frequency distributions for Fig. 1a, b. Note that the good thresholds (28 and 37) are in the middles of the valleys.

```
 0  2048   4
 2   949   6
 4  1052   9
 6   968  11
 8   976  13
10   960  15
12  1025  18
14   986  20
16  1034  22
18  1117  25
20  1164  27
22  1136  30
24  1146  33
26  1228  35
28  1324  39
30  1560  42
32  1724  46
34  1992  51
36  2692  57
38  3557  65
40  4220  74
42  4660  85
44  2437  93
46  1569  96
48   622  98
50   249  98
52   153  99
54   126  99
56    97  99
58    46  99
60    21  99
62     2 100
```

(c)

Fig. 6   Threshold selection by the mode method. (c) Gray level frequency distribution for Fig. 1c. Note that in this case there is no valley.

however, be available. To take another example from the printed page case, we may know that the characters should be composed of strokes having a certain width; if so, we can threshold at that value of t which yields objects of the desired width. (Measurement of object widths will be discussed in Section 9.4.) Analogous examples can be given for other classes of pictures.

The p-tile method is one way of using f's gray level histogram $p_f(z)$ as a guide in threshold selection. Another approach that makes use of the histogram, which we shall call the *mode method*, can be motivated as follows: Suppose that f contains, say, large dark objects on a light background. Then p_f should contain a peak (or "mode") corresponding to the gray levels in the objects, and another peak corresponding to the background levels. (In other words, p_f should be *bimodal*.) Moreover, if gray levels intermediate between those of the objects and the background are relatively uncommon in f, then p_f will have a valley between these two peaks. It seems reasonable, if we desire to separate the objects from the background, to threshold f at the gray level corresponding to the bottom of this valley. This method, which frequently gives good results, is illustrated in Fig. 6.

The choice of the valley bottom as a threshold level in the mode method can be justified on several grounds. Intuitively, this choice should minimize the probability of misclassifying an object point as background or vice versa. (This justification will be examined more closely in Section 8.1.3.) In addition, when we threshold at the bottom of a valley, the results are relatively insensitive to our exact choice of the threshold level, since the gray levels at a valley bottom are relatively unpopulated. In general, if f's histogram contains several major peaks separated by valleys, one can use thresholds at the bottoms of all the valleys to define several segmentations of f.

Determining that $p_f(z)$ is bimodal (or more generally, that it contains significant peaks and valleys), and locating these peaks and valleys, are nontrivial tasks in themselves. One possible approach is as follows:

(a) Find the two highest local maxima on p_f that are at least some minimum distance apart. (The second condition is intended to prevent local irregularities near the top of a single major peak from being detected as different major peaks.) Let the locations of these maxima be z_i and z_j.

(b) Find a lowest point of p_f between z_i and z_j, i.e., a point z_k such that $p_f(z_k) \leqslant p_f(z)$ for all $z_i \leqslant z \leqslant z_j$.

(c) The flatness of the histogram can be measured by, e.g., $p_f(z_k)/\min(p_f(z_i), p_f(z_j))$. If this is small, the histogram should be strongly bimodal, since the peaks are much higher than the valley, and thus z_k should be a useful threshold for segmenting f into significant parts.

Another approach is to find two unimodal (= one-peak) density functions

$p(z)$ and $q(z)$, e.g., normal density functions, whose weighted sum best approximates $p_f(z)$. We can then take, as the bottom of the valley, the lowest point of $p_f(z)$ that lies between the peaks of $p(z)$ and $q(z)$. (The case of a sum of normal density functions will be studied in greater detail in Section 8.1.3.)

If the given picture actually does contain objects on a contrasting background, then the mode method is a reasonable way to find a threshold that separates the objects from the background. It should be realized, however, that even if a picture's histogram is strongly bimodal, the picture itself need not consist of objects and a background. The histogram tells us only how many points of the picture have each gray level, but not how these points are arranged in the picture. A picture whose left half is black and whose right half is white has exactly the same histogram as a picture that consists of random salt-and-pepper noise. Thus analysis of the histogram alone, in the absence of any knowledge about the picture, is not an adequate guide for making decisions about how to threshold the picture. If a threshold is selected based on the histogram alone, it is important to examine the resulting thresholded picture to see whether objects of the desired types have indeed been extracted.

A somewhat more general basis for segmenting a picture can be defined by studying the tendencies of the picture's gray levels to occur as neighbors of one another. Specifically, let $p_f(z, w)$ denote the relative frequency with which points having gray level w occur in some neighborhood of points having gray level z (or vice versa); this is a type of joint probability density for pairs of gray levels of f. If f is a quantized picture having K gray levels, $p_f(z, w)$ can be represented as a symmetric $K \times K$ matrix, whose (i, j) entry, call it p_{ij}, is the number of points having level z_j that occur as neighbors of points having level z_i. (Such gray level cooccurrence matrices will be studied further when we deal with textural properties in Section 10.2.1.) For each i, the vector $(p_{i1}, \ldots, p_{iK})$ characterizes the types of neighbors that points of gray level z_i have in f. We can use these vectors to measure the "closeness" between pairs of gray levels, e.g., we can say that levels z_i and z_j are closely related if the normalized cross correlation

$$\sum_{k=1}^{K} p_{ik} p_{jk} \bigg/ \left(\sum_{k=1}^{K} p_{ik}^2 \sum_{k=1}^{K} p_{jk}^2 \right)^{1/2}$$

is close to 1. (On this cross correlation and its properties see Section 8.3.1.) If we can find a "cluster" of gray levels with respect to this closeness measure, i.e., a set of levels that are all mutually close, then it is possible that this cluster results from the occurrence of objects or regions of a particular type in the picture. Such an object would be characterized by having particular neighbor associations among gray levels, even though it might not occupy any particular gray level range. We can segment the picture to extract such objects by singling out those points whose neighbor frequency vectors are close to (i.e., correlate highly with) the cluster.

It should be noted that the segmentation method just described is not a thresholding operation, since the selection of a point depends not on its own gray level, but on the frequency vector of its neighbors' gray levels. Note also that this method would distinguish quite well between a half black, half white picture and a picture of salt-and-pepper noise. In the latter case, all points would have the same neighbor frequency vector; while in the former case, black points' neighbors would always be black, and white points' neighbors white, except at the border between the black and white regions. Thus in the salt-and-pepper case, we would get no segmentation; but in the other case, the interiors of the black and white regions would be found as clusters.

The methods described in this section can all be applied to thresholding with respect to the value, or average value, of some local property other than gray level. Here too, thresholds can be selected at p-tiles of the local property value, or at valleys between peaks in the probability density of this value; or we can find clusters with respect to the frequencies with which neighboring points have various values of the local property.

8.1.3 Minimum-Error Thresholding

In this section we consider an example in which the p-tile and mode methods of threshold selection, described in Section 8.1.2, can be shown to give optimal results.

Let the pictures of interest consist of objects on a background, where the gray levels of points in the objects have a normal[†] probability density $p(z)$ with mean μ and standard deviation σ, while the gray levels of background points have a normal density $q(z)$ with mean ν and standard deviation τ. We shall assume, without loss of generality, that $\mu < \nu$. Suppose further that the objects occupy fraction θ of the picture area, so that the background occupies fraction $1 - \theta$. Thus such a picture has overall gray level probability density $\theta p(z) + (1 - \theta) q(z)$.

Exercise 1. Prove that $\theta p(z) + (1 - \theta) q(z)$ has mean $\theta\mu + (1 - \theta)\nu$, and variance $\theta\sigma^2 + (1 - \theta)\tau^2 + \theta(1 - \theta)(\mu - \nu)^2$. ∎

Suppose that we threshold the picture at t, and call all points with gray level $< t$ object points, and all points with level $\geq t$ background points. The

---

[†] Actual gray level probability densities cannot be normal, since gray levels are non-negative and occupy only a finite range. If, however, the mean gray level is well inside the range, and the standard deviation is small relative to the size of the range, the normal density can be a reasonable approximation to the actual density. We also treat the gray scale here as though it were continuous.

probability of misclassifying an actual background point as an object point is $Q(t) \equiv \int_{-\infty}^{t} q(z)\, dz$. Similarly, the probability of misclassifying an object point as background is $1 - P(t) \equiv \int_{t}^{\infty} p(z)\, dz$. The overall misclassification probability when we threshold at t is thus

$$\theta(1 - P(t)) + (1 - \theta)\, Q(t)$$

To find the threshold value for which this is a minimum, we can differentiate this with respect to t and set the result equal to zero, obtaining

$$(1 - \theta)\, q(t) = \theta p(t)$$

Now

$$p(t) = \frac{1}{\sqrt{2\pi}\sigma} \exp\left[\frac{-(t - \mu)^2}{2\sigma^2}\right]; \qquad q(t) = \frac{1}{\sqrt{2\pi}\tau} \exp\left[\frac{-(t - v)^2}{2\tau^2}\right]$$

If we substitute these in the previous equation and take logarithms of both sides, we obtain

$$\ln \sigma + \ln(1 - \theta) - \frac{(t - v)^2}{2\tau^2} = \ln \tau + \ln \theta - \frac{(t - \mu)^2}{2\sigma^2}$$

or

$$\tau^2 (t - \mu)^2 - \sigma^2 (t - v)^2 = 2\sigma^2 \tau^2 \ln \frac{\tau \theta}{\sigma(1 - \theta)}$$

This quadratic equation can be solved for t in terms of μ, σ, v, τ, and θ. Note in particular that if $\theta = \frac{1}{2}$ and $\tau = \sigma$, the equation simplifies to $(t - \mu)^2 = (t - v)^2$, so that $t = (\mu + v)/2$.

Exercise 2. Prove that if $\sigma = \tau$, but θ is arbitrary, the threshold that minimizes misclassification probability is

$$\frac{\mu + v}{2} + \frac{\sigma^2}{v - \mu} \ln \frac{\theta}{1 - \theta} \quad \blacksquare$$

We shall now verify that, in the special case $\theta = \frac{1}{2}$ and $\tau = \sigma$, the p-tile and mode methods both give the same result as that just obtained—namely, that the threshold should be midway between the means of the object and background gray level density functions.

If we know that fraction θ of the picture points are object points, then according to the p-tile method, we should threshold at the θ-tile, i.e., at that t such that fraction θ of the picture points have gray level less than t, so that

$$\int_{-\infty}^{t} [\theta p(z) + (1 - \theta)\, q(z)]\, dz = \theta$$

In general, this equation can be solved for t numerically. In the special case where $\theta = \frac{1}{2}$ and $\tau = \sigma$, however, we can solve it directly by observing that the density function

$$d(z) \equiv \frac{1}{\sqrt{2\pi}\sigma}\left\{\exp\left[\frac{-(z-\mu)^2}{2\sigma^2}\right] + \exp\left[\frac{-(z-v)^2}{2\sigma^2}\right]\right\}$$

is symmetrical around the point $z = (\mu+v)/2$; i.e., for any w we have

$$d\left(\frac{\mu+v}{2}+w\right) = d\left(\frac{\mu+v}{2}-w\right)$$

Thus $(\mu+v)/2$ is the median ($\equiv$ the "$\frac{1}{2}$-tile") of $d(z)$.

The peaks and valleys of $\theta p(z)+(1-\theta)q(z)$ can be found by differentiating it with respect to z and equating the result to zero. This gives

$$\theta\tau^3 \exp[-(z-\mu)^2/2\sigma^2](z-\mu) + (1-\theta)\sigma^3 \exp[-(z-v)^2/2\tau^2](z-v) = 0$$

In general, this too would have to be solved numerically. In our special case, it simplifies to

$$\exp[-(z-\mu)^2/2\sigma^2](z-\mu) + \exp[-(z-v)^2/2\sigma^2](z-v) = 0$$

It is easily verified that this has a root at $z = (\mu+v)/2$. To see whether this root corresponds to a maximum or a minimum, let us examine the derivatives of $d(z)$. Ignoring common positive constant factors, its first derivative is

$$d'(z) = -(z-\mu)\exp[-(z-\mu)^2/2\sigma^2] - (z-v)\exp[-(z-v)^2/2\sigma^2]$$

For $z < \mu$, both terms of this are positive, and for $z > v$, both terms are negative; thus $d(z)$ is strictly increasing for $z < \mu$ and strictly decreasing for $z > v$. We also have

$$d''(z) = \left(\frac{(z-\mu)^2}{\sigma^2}-1\right)\exp\left[\frac{-(z-\mu)^2}{2\sigma^2}\right] + \left(\frac{(z-v)^2}{\sigma^2}-1\right)\exp\left[\frac{-(z-v)^2}{2\sigma^2}\right]$$

where we have also ignored common positive constant factors. At $z = (\mu+v)/2$, this has value

$$2\left(\frac{(v-\mu)^2}{4\sigma^2}-1\right)\exp\left[\frac{-(v-\mu)^2}{8\sigma^2}\right]$$

Thus $d''((\mu+v)/2)$ is positive provided that $(v-\mu)^2 > 4\sigma^2$, i.e., $v-\mu > 2\sigma$. In this case $d(z)$ has a minimum at $(\mu+v)/2$; and since it is increasing for $z < \mu$ and decreasing for $z > v$, it must also have maxima between μ and $(\mu+v)/2$ and between $(\mu+v)/2$ and v. Thus the threshold $(\mu+v)/2$ corresponds to a valley between two peaks on $d(z)$, provided that $v-\mu > 2\sigma$.

Exercise 3. If $v-\mu < 2\sigma$, $d(z)$ has a maximum, rather than a minimum, at $(\mu+v)/2$. Does it have any minima at all in this case (i.e., is $(\mu+v)/2$ its only maximum)? ▌

Exercise 4.   Let f be a picture, let f_t be the two-valued result of threshold-ing f at t, and let the two gray levels in f_t be B and W. Prove that, no matter what the probability density of the gray levels in f, the value of t that mini-mizes $\iint (f-f_t)^2 \, dx \, dy$ is always $t = (B+W)/2$. ∎

Even if the probability densities $p(z)$ and $q(z)$ are not normal, the equation $(1 - \theta) q(t) = \theta p(t)$ can still be used to determine the minimum-error thresh-old t; as long as p and q are known functions, we can solve this equation for t using numerical methods. It should be realized, however, that for arbitrary p and q, simple thresholding is not necessarily an appropriate basis for deciding whether a given point of the picture belongs to an object or to the back-ground. Rather, we want to determine the set Z of gray levels such that, if a point's gray level is in Z, the point is more likely to belong to the objects than to the background. We shall now show how to do this whenever $p(z)$, $q(z)$, and θ are known.

Let $p(o, z)$ be the joint probability that a point belongs to an object and has gray level z, and let $p(b, z)$ be the joint probability that a point belongs to the background and has gray level z. Let $p(o)$ and $p(b)$ be the _a priori_ probabi-lities of a point belonging to an object and to the background, respectively; in our earlier notation, we have $p(o) = \theta$ and $p(b) = 1 - \theta$. Let $p(z \mid o)$ and $p(z \mid b)$ be the conditional probabilities that a point has gray level z, given that it belongs to an object and to the background, respectively; these are just our probability densities $p(z)$ and $q(z)$. Finally, let $p(o \mid z)$ and $p(b \mid z)$ be the conditional probabilities that a point belongs to the object or to the background, respectively, given that it has gray level z. Now

$$p(o, z) = p(o) \, p(z \mid o) = \theta p(z)$$

and

$$p(b, z) = p(b) \, p(z \mid b) = (1 - \theta) \, q(z)$$

On the other hand, if we let $\pi(z)$ be the _a priori_ probability that an arbitrary point has gray level z, we have

$$p(o, z) = \pi(z) \, p(o \mid z) \qquad \text{and} \qquad p(b, z) = \pi(z) \, p(b \mid z)$$

Using these four equations, we can express $p(o \mid z)$ and $p(b \mid z)$ in terms of θ, $p(z)$, $q(z)$, and $\pi(z)$:

$$p(o \mid z) = \theta p(z)/\pi(z); \qquad p(b \mid z) = (1 - \theta) q(z)/\pi(z)$$

If a point has gray level z, it is more likely to belong to an object than to the background if $p(o \mid z) > p(b \mid z)$, i.e., if

$$\theta p(z) > (1 - \theta) q(z)$$

Thus if we know θ, p, and q, we can determine which gray levels z belong to the set Z just defined.

We can generalize this discussion to the case where there are k possible classes of picture points, rather than just the two classes "object" and "background." Let $p(i, z)$ be the joint probability that a point belongs to the ith class $(1 \leqslant i \leqslant k)$ and has gray level z; let $p(i)$ be the *a priori* probability of a point belonging to the ith class, and let $p(z|i)$ and $p(i|z)$ be conditonal probabilities defined analogously to those in the last paragraph. Note that

$$\sum_{i=1}^{k} p(i) = 1; \qquad \sum_{i=1}^{k} p(i|z) = 1 \qquad \text{for any } z$$

and

$$\pi(z) = \sum_{i=1}^{k} p(i) p(z|i)$$

We have

$$p(i, z) = p(i) p(z|i) = \pi(z) p(i|z)$$

so that

$$p(i|z) = p(i) p(z|i)/\pi(z)$$

and if the $p(i)$'s and $p(z|i)$'s are known, we can determine the i for which $p(i|z)$ is greatest. In practice, unfortunately, the $p(i)$'s and $p(z|i)$'s are usually not known; but we can estimate them if we are given a set of samples, from the given ensemble of pictures, in which the points belonging to the various classes have been identified.

Still more generally, we may want to classify picture points on the basis of a set of local property values (e.g., color parameters, values of various derivative operators at the point, etc.), rather than just on the basis of gray level. To generalize our discussion to this case, let $\vec{z}$ be the vector whose components are the property values in question. Then we can define $p(i, \vec{z})$, $p(\vec{z}|i)$, and $p(i|\vec{z})$ just as above, and can determine the i for which $p(i|\vec{z})$ is greatest. Of course, the needed probability densities are now even harder to estimate. Classification on the basis of a set of property values is discussed extensively in books on pattern recognition, to which the reader is referred for further details (see, e.g., reference [4] in Section 1.3). If we are also given a set of numbers $\lambda(i|j)$ which represent the *costs* of assigning a point to class i when it really belongs to class j $(1 \leqslant i, j \leqslant n)$, then we can determine the class i to which a given $\vec{z}$ should be assigned so as to minimize the expected cost; this is not necessarily the same as the class i to which $\vec{z}$ has maximum probability of belonging.

8.1.4 Variable Thresholding

In many cases, no single threshold gives good segmentation results over an entire picture. Suppose, for example, that the picture shows dark objects on a

light background, but that it was made under conditions of uneven illumination. The objects will still contrast with the background throughout, but both
background and objects may be much lighter on one side of the picture than
on the other. Thus a threshold that nicely separates objects from background
on one side may accept too much of the background as belonging to the
objects, or reject too much of the objects as belonging to the background, on
the other side.

   If the uneven illumination is described by some known function of position
in the picture, one could attempt to correct for it using gray level correction
techniques (Section 6.2.1), after which a single threshold should work for the
whole picture. If this information is not available, one can divide the picture
into pieces and apply threshold selection techniques (Section 8.1.2) to each
piece. If a piece contains both objects and background, its histogram should
be bimodal, and the mode method should yield a *local threshold*, suitable for
separating objects from background in that part of the picture. If a piece
contains objects only, or background only, it will not have a bimodal histogram, and no such threshold will be found for it; but a threshold can still be
assigned to it by interpolation from the local thresholds that were found for
nearby bimodal pieces. (Some smoothing of the resulting thresholds may be
necessary, since if a threshold changes abruptly from one piece to the next,
artifacts may result.)

   The results of thresholding are often noisy; if we threshold high, occasional
object points will be called background, while if we threshold low, the opposite will happen. One way to overcome this problem is to use two thresholds,
say $t_1 < t_2$. Points whose gray levels exceed t_2 are classified as "core" object
points; while points whose levels exceed t_1 are called object points only if
they are close to core object points. (This method is a simple example of
tracking or region growing; such methods will be discussed in greater detail
in Section 8.4.) Here t_2 should be chosen so that every object contains some
points with levels higher than t_2, but the background contains no such points;
while t_1 should be chosen so that every object point has level higher than t_1.
If we used t_2 alone, the objects would be incomplete; if we used t_1 alone,
many background points would be misclassified as object points; but when
we use them together, good separation of the objects from the background
may be possible. This method should not be used unless the contrast between
objects and background is relatively sharp; otherwise, points of the background that are near the objects may have levels higher than t_1.

   Alternatively, if there exists a threshold t_2 such that every object contains
points with levels higher than t_2, but the background contains no such points,
we can threshold the picture at t_2 and then examine the neighborhoods of the
above-threshold points, with the aim of finding a local threshold that will
separate object from background in each of these neighborhoods. (This is

evidently more general than using the same threshold t_1 in each neighborhood.) This method is appropriate if the objects are relatively small and do not occur too close together. The neighborhoods used should be large enough to ensure that they contain both object and background points, so that the histograms of the gray levels in the neighborhoods will be bimodal.

One sometimes wants to find *local maxima* in a picture, i.e., to extract points which have higher value, with respect to some local property, than the nearby points. Typically, one would also require these points to have values above some low threshold t_1; but once t_1 is exceeded, all relative maxima are accepted, no matter what their absolute sizes. Local maximum finding can thus be regarded as an extreme case of local thresholding. The following are some cases in which local maxima are of interest:

(a)   When a matching operation has been applied to a picture, we may want to find the best matches, even though they are far from perfect matches, as long as the degree of match exceeds some low threshold. If, however, two or more high match values occur too close together, we would normally want to keep only the highest of them, representing the position of best match. This is because if the shape or pattern that is being matched occurs at position (x, y), say, we do not want to detect it also at nearby positions where it would have to overlap its occurrence at (x, y). For examples of matching operation outputs, which show local peaks at the positions of best match, see the figures in Section 8.3.

(b)   When edges (or lines, etc.) are detected in a picture, a given edge may be detected in more than one position (see Section 8.2); if we kept all of these, the resulting edge would be thick. We can keep a detected edge thin by rejecting nonmaxima in the direction across the edge, i.e., we ignore an edge detection value if there exists a higher value nearby in that direction. (On the other hand, we should not suppress an edge value if there is a higher value nearby along the edge, since we do not want the points along the edge to compete with one another.) Examples of nonmaximum suppression in edge detection will be given in Section 8.2.4.

Still other types of local maxima are of interest in the analysis of shapes, e.g., points of an object at which the distance to the background is a local maximum constitute the "skeleton" of the object; see Section 9.2.3.

8.2   EDGE DETECTION

The thresholding techniques described in Section 8.1 are primarily designed to extract objects that have characteristic gray level ranges or textures; in other words, they yield objects that have some type of *uniformity*. Another important approach to picture segmentation is based on the detection of *discontinuity*, i.e., of places where there is a more or less abrupt change in gray level or in texture, indicating the end of one region and the beginning of another. Such a discontinuity is called an *edge*. This section reviews a variety of edge detection techniques.

Biological visual systems appear to make use of edge detection, but not of thresholding. When a person looks at a region across which there is a gradual brightness change, he cannot see the region as broken up into two clear-cut parts, no matter how hard he tries to threshold it mentally. If there is an abrupt brightness change cutting across the region, on the other hand, the person immediately sees an edge there which breaks the region apart.

Abrupt changes in gray level can take several different forms. The most common is the *step edge*, illustrated by the cross section in Fig. 7a. This, of course, is an ideal example; the presence of blur and noise turns steps into noisy ramps such as that in Fig. 7b. The discussion in this section will deal primarily with step edges, but other types can occur, e.g., abrupt changes in

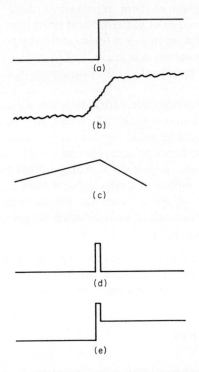

Fig. 7 Idealized edge cross sections: how the gray level changes as we move across the edge. (a) Perfect step edge. (b) Noisy, blurred step edge. (c) "Roof" edge: abrupt change in rate of change of gray level. (d) Perfect "spike" line. (e) Line combined with step edge.

the rate of change of gray level (Fig. 7c). Edges at which there is an abrupt change in texture will be discussed in Section 8.2.4.

A step edge separates two regions in each of which the gray level (or texture) is relatively uniform, with different values on the two sides of the edge. Another important type of gray level discontinuity is the *line*, which is a thin strip that differs from the regions on both sides of it; it has the spike-like cross section shown in Fig. 7d. Line detection will be discussed in Section 8.3. It should be mentioned that lines often occur in association with edges (highlights on edges of blocks; membrane separating parts of a cell; roads running between fields bearing different crop types); an idealized version of the resulting combined cross section is shown in Fig. 7e.

A difficulty with edge detection as an approach to picture segmentation is that the detected edges often have gaps in them, at places where the transitions between regions are not abrupt enough. Moreover, edges may be detected at points that are not part of region boundaries, if the given picture is noisy. Thus the detected edges will not necessarily form a set of closed, connected curves that surround closed, connected regions. One way to overcome these problems is to use tracking techniques (Section 8.4) to follow the edges around the regions; such techniques can be designed to tolerate gaps in edges which do lie on a curve, and to reject as noise edges which do not. Another possibility is to apply curve detection operations (Section 8.3) to the edge detector output; this rejects edge points that do not lie on curves, and can also be designed to fill gaps in edges that do. A method of "filling in" regions that are surrounded by broken edges, using a propagation process, will be described in Section 9.2.4. The tracking approach is usually the best, because of its great flexibility; in particular, it can be designed to take into account any information that may be available about the shapes of the regions whose boundaries are being tracked.

8.2.1 Differentiation

Derivative operators, which give high values at points where the gray level of the picture is changing rapidly, were discussed in Section 6.4. Evidently, any such operator can be used as an edge detector; its value at a point represents the "edge strength" at that point, and we can explicitly extract sets of edge points from the picture by thresholding these values.

The simplest derivative operators are the first partial derivatives $\partial f/\partial x$ and $\partial f/\partial y$, which give the rates of change of gray level in the x- and y-directions. As we saw in Section 6.4.1, the rate of change in any direction θ is a linear combination of these:

$$\frac{\partial f}{\partial x'} = \frac{\partial f}{\partial x} \cos \theta + \frac{\partial f}{\partial y} \sin \theta$$

For digital pictures, we use differences instead of derivatives:

$$\Delta_x f(i,j) = f(i,j) - f(i-1,j)$$

$$\Delta_y f(i,j) = f(i,j) - f(i,j-1)$$

$$\Delta_\theta f(i,j) = \Delta_x f(i,j) \cos\theta + \Delta_y f(i,j) \sin\theta$$

It should be noted that the values of these operations can be either positive or negative, depending on whether the gray level goes upward or downward as one moves in the positive x- (or y-, or θ-) direction. If one wants operations that always have nonnegative values at edges, absolute values of derivatives or differences can be used.

A directional derivative (or difference) measures only the component of the rate of change of gray level in one particular direction. Suppose, for example, that the gray level is given by the linear ramp function $f(x,y) = a(x\cos\varphi + y\sin\varphi) + b$ over some portion of the picture. Here the x-component of the rate of change is $\partial f/\partial x = a\cos\varphi$; the y-component is $\partial f/\partial y = a\sin\varphi$; the component in direction φ is

$$\frac{\partial f}{\partial x}\cos\varphi + \frac{\partial f}{\partial y}\sin\varphi = a(\cos^2\varphi + \sin^2\varphi) = a$$

and the component in direction $\varphi + (\pi/2)$ is

$$\frac{\partial f}{\partial x}\cos\left(\varphi + \frac{\pi}{2}\right) + \frac{\partial f}{\partial y}\sin\left(\varphi + \frac{\pi}{2}\right) = a(-\cos\varphi\sin\varphi + \sin\varphi\cos\varphi) = 0$$

Similarly, the components in directions $\varphi + \pi$ and $\varphi + (3\pi/2)$ are $-a$ and 0, respectively. Thus the partial derivatives in various directions give responses to this ramp edge that vary from 0 to $\pm a$, depending on their orientations relative to the ramp direction. The effects of edge orientation on the response of a directional difference operator are illustrated in Figs. 8a–e.

As we recall, the direction θ_n at a given point in which the partial derivative has greatest magnitude is $\tan^{-1}((\partial f/\partial y)/(\partial f/\partial x))$, and this magnitude is $\sqrt{(\partial f/\partial x)^2 + (\partial f/\partial y)^2}$. The vector having this magnitude and direction is called the *gradient* of f. Thus in the previous ramp example the gradient direction is φ, and its magnitude is a. (Directional derivatives in any pair of perpendicular directions could be used, in place of $\partial/\partial x$ and $\partial/\partial y$, to compute the gradient.) If we want a simple operator that detects edges in all orientations with equal sensitivity, we can use the magnitude of the gradient (see Fig. 8f).

For digital pictures, we can use differences in place of derivatives in the previous definitions. Thus the magnitude of the "digital gradient" of f at (i,j) is $\sqrt{\Delta_x f(i,j)^2 + \Delta_y f(i,j)^2}$. It is common practice to approximate this

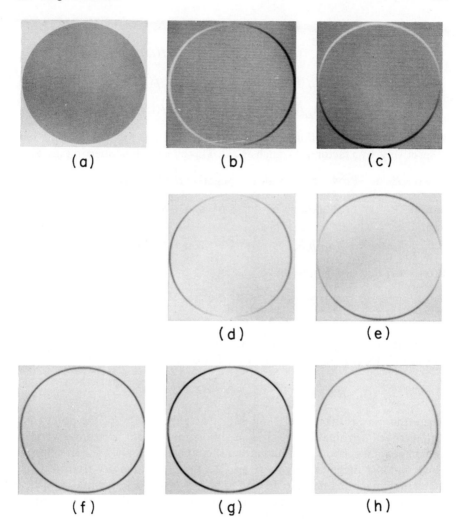

Fig. 8 Directional sensitivity of difference operators. (a) Input picture f (gray levels 32 and 0). (b) $\Delta_x f + 32$. (c) $\Delta_y f + 32$. (d) $|\Delta_x f|$. (e) $|\Delta_y f|$. (f) $\sqrt{(\Delta_x f)^2 + (\Delta_y f)^2}$. (g) $|\Delta_x f| + |\Delta_y f|$. (h) $\max(|\Delta_x f|, |\Delta_y f|)$.

expression, either by $|\Delta_x f(i,j)| + |\Delta_y f(i,j)|$ or by

$$\max(|\Delta_x f(i,j)|, |\Delta_y f(i,j)|)$$

Note, however, that these approximations are no longer equally sensitive to edges in all orientations. They agree with the exact expression for horizontal or vertical edges, e.g., if $\Delta_y f(i,j) = 0$, all three expressions have value

$\Delta_x f(i, j)$. For a $45°$ edge, however, where $\Delta_x f = \Delta_y f$, we have

$$\sqrt{\Delta_x f(i, j)^2 + \Delta_y f(i, j)^2} = \Delta_x f(i, j)\sqrt{2}$$

$$|\Delta_x f(i, j)| + |\Delta_y f(i, j)| = 2\Delta_x f(i, j)$$

$$\max(|\Delta_x f(i, j)|, |\Delta_y f(i, j)|) = \Delta_x f(i, j)$$

so that the approximations can give values that are too high or too low, respectively, by a factor of as much as $\sqrt{2}$. This effect is illustrated in Figs. 8f–h.

Exercise 5.   Prove that for all nonnegative a, b we have

$$\max(a, b) \leqslant \sqrt{a^2 + b^2} \leqslant a + b$$

and

$$(a+b)/\sqrt{2} \leqslant \sqrt{a^2 + b^2} \leqslant \max(a, b)\sqrt{2}$$

Prove also that if $a > b$, we have

$$\sqrt{a^2 + b^2} < a + \frac{b}{2} \leqslant \frac{\sqrt{5}}{2}\sqrt{a^2 + b^2} \quad \blacksquare$$

Various other approximations to the digital gradient (we shall omit "magnitude of" from now on) are often used. For example, one can use

$$\max_{u, v} |f(i, j) - f(u, v)|$$

where the max is taken over some set of neighbors (u, v) of the point (i, j), e.g., its four horizontal and vertical neighbors, or its eight horizontal, vertical, and diagonal neighbors. The sum rather than the max can be used, if preferred. Note that the diagonal neighbors are $\sqrt{2}$ times as far away from (i, j), so that a diagonal difference should tend to be larger than a horizontal or vertical difference, given that the slope components in the two directions are equal. If desired, we can compensate for this by weighting the diagonal difference terms by the factor $1/\sqrt{2}$.

Still another popular digital gradient approximation, due to Roberts [34], is

$$\max(|f(i, j) - f(i+1, j+1)|, |f(i+1, j) - f(i, j+1)|)$$

(We recall that differences in any pair of perpendicular directions can be used to compute the gradient.) Here again, the sum rather than the max can be used, if one wishes. Note that the differences used in this expression are symmetrical about the interpolated point $(i+\frac{1}{2}, j+\frac{1}{2})$; thus we should regard the "Roberts gradient" as being an approximation to the continuous gradient at that point, rather than at the point (i, j).

The various digital gradient approximations yield different numerical "edge values" for any given picture. When, however, we display these values in picture form, representing edge values by gray levels, the results all tend to look quite similar. Some examples of edge values obtained using digital gradients are shown in Fig. 9.

Higher-order derivative operators can also be used to detect edges in a picture. The Laplacian $(\partial^2 f/\partial x^2)+(\partial^2 f/\partial y^2)$ (Section 6.4.2) is an example of such an operator that is orientation insensitive. In the digital case, it is approximated by

$$\nabla^2 f(i,j) = f(i+1,j) + f(i-1,j) + f(i,j+1) + f(i,j-1) - 4f(i,j)$$

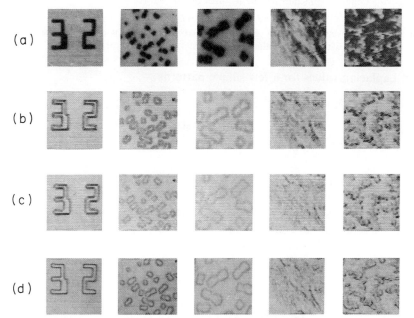

Fig. 9 Digital gradients. (a) Original pictures $(f(i,j))$. (b) Max$|f(i,j)-f(u,v)|$, where (u,v) ranges over the eight neighbors of (i,j). (c) Same using only the four horizontal and vertical neighbors. (d) Max$(|f(i,j)-f(i+1,j+1)|,|f(i+1,j)-f(i,j+1)|)$.

Note that this can take on either positive or negative values; to guarantee nonnegative values, its absolute value can be used. One can also use sums involving more than just four neighbors of (i,j), e.g.,

$$\sum_{u,v} (f(u,v)-f(i,j))$$

where the sum is taken over all eight horizontal, vertical, and diagonal neighbors of (i,j).

Digital Laplacian values, for the same set of pictures used in Fig. 9, are shown in Fig. 10. These results do not seem to be as good as the gradient results. To see one reason why this might be so, let us compute the gradient

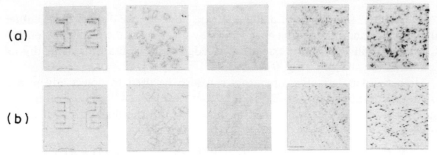

Fig. 10 Digital Laplacians for the pictures of Fig. 9a.

(a) $|\nabla^2 f| = |f(i+1, j) + f(i-1, j) + f(i, j+1) + f(i, j-1) - 4f(i, j)|$

where values above 63 have been set to 63. (b) Same as (a), but without the absolute value being taken, and with negative values set to zero.

and Laplacian values for a few simple patterns:

$$
\begin{array}{cccc}
 & & \vdots & \vdots \;\; \vdots \\
 & & 1 & 1 \;\; 1 \;\;\cdots \\
\text{(a)} \;\; \underline{1}, & \text{(b)} \;\; \underline{1}, & \text{(c)} \;\; \underline{1}, & \text{(d)} \;\; \underline{1} \;\; 1 \;\;\cdots \\
 & 1 & 1 & 1 \;\; 1 \;\;\cdots \\
 & \vdots & \vdots & \vdots \;\; \vdots
\end{array}
$$

Here the unmarked points all have gray level 0; one point is underlined to serve as a reference for comparing the input and output patterns. The gradient values $\max(|\Delta_x f(i, j)|, |\Delta_y f(i, j)|)$ are

$$
\begin{array}{cccc}
 & & \vdots & \vdots \\
1 & 1 & 1 \;\; 1 & 1 \\
\text{(a)} \;\; \underline{1} \;\; 1, & \text{(b)} \;\; \underline{1} \;\; 1, & \text{(c)} \;\; \underline{1} \;\; 1, & \text{(d)} \;\; \underline{1} \\
 & 1 \;\; 1 & 1 \;\; 1 & 1 \\
 & \vdots & \vdots & \vdots
\end{array}
$$

The Laplacian values $|\nabla^2 f(i, j)|$ are

$$
\begin{array}{cccc}
 & & \vdots & \vdots \\
1 & 1 & 1 \;\; 2 \;\; 1 & 1 \;\; 1 \\
\text{(a)} \;\; 1 \;\; \underline{4} \;\; 1, & \text{(b)} \;\; 1 \;\; \underline{3} \;\; 1, & \text{(c)} \;\; 1 \;\; \underline{2} \;\; 1, & \text{(d)} \;\; 1 \;\; \underline{1} \\
1 & 1 \;\; 2 \;\; 1 & 1 \;\; 2 \;\; 1 & 1 \;\; 1 \\
 & \vdots & \vdots & \vdots
\end{array}
$$

For both the gradient and the Laplacian, the output for the isolated point (a) is a slightly thickened or smeared point; for the line end (b) and the line (c), we obtain a thickened line (end); for the step edge (d), we also obtain a line. Note also that the gradient gives the same maximum output value 1 for all four patterns. The Laplacian, on the other hand, detects the point four times as strongly as it does the edge; the line end, three times as strongly; and the line, twice as strongly. Thus the Laplacians favor lines, line ends, and points over edges. (Methods of detecting points and lines in a picture will be discussed in Section 8.3.) In a noisy picture, the high Laplacian values at points and lines may overshadow the lower values at edges.

As we saw in Section 6.4.3, when we apply the Laplacian to a picture, low spatial frequencies are weakened, while high ones remain relatively intact. Thus high-pass spatial frequency filtering (suppressing low frequencies, and retaining high ones) should have effects similar to those of applying the Laplacian. These effects are illustrated, for the same set of pictures, in Fig. 11. Again, the results do not seem to be as useful as those obtained with gradient operators.

If an edge is blurred rather than abrupt, the gradient should have its highest value at the steepest point, which would normally be a good location to assign to the edge; see the cross-sectional sketch in Fig. 12a. If, however, the edge involves a long ramp, the gradient will be relatively constant on the ramp, and the Laplacian will have low or zero values, as shown schematically in Fig. 12b. Thus the center of the ramp is not given any uniquely distinguishable value by these derivative operators. We can, however, detect the "shoulders" of the ramp (where the gradient rises to its plateau value, and where the Laplacian has its peaks); and we can locate the edge as lying midway between these shoulders.

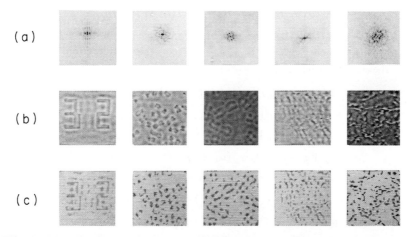

(a)

(b)

(c)

Fig. 11 Edge detection by high-pass spatial frequency filtering, applied to the pictures of Fig. 9a. (a) Power spectra (64×64). (b) Reconstructed pictures with spatial frequency range 1–10 (in all directions from the origin) suppressed. (c) Results of setting all but the darkest 15% of the points in (b) to zero.

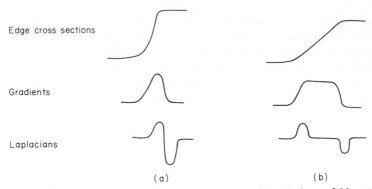

Edge cross sections

Gradients

Laplacians

(a)   (b)

Fig. 12   Idealized cross sections showing gradients and Laplacians of blurred edges.
(a) Smoothly blurred edge. (b) Ramp edge.

8.2.2   Best-Fit Edge Detection

This section describes two approaches to edge detection which involve
"best fits" of functions to the given picture. Let $f(i, j)$, $f(i+1, j)$, $f(i, j+1)$,
and $f(i+1, j+1)$ be the gray levels at four neighboring points of a picture.
Suppose that we want to fit a plane $z = ax + by + c$ to these four values.
This cannot be done exactly, but we can find a plane for which some measure
of the error in fit, e.g.,

$$(ai + bj + c - f(i, j))^2 + (a(i+1) + bj + c - f(i+1, j))^2$$
$$+ (ai + b(j+1) + c - f(i, j+1))^2$$
$$+ (a(i+1) + b(j+1) + c - f(i+1, j+1))^2$$

is a minimum. To find this best-fitting plane, we differentiate the error expression with respect to each of a, b, and c, set the results equal to zero, and
solve these equations for a, b, and c:

$$i(ai + bj + c - f(i, j)) + (i+1)(a(i+1) + bj + c - f(i+1, j))$$
$$+ i(ai + b(j+1) + c - f(i, j+1))$$
$$+ (i+1)(a(i+1) + b(j+1) + c - f(i+1, j+1)) = 0$$
$$j(ai + bj + c - f(i, j)) + j(a(i+1) + bj + c - f(i+1, j))$$
$$+ (j+1)(ai + b(j+1) + c - f(i, j+1))$$
$$+ (j+1)(a(i+1) + b(j+1) + c - f(i+1, j+1)) = 0$$
$$(ai + bj + c - f(i, j)) + (a(i+1) + bj + c - f(i+1, j))$$
$$+ (ai + b(j+1) + c - f(i, j+1))$$
$$+ (a(i+1) + b(j+1) + c - f(i+1, j+1)) = 0$$

It is easily verified that the solution is

$$a = \frac{f(i+1,j)+f(i+1,j+1)}{2} - \frac{f(i,j)+f(i,j+1)}{2}$$

$$b = \frac{f(i,j+1)+f(i+1,j+1)}{2} - \frac{f(i,j)+f(i+1,j)}{2}$$

$$c = \tfrac{1}{4}(3f(i,j)+f(i+1,j)+f(i,j+1)-f(i+1,j+1)-ia-jb)$$

The gradient of the plane $z = ax+by+c$ has magnitude $\sqrt{a^2+b^2}$, which we can approximate by $|a|+|b|$ or by $\max(|a|,|b|)$, if desired. In any case, the "components" $|a|$ and $|b|$ of this gradient are simply absolute differences of average gray levels computed within the original 2×2 picture neighborhood. Graphically, the neighborhood is

$$(i,j+1) \quad (i+1,j+1)$$
$$(i,j) \qquad (i+1,j)$$

and we see that a is the difference between the averages of the two columns, while b is the difference between the averages of the two rows. In other words, a and b are, respectively, horizontal and vertical differences, where the differencing has been preceded by smoothing (i.e., averaging).

Since the plane is a best approximation to the picture gray levels in the given 2×2 neighborhood, it is reasonable to take the gradient of the plane as a useful approximation to the gradient of the picture at the center $(i+\tfrac{1}{2},j+\tfrac{1}{2})$ of that neighborhood. As in Section 8.2.1, the approximation that we obtain in this way is indeed a plausible "digital gradient," since it is composed of horizontal and vertical differences. It is more complicated than the "Roberts gradient" defined in the last section, but it should be less sensitive to noise, since it involves averaging prior to differencing. Its output, for the same set of pictures used in the last section, is shown in Fig. 13.

The method of defining a digital gradient just described can be generalized as follows: Given an $m\times m$ neighborhood N on the picture, fit a polynomial

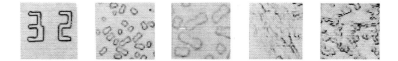

Fig. 13   Digital gradient defined by fitting a plane to the four gray levels in a 2×2 neighborhood (see text). It is easily shown that this gradient responds only half as strongly to a diagonal step edge as do those used in Fig. 9. The results (shown for the pictures in Fig. 9a) have been multiplied by 2 to make the information in them more easily visible; they would otherwise have been much poorer than the results in Fig. 9.

surface to the gray levels of the points in N. (The degree of this surface should be such that the number of coefficients is less than m^2.) Take the gradient of this surface at the center (x, y) of N. (In the previous example the surface was a plane, so that its gradient was constant.) This gradient, which can be expressed in terms of the gray levels of the points in N, can be regarded as a "best approximation" to the gradient of the given picture at (x, y), since it was computed using a locally best-fitting surface.

Exercise 6.   Show that if we least squares fit a second-degree surface to a 3×3 picture neighborhood centered at (i, j), then the gradient of this surface at (i, j) has x- and y-components proportional to

$$|[f(i-1, j+1) + f(i-1, j) + f(i-1, j-1)]$$
$$- [f(i+1, j+1) + f(i+1, j) + f(i+1, j-1)]|$$

and

$$|[f(i-1, j+1) + f(i, j+1) + f(i+1, j+1)]$$
$$- [f(i-1, j-1) + f(i, j-1) + f(i+1, j-1)]| \quad \blacksquare$$

Exercise 7.   What is the Laplacian of the surface in Exercise 6 at (i, j)? What happens if we fit our second-degree surface to a 4×4 neighborhood? $\blacksquare$

In these examples we defined useful picture derivative operators by locally fitting smooth surfaces to the picture; these operators can then be used for edge detection. We conclude by mentioning a rather different approach, in which edges are detected by locally fitting ideal step edges to the picture.

An ideal step edge is described by a function of the form

$$e(x, y) = \begin{cases} a, & \alpha x + \beta y \leqslant \gamma \\ a + h, & \alpha x + \beta y > \gamma \end{cases}$$

Here the lower and upper levels of the step have values a and $a+h$, respectively, so that the step has height h; and the step occurs along the line $\alpha x + \beta y = \gamma$. Given a picture $f(x, y)$ and a neighborhood N in it, we would like to find the step edge that has least mean squared difference from the actual gray levels in N. In other words, we want to choose the parameters a, h, α, β, and γ so as to minimize

$$\int \int_N [e(x, y) - f(x, y)]^2 \, dx \, dy$$

A solution to this minimization problem was derived by Hueckel [18] for the case of a circular neighborhood D of unit radius, using a truncated series expansion in terms of a set of basis functions defined on D. This set of functions was determined as the unique solution to a set of functional equations expressing the requirements that the functions have decreasing weight as one moves away from the center of N, and that they be insensitive to high-frequency noise. The reader is referred to Hueckel's papers [18, 19] for the details, which are somewhat complicated and will not be given here.

8.2.3 Minimum-Error Edge Detection

Suppose that we are dealing with a class of pictures that consist of objects on a background, where the probability density of gray levels in the objects is $p(z)$, and that in the background is $q(z)$, as in Section 8.1.3. Let f be such a picture, and let f' be the result of applying some edge detection operation, such as a gradient or Laplacian, at every point of f. In terms of $p(z)$ and $q(z)$, we can compute the probability density of the gray levels in f' at points interior to the objects or to the background, and at points just on the border between objects and background. If we want to detect edges in f by thresholding f', these probability densities give us the probabilities of incorrectly detecting an edge point inside the objects or the background, as well as the probability of failing to detect an edge point on the border between them. If the costs of such false alarms and false dismissals are known, we can thus determine the expected cost when a particular threshold t is used; and we can attempt to choose t so as to minimize this cost.

To illustrate these ideas, suppose that f is a digital picture, and that we are trying to detect vertical edges in f by computing the first difference $f'(i, j) = \Delta_x f(i, j) = f(i, j) - f(i-1, j)$. Now f' has value ζ whenever the values of f at the two points differ by ζ. If we assume that the values of f at adjacent points are independent, and that both points are object points, then the probability that f' has value ζ is

$$p_o'(\zeta) = \int_{-\infty}^{\infty} p(z)\, p(z+\zeta)\, dz$$

(This is obtained by summing the probabilities that the two points have all possible pairs of values that differ by ζ.) In other words, the probability density $p_o'(\zeta)$ of f' interior to the objects is given by the autocorrelation of the objects' gray level probability density $p(z)$. Similarly, the probability density of f' interior to the background is just the autocorrelation of $q(z)$, i.e.,

$$p_b'(\zeta) = \int_{-\infty}^{\infty} q(z)\, q(z+\zeta)\, dz$$

Finally, the probability density of f' when one of the points is in an object and the other is in the background is just the cross correlation of $p(z)$ and $q(z)$, i.e.,

$$p_e'(\zeta) = \int_{-\infty}^{\infty} p(z)\, q(z \pm \zeta)\, dz$$

where we use the plus sign if the first point is in the background and the second point in an object, and the minus sign if the reverse is true.

Exercise 8. What are the probability densities of the absolute difference $|\Delta_x f|$ inside the objects, inside the background, and on the border between them? ∎

Exercise 9. What are the probability densities for the digital gradient $|\Delta_x f| + |\Delta_y f|$? (Here you must consider a few additional cases, since when (i, j) is in an object, one or both of $(i-1, j)$ and $(i, j-1)$ can be in the background, and vice versa.) ∎

Exercise 10. What are the probability densities for the digital Laplacian

$$f(i+1, j) + f(i-1, j) + f(i, j+1) + f(i, j-1) - 4f(i, j)?$$ ∎

If we want to choose an edge detection threshold t to minimize the probability of making an error, we can now proceed exactly as in Section 8.1.3. The probabilities of misclassifying an object or background point as an edge point are

$$1 - P_o'(t) \equiv \int_t^\infty p_o'(\zeta)\, d\zeta \quad \text{and} \quad 1 - P_b'(t) \equiv \int_t^\infty p_b'(\zeta)\, d\zeta$$

respectively. Similarly, the probability of misclassifying a border point as a nonedge point is

$$P_e'(t) \equiv \int_{-\infty}^t p_e'(\zeta)\, d\zeta$$

If fraction θ of the picture points are object points, and fraction ε are border points, the overall error probability when we threshold at t is thus

$$\theta(1 - P_o'(t)) + \varepsilon P_e'(t) + (1 - \theta - \varepsilon)(1 - P_b'(t))$$

We can find the t that minimizes this probability by taking its derivative with respect to t and equating the result to zero, which yields

$$\varepsilon p_e'(t) = \theta p_o'(t) + (1 - \theta - \varepsilon) p_b'(t)$$

This equation can be solved for t using numerical methods.

More generally, as in Section 8.1.3, we can determine the set Z of values of f' for which it is more probable than not that a point is an edge point. A value ζ belongs to this set provided that

$$\varepsilon p_e'(\zeta) > \theta p_o'(\zeta) \quad \text{and} \quad \varepsilon p_e'(\zeta) > (1 - \theta - \varepsilon) p_b'(\zeta)$$

This result can be generalized to the case where there are k possible classes of picture points, rather than just the "object" and "background" classes; and to the case where we make the classification on the basis of a set of property values measured at the given point, rather than just the single value ζ. In practice, of course, this approach depends on our knowing, or being able to estimate, the probabilities ε, θ, $p_e'(\zeta)$, $p_o'(\zeta)$, and $p_b'(\zeta)$.

The maximum-probability approach can also be used to detect edges based on measurements made at a set of points rather than a single point. For example, suppose that we are trying to decide whether m points that lie along a line, say $(i, j), (i+1, j), ..., (i+m-1, j)$, are points of an edge, based on measuring some property value $f'(i+h, j)$, $0 \leqslant h < m$, at each of the points. If we know the probability densities $p_e'(\zeta)$ and $p_n'(\zeta)$ of the values of f' at edge and at nonedge points, and if the values of f' at distinct points are independent, then the probability of obtaining a given m-tuple of values $(\zeta_1, ..., \zeta_m)$ at the m points, when an edge is present, is just the product

$$\prod_{i=1}^{m} p_e'(\zeta_i)$$

The probabilities of other alternatives (edge totally absent, edge partially present, etc.), can be computed similarly, and we can thus determine conditions under which an edge is more likely than not. A special case of this problem, involving normal probability densities, has been treated by Griffith, to whose paper [13] the reader is referred for the details.

As another example, suppose we suspect that (i, j) is in an object and $(i+m-1, j)$ is in the background, and we want to determine the most probable location of the edge between object and background along the line $(i, j), ..., (i+m-1, j)$. Suppose that we know the conditional gray level probability densities $p(z)$ and $q(z)$ for the objects and the background, and that the gray levels of distinct points are independent. Given the m-tuple of gray levels $(z_{i_1}, ..., z_{i_m})$ at the points $(i, j), ..., (i+m-1, j)$, the probability that points $(i, j), ..., (i+h-1, j)$ are in the object and points $(i+h, j), ..., (i+m-1, j)$ are in the background is

$$p(o \mid z_{i_1}) \cdots p(o \mid z_{i_h}) p(b \mid z_{i_{h+1}}) \cdots p(b \mid z_{i_m})$$

As in Section 8.1.3, this expression is equal to

$$\theta^h (1-\theta)^{m-h} p(z_{i_1}) \cdots p(z_{i_h}) q(z_{i_{h+1}}) \cdots q(z_{i_m})/\pi(z_{i_1}) \cdots \pi(z_{i_m})$$

where the denominator is the same for all h. We can evaluate the numerator for each h and determine which h yields the maximum value.

Exercise 11. Suppose we know the *a priori* probability $p_o(h)$ that the object extends as far as point $(i+h, j)$, for each $0 \leqslant h < m$, and we also know the conditional densities $p(z)$ and $q(z)$. Assuming independence, how do we determine the most probable extent of the object when the gray levels $(z_{i_1}, ..., z_{i_m})$ are given? ▊

8.2.4 Texture Edge Detection

Up to now we have been concerned with detecting edges at which there is an abrupt change (e.g., a "step") in gray level. In this section we consider edges at which there is an abrupt change in *texture*. We saw a few examples of such edges in Figs. 4 and 5 (Section 8.1.1), involving objects composed of salt-and-pepper noise. In cases like these, conventional edge detection operators such as gradients or Laplacians would detect the individual noise dots, but the values of these operators would be no higher on the borders of the objects than in their interiors.

The most common type of "texture edge" is one at which there is an abrupt change in *average gray level*. Such edges result when noise is added to a perfect step edge, as shown in Figs. 7a and 7b. An extreme case is the "pure noise" edge of Fig. 4, where the gray levels on both sides of the edge are the same, but they have different probabilities. Note that in the case of Fig. 7b, we can detect the presence of the edge by looking for a piece of the picture that has a bimodal histogram; but this method would not work in the case of Fig. 4.

A straightforward way to detect abrupt changes in average gray level in a picture f is to average f, and then compute differences of averages taken over adjacent, nonoverlapping neighborhoods. For example, let $f^{(r)}(u, v)$ denote the average gray level of f in a circular neighborhood of radius r centered at (u, v). Then the absolute difference of such averages for a pair of horizontally adjacent, nonoverlapping circular neighborhoods that touch at the point (x, y) is $|f^{(r)}(x-r, y) - f^{(r)}(x+r, y)|$. More generally, if the centers of the two touching neighborhoods lie on a line of slope θ, and the neighborhoods touch at (x, y), the absolute difference is

$$e^{(r, \theta)}(x, y) = |f^{(r)}(x - r \cos \theta, y - r \sin \theta) - f^{(r)}(x + r \cos \theta, y + r \sin \theta)|$$

This difference will be high if the averages over the two neighborhoods are very different; in particular, it will be high if the neighborhoods lie just on opposite sides of an edge of the desired type. The question of how to choose r, the size of the averaging neighborhood, will be discussed later.

If r is large, $e^{(r, \theta)}$ will be high not only just at an edge, but over a range of positions on either side of it, as illustrated in Fig. 14a. This is because the averages over the two touching neighborhoods begin to differ as soon as one of the neighborhoods overlaps both sides of the edge, while the other neighborhood lies entirely on one side. However, $e^{(r, \theta)}$ should be a local maximum when the neighborhoods touch just at the edge, and each neighborhood lies entirely on one side, since the averages are then as different as possible (Fig. 14b). Thus if we want to determine the location of an edge using $e^{(r, \theta)}$, we should ignore high values if there are higher ones nearby in the θ-direction;

Inputs   Outputs

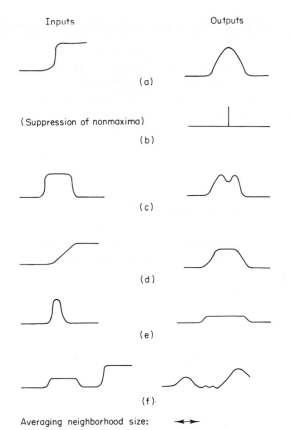

(a)

(Suppression of nonmaxima)

(b)

(c)

(d)

(e)

(f)

Averaging neighborhood size: ◄──►

Fig. 14   Idealized cross sections illustrating input edges and the corresponding outputs for an edge detector based on differences of averages; the size of the averaging neighborhood is indicated by the arrow below the figure. (a) Step edge. (b) Nonmaximum suppression. (c) Interference by opposite edges; one of the peaks will be suppressed as a nonmaximum unless they are exactly equal. (d) Ramp edge: the peak becomes a plateau. (e) When the object is too small, the edge output is weakened. (f) A high-contrast object nearby can interfere with detecting the edge of a low-contrast object.

in other words, we should keep only the local maxima in the direction across the edge, as pointed out at the end of Section 8.1. Here "nearby" could be defined as "within distance r"; if we made it much larger than this, e.g., as large as $2r$, then the opposite edges of an object would compete with one another, as shown in Fig. 14c. If the edge being detected is ramplike rather than steplike, it may give rise to a plateau of highest values of $e^{(r,\theta)}$, rather than a peak (Fig. 14d). In such a case, one can estimate the location of the edge as midway between the "shoulders" in the values of $e^{(r,\theta)}$, as suggested at the end of Section 8.2.1.

The difference of averages $e^{(r, \theta)}$ was designed to be sensitive to edges that run perpendicular to θ; it will be less sensitive to edges in other directions, and it will not be sensitive at all to edges that run parallel to θ, since in this case equal fractions of both neighborhoods will lie on each side of the edge, so that the averages will be equal. To detect edges in all directions, one could use difference operators $e^{(r, \theta)}$ and $e^{(r, \theta + \pi/2)}$ in a pair of perpendicular directions, and combine the results to form a "gradient"

$$e^{(r)} = \sqrt{(e^{(r, \theta)})^2 + (e^{(r, \theta + \pi/2)})^2}$$

(or the sum or max of the absolute values, if preferred).

We have assumed here that the averages $f^{(r)}$ are unweighted. One could also design difference operators based on weighted averages. For example, if we are trying to detect ramplike edges, we might use weights that are high at the centers of the two touching neighborhoods, and taper off away from the centers, so that in particular the weights near where the neighborhoods touch are low. On the other hand, if we are interested in abrupt steps, weights that are highest where the neighborhoods touch might be more appropriate.

We must now discuss how to choose the radius r in designing the difference operators $e^{(r, \theta)}$. If r is too small, the averaging process may not smooth out the noise to a sufficient degree, and there will be too many false-alarm edge detections. On the other hand, if r is too large, we lose our ability to detect the edges of objects that are small relative to r. If we know in advance what the smallest object size will be, we should evidently choose the largest possible r such that objects of this size can just be detected. If we do not know this in advance, a possible approach is to compute $e^{(r)}$ values for a range of r's, and use some combination of these values as our final edge value at each point, or pick a "best size" at each point of the picture by comparing these values. For example, if we multiply several $e^{(r)}$'s together, the result will be large only at points where the change in average gray level is not only abrupt, but also extends over a significant region on each side of the edge; this method has been used to both detect and sharply localize steps in average gray level. Another possibility is to look for the smallest r such that $e^{(r)} > e^{(s)}$ for all $s > r$. This can be justified as follows: If r is just small enough to fit the object whose edge is being detected, then for $s > r$, at least one of the touching neighborhoods will not fit (i.e., it will overlap both the object and its background; see Fig. 14e), so that we will have $e^{(s)} < e^{(r)}$. (This argument assumes that the background at the point where the edge is being detected differs from the object in the same way as the background which the too large neighborhood overlaps, i.e., both are lighter, or both darker, than the object. If they differ from the object in opposite ways, the argument breaks down, as shown in Fig. 14f.)

For a digital picture, computing averages over circular neighborhoods is expensive. It turns out, fortunately, that one can obtain useful results with square neighborhoods, e.g., with

$$f^{(r)}(u,v) = \left(\sum_{i=u-r}^{u+r} \sum_{j=v-r}^{v+r} f(i,j) \right) \Big/ (2r+1)^2$$

In this case it is convenient to use difference operators based on horizontally and vertically adjacent pairs of neighborhoods, i.e.,

$$e^{(r,\,\mathrm{H})}(x,y) = |f^{(r)}(x+r,y) + f^{(r)}(x-r-1,y)|$$

$$e^{(r,\,\mathrm{V})}(x,y) = |f^{(r)}(x,y+r) - f^{(r)}(x,y-r-1)|$$

These can then be combined to yield a "gradient" difference at (x, y). The results obtained by this method for the "pure noise" edge of Fig. 4 are shown in Fig. 15.

Recall that the digital difference operators described in Section 8.2.2,

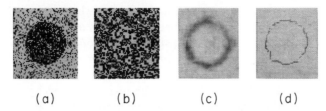

(a)   (b)   (c)   (d)

Fig. 15 "Gradient" based on differences of averages. (a) Input picture (same as Fig. 4a). (b) Roberts gradient of (a); this is of little use in detecting the edge of the densely dotted region. (c) Gradient of (a) based on averages over 8×8 square neighborhoods (see text). (d) Results of suppressing nonmaxima in (c); we keep only horizontal maxima of the vertical differences, and vertical maxima of the horizontal differences.

which were designed by fitting surfaces to the digital gray levels of a point and its neighbors, also turned out to involve differences of averages, taken over neighborhoods that are thin rectangles rather than squares. The rectangles are elongated in the direction parallel to the edge, but are thin in the across direction; they thus permit some smoothing of noise, while not degrading the sharpness with which the edge is located. (On smoothing a picture without blurring edges see also Section 6.5.2.)

We have assumed up to now that the differences of averages are computed at every point of the given picture. If all we want to do is detect the presence of edges, but not locate them accurately, this is unnecessary. In fact, it would be sufficient to compute the differences at points spaced (say) $r/2$ apart. This would imply that for any edge, there exists a pair of the touching neighborhoods one of which lies on one side of the edge, while the other overlaps the other side of the edge by at least 50%. We would thus be assured of obtaining an $e^{(r)}$ value equal to at least 50% of the difference between the average gray levels on the two sides of the edge. Given such a value at the point (x, y), we know that there must exist an edge of at least that strength somewhere within distance $r/2$ of (x, y). Once an edge has been detected in this way, using an $e^{(r)}$ operator with a large radius r, we can proceed to locate it more accurately using operators with smaller radii; or one could use sequential detection and tracking methods to locate it (see Section 8.4). The use of coarse, low-resolution operators to guide the application of finer operators is sometimes called "planning"; it is a basic idea that has many other applications in picture analysis.

The maximum probability approach to edge detection discussed in Section 8.2.3 can also be applied to edge detectors that use differences of averages. Given the probability densities of the gray levels in objects and background, one can compute densities for average gray levels (see Section 6.5.2), and hence for differences of these averages. This can be done for points interior to the objects or background, as well as for points where the touching neighborhoods overlap both objects and background, as a function of the degree of overlap (compare the case in the last paragraph of Section 8.2.3). One can thus determine which values of the difference of averages are more likely than not to be due to edges.

If the gray level probability densities for objects and background are not known, but we know (say) their means and standard deviations (call them μ_o and μ_b, σ_o and σ_b, respectively) we can define edge detection criteria in terms of this information; for example, we can consider an edge to have been detected if the average gray level in one of the touching regions differs from μ_o by less than some multiple $\lambda\sigma_o$ of σ_o, while the average in the other region differs from μ_b by less than $\lambda\sigma_b$. Even if we know nothing about the object and background densities, we can define detection criteria in terms of statistics of the gray levels in the two touching regions themselves. For example, let μ_1, μ_2, σ_1, σ_2 be the means and standard deviations of the gray levels in the two regions; then we can consider an edge to be present if the "Fisher distance" $|\mu_1 - \mu_2|/(\sigma_1 + \sigma_2)$ is greater than some threshold. (A criterion of this sort should be used only if σ_1 and σ_2 are never both zero; if $\sigma_1 = \sigma_2 = 0$, the Fisher distance becomes infinite, no matter how small the difference between μ_1 and μ_2.)

We conclude this section by considering more general types of texture edges, such as those shown in Fig. 5. As pointed out in Section 8.1.1, these cases involve objects which differ from their backgrounds with respect to the average value of some local property, rather than with respect to average gray level. Thus at the edges of these objects, there are abrupt changes in *average local property value*. A straightforward extension of our method of detecting abrupt changes in average gray level can be used to handle this larger class of edges, just as in Section 8.1.1. If we compute the local property at every point of the given picture f, we obtain a new picture f' in which the objects now differ from the background in average gray level. Our method of edge detection by taking differences of averages can now be applied to f'. This approach is illustrated in Fig. 16.

Differences in averages of local property values may also be useful in de-

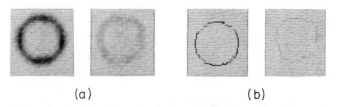

(a)   (b)

Fig. 16 Texture "gradients". (a) Results of applying the gradient used in Fig. 15c to the pictures in Fig. 5b. (b) Results of suppressing nonmaxima in (a).

tecting ordinary (noisy) step or ramp edges. For example, a second derivative (or difference) is nonzero on the "shoulders" of a ramp; in fact, it is positive on the top shoulder and negative on the bottom shoulder, as seen from the cross section in Fig. 14d. To detect the presence of the pair of shoulders, we can examine the average values of the second derivative (or of its sign, if desired) in a pair of adjacent, nonoverlapping neighborhoods. If the difference of these averages has a large absolute value, an edge is probably present.

Still more generally, suppose that the *probability density of the values of some local property* is different in the objects than in the background (but the average values of this local property need not be different). In Fig. 5, for example, the objects and background have different gray level probability densities (i.e., different histograms), even though they have the same average gray level. We can detect such edges by comparing histograms (or probability densities) taken over pairs of adjacent, nonoverlapping neighborhoods. For example, let $p^{(r)}(u, v)$ denote the histogram for a circular neighborhood of radius r centered at (u, v). Then one simple measure of the difference between histograms in a pair of horizontally adjacent circular neighborhoods that touch at (x, y) is

$$\int_{z_1}^{z_n} |p^{(r)}(x-r, y) - p^{(r)}(x+r, y)|,$$

where $[z_1, z_n]$ is the gray level range; and similarly for pairs of neighborhoods having other orientations. Here again, in order to locate edges exactly, we would have to suppress nonmaxima, just as we did for the differences of averages. The neighborhood size selection problem is especially critical when we compare histograms rather than averages, since a histogram obtained from a small neighborhood may be a very unreliable estimate of the actual probability density in that region of the given picture.

8.3 MATCHING

Edge detection can be regarded as finding points of the given picture at which the local pattern of gray levels is steplike. In this section we treat the general problem of finding points at which an arbitrary given pattern of gray levels is present, i.e., finding points where some given pattern *matches* the picture.

Matching of patterns against pictures is of importance in many types of applications; the following are a few examples:

(a) The pattern to be matched may be a simple pattern such as a step or ramp, i.e., an edge. Other examples of such simple patterns are lines, spots, etc; methods for detecting such patterns will be discussed in Sections 8.3.3 and 8.3.4.

(b) The pattern may be a "template" representing a known object. For example, we can match templates of characters against pictures of printed pages; templates of targets against pictures obtained by reconnaissance systems; or templates of landmarks against pictures obtained by navigation systems, e.g., templates of star patterns against pictures of the sky.

(c) The pattern may itself be a piece of picture, i.e., we may want to match a piece of one picture against another picture. For example, if we have two pictures of the same scene taken from different viewpoints, and we can identify the parts of the two pictures that show the same piece of the scene, we can then measure steroscopic parallax and determine the heights or depths of objects in the scene. Similarly, if we have two pictures taken at different times, we can use this information to measure the relative motions of objects in the scene (cars on a road, clouds in the sky, etc.). If we have two differently distorted pictures of the scene, we can use such piecewise matching to determine their relative distortion, so that it can be corrected if desired (Section 6.3).

Matching would be a trivial task if we could expect an exact copy of the pattern (or template, or piece of picture) to be present in the picture being analyzed. In practice, however, the picture will be noisy, and we can at best hope to maximize some measure of the degree of match between pattern and picture. The use of cross-correlational measures of degree of match will be discussed in Section 8.3.1. We can, in particular, try to find a linear shift-invariant filter whose output maximizes some match criterion, usually expressed as a signal-to-noise ratio; this problem is treated in Section 8.3.2.

Of course, nonlinear matching operations may perform better than linear ones; this is illustrated in Sections 8.3.3 and 8.3.4, where we consider the detection of simple patterns such as lines and spots in a picture. Finally, in Section 8.3.5 we discuss the use of search techniques in matching.

8.3.1 Matching by Cross Correlation

There are many possible ways of measuring the degree of match or mismatch between two functions f and g over a region $\mathscr{S}$. For example, one can use as mismatch measures such expressions as

$$\max_{\mathscr{S}} |f-g| \quad \text{or} \quad \iint_{\mathscr{S}} |f-g| \quad \text{or} \quad \iint_{\mathscr{S}} (f-g)^2$$

and so on (where the integrals become sums in the digital case). It is easily verified that these expressions are all "distance measures" or *metrics* (see Section 9.2.2).

If we use $\iint (f-g)^2$ as a measure of mismatch, we can derive a useful measure of match from it. Note, in fact, that

$$\iint (f-g)^2 = \iint f^2 + \iint g^2 - 2 \iint fg$$

Thus if $\iint f^2$ and $\iint g^2$ are fixed, the mismatch measure $\iint (f-g)^2$ is large if and only if $\iint fg$ is small. In other words, for given $\iint f^2$ and $\iint g^2$, we can use $\iint fg$ as a measure of match.

The same conclusion can be reached by making use of the well-known Cauchy–Schwarz inequality, which states that for f and g nonnegative we always have

$$\iint fg \leqslant \sqrt{\iint f^2 \iint g^2}$$

with equality holding if and only if $g = cf$ for some constant c.[†] (The analogous result in the digital case is

$$\sum_{i,j} f(i,j) g(i,j) \leqslant \sqrt{\sum_{i,j} f(i,j)^2 \cdot \sum_{i,j} g(i,j)^2}$$

[†] To prove the Cauchy–Schwarz inequality, consider the polynomial

$$P(z) = \left(\iint f^2 \right) z^2 + 2 \left(\iint fg \right) z + \iint g^2$$

Now $P(z) = \iint (fz+g)^2 \geqslant 0$ for all z. Since P is a quadratic polynomial in z, this implies that it either has no real roots or has just one ("double") root; in other words, the discriminant of P must be negative or zero, i.e., $(\iint fg)^2 - (\iint f^2)(\iint g^2) \leqslant 0$, which immediately gives the desired inequality. Moreover, if equality holds, P does have a root, say, z_0; but then $P(z_0) = \iint (fz_0+g)^2 = 0$ evidently implies $fz_0+g = 0$ for all x, y, so that $g = (-z_0)f$.

with equality holding if and only if $g(i, j) = cf(i, j)$ for all i, j.) Thus when $\iint f^2$ and $\iint g^2$ are given, the size of $\iint fg$ is a measure of the degree of match between f and g (up to a constant factor).

Suppose now that f is a template, g a picture, and we want to find pieces of g that match f. (We are tacitly assuming that f is small compared to g, i.e., f is zero outside a small region $\mathscr{S}$, and we are interested only in matching the nonzero part of f against g.) We can do this by shifting f into all possible positions relative to g, and computing $\iint fg$ for each such shift (u, v). By the Cauchy–Schwarz inequality, we have

$$\int\int_{\mathscr{S}} f(x, y)\, g(x+u, y+v)\, dx\, dy$$

$$\leqslant \sqrt{\int\int_{\mathscr{S}} f^2(x, y)\, dx\, dy \int\int_{\mathscr{S}} g^2(x+u, y+v)\, dx\, dy}$$

Since f is zero outside $\mathscr{S}$, the left-hand side is equal to

$$\int\int_{-\infty}^{\infty} f(x, y)\, g(x+u, y+v)\, dx\, dy$$

which is just the *cross correlation* C_{fg} of f and g (Section 2.1.4). Note that on the right-hand side, while $\iint f^2$ is a constant, $\iint g^2$ is not, since it depends on u and v. Thus we cannot simply use C_{fg} as a measure of match; but we can use instead the *normalized cross correlation*

$$C_{fg} \Big/ \sqrt{\int\int_{\mathscr{S}} g^2(x+u, y+v)\, dx\, dy}$$

This quotient takes on its maximum possible value (namely, $\sqrt{\iint_{\mathscr{S}} f^2}$) for displacements (u, v) at which $g = cf$. Some simple examples of the use of normalized cross correlation to find places where a template matches a picture are shown in Fig. 17; a real example is shown in Fig. 18.

							Template		
			1	1	1		0	0	0
1	1	1	1	1	1		1	1	1
			1	1	1		0	0	0

1		1	1	1
1		1		1
1		1	1	1

(a)

```
                               1  2  3  2  1
           1  2  3  2  1       1  2  3  2  1
                               1  2  3  2  1


              1  1  1          1  2  3  2  1
              1  1  1          1  1  2  1  1
              1  1  1          1  2  3  2  1
                         (b)


                               1  2  3  2  1
           1  2  3  2  1       2  4  6  4  2
           1  2  3  2  1       3  6  9  6  3
           1  2  3  2  1       2  4  6  4  2
                               1  2  3  2  1

              1  1  1          1  2  3  2  1
              2  2  2          2  3  5  3  2
              3  3  3          3  5  8  5  3
              2  2  2          2  3  5  3  2
              1  1  1          1  2  3  2  1
                         (c)
```

$$1 \ \sqrt{3}/2 \ 1$$

$$1 \ \sqrt{2} \ \sqrt{3} \ \sqrt{2} \ 1 \qquad 1$$

$$1 \ \sqrt{3}/2 \ 1$$

A B C

D E F*

(d)

Fig. 17 Matching by normalized cross correlation: some simple examples. (a) Picture g and template f. (b) Cross-correlation C_{fg}. (c) $\sum\sum g^2$. (d) $C_{fg}/\sqrt{\sum\sum g^2}$; values less than 1 have been discarded. Note that the perfect match value ($\sqrt{3}$) is not much better than the near misses in position ($\sqrt{2}$'s) and in shape ($3/\sqrt{5}$'s). All points not shown are 0's.

*A = C = D = F = $2/\sqrt{3}$; B = E = $3/\sqrt{5}$.

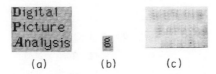

(a)   (b)   (c)

Fig. 18   Matching by normalized cross correlation: a real example. (a) Picture. (b) Template. (c) Their normalized cross correlation. Note the many near misses.

In many cases of interest (character recognition, for example), the templates and picture are—at least approximately—two-valued, i.e., "black and white." In such cases, the process of finding matches can be greatly simplified. Specifically, let the two values be 0 and 1; then $\iint fg$ is just the area of the set of points where f and g are both 1. Let f' be the "negative template" which is 0 when f is 1, and vice versa; then $\iint f'g$ is just the area of the set of points where f' and g are both 1, i.e., where f is 0 and g is 1. Now consider

$$\iint (f-f')g = \iint fg - \iint f'g$$

This takes on its greatest possible value when

(a) $g = 1$ wherever $f = 1$, so that the positive term is as large as possible, namely, equal to the area of the set of points where $f = 1$;

(b) $g = 0$ wherever $f = 0$ (i.e., g is never 1 when $f' = 1$), so that the negative term is zero.

Thus $\iint (f-f')g$ is maximized when f and g match exactly.

It follows from these remarks that we can find matches to f in g by cross-correlating $f-f'$ with g and looking for maxima. Note that $f-f'$ is just the two-valued "picture" that has value 1 where $f = 1$, and value -1 where $f = 0$. Note also that in this two-valued case we can detect matches without the need to "normalize" the cross correlation. Simple examples of this method are shown in Fig. 19.

Exercise 12.   Prove that $\iint (f-f')g = 2\iint fg - \iint g$. ∎

For digital pictures, cross correlations would normally be computed point by point; for each relative shift (u, v) of f and g, we multiply them pointwise and sum the results to obtain $C_{fg}(u, v)$. If f is $m \times m$ and g is $n \times n$, where m is much smaller than n, this requires m^2 multiplications to be performed for each of n^2 relative shifts. If a parallel array-processing computer were available, the sets of multiplications could presumably be done as single operations, but we would still have to go through all n^2 shifts, and for each shift, we must add up the n^2 terms of the resulting product of arrays, which may not be a natural operation for an array processing machine. For the parallel computer, a different method of computing the cross correlation might be more

Template

1	1	1
0	0	0
1	1	1

(a)

```
                                      1  2  3  2  1
        1  2  3  2  1                 1  2  3  2  1
                                      2  4  6  4  2
        1  2  3  2  1                 1  2  3  2  1
                                      1  2  3  2  1
```

```
     1  1  1                 1  2  3  2  1
     1  1  1                 1  1  2  1  1
     2  2  2                 2  4  6  4  2
     1  1  1                 1  1  2  1  1
     1  1  1                 1  2  3  2  1
```

(b)

```
  1  2  3  2  1
```

```
              1  1  1

              1  1  1
```

(c)

Fig. 19 Matching by subtracting cross correlations in the two-valued case. (a) Negative template f'; the picture and positive template are the same as in Fig. 17a. (b) $C_{f'g}$. (c) $C_{fg} - C_{f'g}$; negative values have been discarded.

efficient: We multiply g by each of the m^2 elements in the array f; shift the resulting m^2 arrays relative to one another, by amounts corresponding to the positions of the elements in f; and add them pointwise to obtain the array C_{fg}. This requires only m^2 multiplications and shifts (each carried out in parallel on the array g), and addition of the resulting m^2 arrays (note that this is pointwise addition of arrays, rather than addition of the elements of an array).

By the convolution theorem, cross-correlating a template f with a picture g is equivalent to pointwise multiplying the Fourier transforms F^* and G and then taking the inverse transform. This method can be faster than direct cross correlation, if an efficient Fourier transform algorithm is used. Recall that convolutions obtained using the discrete Fourier transform are *cyclic*, since the transform treats the pictures as though they were periodic (see Section 2.2.1). Thus these convolutions will have values even for shifts such that the template is no longer entirely inside the picture. For most purposes, such values would not be useful. Note also that the Fourier transform arrays that we multiply pointwise should be of the same size; to achieve this when the template is smaller than the picture, we can fill out the template to the size of the picture by adding rows and columns of zeros, or of some other constant value (the average gray level of the picture is sometimes a good choice).

It is sometimes advantageous to work with pictures whose gray scales have been normalized (see Section 10.1.2) by subtracting the average gray level of the picture from the gray level of each point. Given two pictures f and g, whose average gray levels are μ and v, respectively, the normalized pictures are then $f - \mu$ and $g - v$. Our normalized match measure for $f - \mu$ and $g - v$ is then

$$\int\int (f-\mu)(g-v) \bigg/ \sqrt{\int\int (f-\mu)^2 \int\int (g-v)^2} = \left(\int\int fg - \mu v\right) \bigg/ \sigma\tau$$

where σ and τ are the standard deviations of the gray levels of f and g, respectively.

8.3.2 Linear Matched Filtering

Cross-correlating f and g is a linear shift-invariant operation, since it is the same as convolving g with f rotated 180°. In this section we consider the question of finding matches between a template f and a picture g by cross-correlating an arbitrary filter with g. We shall show that, under certain assumptions, the best filter to use for this purpose is f itself; this result is known as the *matched filter* theorem. In what follows we use boldface to denote random fields, as in Section 2.4.

Let $\mathbf{g} = f + \mathbf{n}$, where the noise field $\mathbf{n}$ is homogeneous, and suppose that we convolve the filter h with $\mathbf{g}$ to obtain $h * \mathbf{g}$. Let $\mathbf{g}' = h * \mathbf{g}$ and $\mathbf{n}' = h * \mathbf{n}$. We have seen in Section 2.4.5 that $S_{n'n'} = S_{nn}|H|^2$. Now the expected noise power, i.e., the expected value $E\{|h * \mathbf{n}|^2\}$ at any point, by homogeneity, is $R_{n'n'}(0,0)$. This, however, is just the inverse Fourier transform of $S_{n'n'}$ evaluated at $(0,0)$, i.e., it is $\int\int S_{n'n'} = \int\int S_{nn}|H|^2$. If the noise is "white," i.e., S_{nn} is constant, the expected noise power is thus $S_{nn}\int\int|H|^2$.

On the other hand, the signal power $|h * f|^2$ at a point (x, y) is equal to

$$|\mathscr{F}^{-1}(HF)|^2 = \left|\int\int HF \exp[2\pi j(ux+vy)]\, du\, dv\right|^2$$

By the Cauchy–Schwarz inequality,[†] the right-hand side is less than or equal

---

[†] We use here the complex form of the Cauchy–Schwarz inequality:

$$\left|\int\int fg\right|^2 \leq \int\int |f|^2 \int\int |g|^2.$$

with equality holding only when $g = cf^*$.

to $\iint|H|^2 \iint|F|^2$. Hence the ratio of signal power to expected noise power in the white noise case is

$$\frac{|h*f|^2}{E\{|h*\mathbf{n}|^2\}} \leqslant \frac{\iint|H|^2 \iint|F|^2}{S_{nn} \iint|H|^2} = \frac{1}{S_{nn}} \int\int |F|^2$$

Moreover, the upper bound is achieved only when

$$H = F^* \exp[-2\pi j(ux+vy)]$$

This is equivalent to

$$h(\alpha, \beta) = f(x-\alpha, y-\beta)$$

In other words, h is f rotated 180° and shifted by the amount (x, y). Thus we see that in this case, the linear shift-invariant filter that maximizes the ratio of signal power to expected noise power at (x, y) is just the template f, rotated 180°. Convolving this filter with $\mathbf{g}$ is thus the same as cross-correlating f with $\mathbf{g}$.

Other signal-to-noise criteria can also be used to define optimum filters. Suppose, for instance, that we want to choose h so as to maximize the ratio

$$(E\{h*\mathbf{g}\})^2/\mathrm{var}(h*\mathbf{g})$$

where the numerator is the expected value of $h*\mathbf{g}$ at the point (x, y), and the denominator is the variance of its values taken over all points. It turns out that this can be done by solving the integral equation

$$R_{gg} * h = f$$

where we assume that f is the expected value of $\mathbf{g}$. If $\mathbf{g}$ is highly uncorrelated, so that R_{gg} approximates a δ function, this gives $h = f$; but in other cases, $h = f$ is no longer the optimum solution. For example, in the one dimensional case, if R_{gg} is exponential, say of the form $R_{gg}(x-u) = e^{-|x-u|/L}$, it can be shown that the solution is

$$h = \frac{1}{2L}f - \frac{L}{2}\frac{\partial^2 f}{\partial x^2}$$

Thus for L small ($\mathbf{g}$ highly uncorrelated) the f term dominates; but for L large, the term $\partial^2 f/\partial x^2$ dominates.

The use of derivatives of f as filters can be justified on other grounds as well. If we are interested only in finding the shape of the template f in g, but we do not care about its gray level, it makes sense to convert f and g into outline form, say by differentiation or high-pass filtering, before we cross correlate them. This type of filtering can be used to maximize the ratio of power in the derivative (gradient, Laplacian, etc.) of the signal to expected power in the derivative of the noise. It should also be noted that matching of outlines tends to yield sharper matches than does matching of solid objects. Some simple examples of this last phenomenon are shown in Fig. 20.

0	0	0	0	0
0	1	1	1	0
0	1	1	1	0
0	1	1	1	0
0	0	0	0	0

1	1	1
1	1	1
1	1	1

(a)

1	2	3	2	1
2	4	6	4	2
3	6	9	6	3
2	4	6	4	2
1	2	3	2	1

(b)

1	2	3	3	3	2	1
2	4	6	6	6	4	2
3	6	9	9	9	6	3
3	6	9	9	9	6	3
3	6	9	9	9	6	3
2	4	6	6	6	4	2
1	2	3	3	3	2	1

(c)

$$
\begin{array}{ccccc}
 & & \sqrt{3/2} & & \\
 & 4/3 & 2 & 4/3 & \\
\sqrt{3/2} & 2 & 3 & 2 & \sqrt{3/2} \\
 & 4/3 & 2 & 4/3 & \\
 & & \sqrt{3/2} & &
\end{array}
$$

(d)

For digital pictures, the role of cross correlation in picture matching can be further justified on grounds of minimum error decision making. Suppose that, in a given region $\mathscr{S}$ of the digital picture g, we are trying to decide between the two equally likely alternatives $g = f_1 + n$ and $g = f_2 + n$, where n is signal-independent Gaussian noise with mean 0 and standard deviation σ. If f_1 is present, the probability of obtaining the observed pattern of gray levels g over $\mathscr{S}$ is proportional to $\prod_{\mathscr{S}} \exp\left[-(g(i,j) - f_1(i,j))^2 / 2\sigma^2\right]$ (where we have ignored a factor which is a power of $1/\sqrt{2\pi}\sigma$). This is because, for each (i,j) in $\mathscr{S}$, the probability of having $n(i,j) = g(i,j) - f_1(i,j)$ is $\exp\left[-n(i,j)^2 / 2\sigma^2\right]$, and these events are independent for the different points

Template

0	0	0	0	0
0	1	1	1	0
0	1	0	1	0
0	1	1	1	0
0	0	0	0	0

1	1	1
1		1
1	1	1

(a')

1	2	3	2	1
2	2	4	2	2
3	4	8	4	3
2	2	4	2	2
1	2	3	2	1

(b')

1	2	3	3	3	2	1
2	3	5	5	5	3	2
3	5	8	8	8	5	3
3	5	8	8	8	5	3
3	5	8	8	8	5	3
2	3	5	5	5	3	2
1	2	3	3	3	2	1

(c')

		$3/\sqrt{5}$		
		$\sqrt{2}$		
$3/\sqrt{5}$	$\sqrt{2}$	$2\sqrt{2}$	$\sqrt{2}$	$3/\sqrt{5}$
		$\sqrt{2}$		
		$3/\sqrt{5}$		

(d')

Fig. 20 Matching of outline figures yields sharper matches than matching of solid figures. (a) Templates f: a solid square and a hollow square. The pictures g are the same as the templates, surrounded by many 0's on all sides. (b) C_{fg}. (c) $\sum\sum g^2$. (d) $C_{fg}/\sqrt{\sum\sum g^2}$; values less than 1 have been discarded.

in $\mathscr{S}$, so that their probabilities multiply. Similarly, if f_2 is present, the probability of observing g is proportional to

$$\prod_{\mathscr{S}} \exp\left[-(g(i,j)-f_2(i,j))^2/2\sigma^2\right].$$

To determine which of f_1 and f_2 is more likely, we can compare these probabilities and decide in favor of the case having the larger probability. Now comparing the probabilities is equivalent to comparing their logarithms

$$-\sum_{\mathscr{S}}[g(i,j)-f_1(i,j)]^2/2\sigma^2 \qquad \text{and} \qquad -\sum_{\mathscr{S}}[g(i,j)-f_2(i,j)]^2/2\sigma^2$$

Here we can cancel the common factor $1/2\sigma^2$, and can also cancel $\sum_{\mathscr{S}} g(i,j)^2$ from both sides, so that it remains to compare

$$\sum[2g(i,j)f_1(i,j)-f_1(i,j)^2] \qquad \text{and} \qquad \sum[2g(i,j)f_2(i,j)-f_2(i,j)^2]$$

If $\sum f_1(i,j)^2 = \sum f_2(i,j)^2$ (i.e., the two patterns have the same "power"), this is equivalent to comparing $\sum g(i,j)f_1(i,j)$ and $\sum g(i,j)f_2(i,j)$; doing this for all regions $\mathscr{S}$ of g is tantamount to cross-correlating f_1 and f_2 with g, and deciding in favor of whichever one yields the higher cross correlation at a given point. Note that if $\sum f_1(i,j)^2 \neq \sum f_2(i,j)^2$, it is no longer sufficient to compare the cross correlations.

8.3.3 Line Detection

In this and the next section we consider, in some detail, methods of detecting simple patterns such as lines (or curves), spots, corners, etc., in a picture. We first discuss a linear matched filtering approach, and compare it with nonlinear filtering.

Let us first consider how to detect thin vertical straight line segments that are darker (i.e., have higher gray level) than their background. Since it is the shape (thin, vertical) that is important here, rather than the specific gray levels of the line and the background, it is reasonable to use a filter function based on a derivative of the line pattern, as indicated at the end of Section 8.3.2. In fact, it is customary to use a filter based on the Laplacian of an ideal line. In a digital picture, a vertical line of 1's on a background of 0's

$$\vdots$$
$$1$$
$$1 \qquad \text{(where all points not marked}$$
$$\qquad \qquad \text{are 0's)}$$
$$1$$
$$\vdots$$

has digital Laplacian proportional to

$$
\begin{array}{ccc}
\vdots & \vdots & \vdots \\
-\tfrac{1}{2} & 1 & -\tfrac{1}{2} \\
-\tfrac{1}{2} & 1 & -\tfrac{1}{2} \\
-\tfrac{1}{2} & 1 & -\tfrac{1}{2} \\
\vdots & \vdots & \vdots
\end{array}
$$

The 3×3 filter h_V defined by

$$
\begin{array}{ccc}
-\tfrac{1}{2} & 1 & -\tfrac{1}{2} \\[4pt]
-\tfrac{1}{2} & 1 & -\tfrac{1}{2} \\[4pt]
-\tfrac{1}{2} & 1 & -\tfrac{1}{2}
\end{array}
$$

is commonly used as a detector for thin vertical lines. Note that this filter gives zero output on a constant or ramp portion of the picture, as well as on a horizontal line (e.g., $\cdots 111 \cdots$ on a background of 0's). On the other hand, h_V has a high positive output value for a dark vertical line on a light background; for example, if the line is

$$
\begin{array}{ccccc}
\vdots & \vdots & \vdots & \vdots & \vdots \\
\cdots\ b & b & a & b & b\ \cdots \\
\cdots\ b & b & a & b & b\ \cdots \\
\cdots\ b & b & a & b & b\ \cdots \\
\vdots & \vdots & \vdots & \vdots & \vdots
\end{array}
\qquad \text{(where } a > b)
$$

the output of h_V is

$$
\begin{array}{ccccc}
\vdots & \vdots & \vdots & \vdots & \vdots \\
\cdots\ 0 & 3(b-a)/2 & 3(a-b) & 3(b-a)/2 & 0\ \cdots \\
\cdots\ 0 & 3(b-a)/2 & 3(a-b) & 3(b-a)/2 & 0\ \cdots \\
\cdots\ 0 & 3(b-a)/2 & 3(a-b) & 3(b-a)/2 & 0\ \cdots \\
\vdots & \vdots & \vdots & \vdots & \vdots
\end{array}
$$

which has the high positive value $3(a-b)$ at points of the line, flanked by negative values $3(b-a)/2$ just alongside the line.

In spite of its plausible behavior in the cases just considered, the filter h_V has serious shortcomings; it responds to many patterns that are not linelike, and it may respond to them more strongly than it does to lines. For example, at a vertical step edge of the form

$$
\begin{array}{cccc}
\vdots & \vdots & \vdots & \vdots \\
\cdots\ a & a & b & b\ \cdots \\
\cdots\ a & a & b & b\ \cdots \\
\cdots\ a & a & b & b\ \cdots \\
\vdots & \vdots & \vdots & \vdots
\end{array}
$$

the output of h_V is

$$
\begin{array}{cccccc}
\vdots & \vdots & & \vdots & \vdots & \\
\cdots\ 0 & 3(a-b)/2 & 3(b-a)/2 & 0 & \cdots \\
\cdots\ 0 & 3(a-b)/2 & 3(b-a)/2 & 0 & \cdots \\
\cdots\ 0 & 3(a-b)/2 & 3(b-a)/2 & 0 & \cdots \\
\vdots & \vdots & & \vdots & \vdots &
\end{array}
$$

Thus its peak positive output $3(a-b)/2$, is only half the peak output from a line of the *same* contrast; but the peak output from a step edge of high contrast can be greater than the peak output from a line of lower contrast. Similarly, at an isolated point, say

$$
\begin{array}{cccccc}
 & \vdots & \vdots & \vdots & & \\
 & b & b & b & b & b \\
\cdots & b & b & b & b & b & \cdots \\
\cdots & b & b & a & b & b & \cdots \\
\cdots & b & b & b & b & b & \cdots \\
 & b & b & b & b & b \\
 & & \vdots & \vdots & \vdots &
\end{array}
$$

the output of h_V is

$$
\begin{array}{cccccc}
 & \vdots & & \vdots & & \vdots & \\
0 & 0 & 0 & 0 & 0 \\
\cdots\ 0 & (b-a)/2 & a-b & (b-a)/2 & 0 & \cdots \\
\cdots\ 0 & (b-a)/2 & a-b & (b-a)/2 & 0 & \cdots \\
\cdots\ 0 & (b-a)/2 & a-b & (b-a)/2 & 0 & \cdots \\
0 & 0 & 0 & 0 & 0 \\
 & \vdots & & \vdots & & \vdots &
\end{array}
$$

which again is weaker than that from a line of the same contrast, but may be stronger than that from a line of lower contrast.

Exercise 13.   If the picture is two-valued (with values a and b, say), verify that the filter h_V takes on its maximum possible positive value $3(a-b)$ at points that lie on vertical lines, and nowhere else. Thus in the two-valued case, h_V detects vertical lines more strongly than anything else. ∎

Exercise 14. It has been suggested that if we want a filter that responds to vertical lines but not to isolated points, we can subtract the output of the horizontal line filter h_H defined by

$$
\begin{array}{ccc}
-\tfrac{1}{2} & -\tfrac{1}{2} & -\tfrac{1}{2} \\[4pt]
1 & 1 & 1 \\[4pt]
-\tfrac{1}{2} & -\tfrac{1}{2} & -\tfrac{1}{2}
\end{array}
$$

pointwise from the output of our vertical line filter. This seems plausible at first glance, since h_H has output zero at a vertical line, so that subtracting it has no effect; while both h_H and h_V have output $a-b$ at an isolated point, so that they cancel when we subtract. What happens, however, at the neighbors of an isolated point? ∎

These remarks suggest that if we want a filter which responds to thin vertical lines, but not to edges or isolated points, we must use something more elaborate than our simple linear filter h_V. We shall now define a nonlinear filter h_V' that has the desired properties. At the point (i, j) of the picture g, the output of h_V' is defined to be the same as that of h_V [i.e., $g(i, j+1) + g(i, j) + g(i, j-1) - \tfrac{1}{2}(g(i-1, j+1) + g(i+1, j+1) + g(i-1, j) + g(i+1, j) + g(i-1, j-1) + g(i+1, j-1))$] whenever

$$g(i, j+1) > g(i-1, j+1), \qquad g(i, j+1) > g(i+1, j+1)$$

$$g(i, j) > g(i-1, j), \qquad\qquad g(i, j) > g(i+1, j)$$

$$g(i, j-1) > g(i-1, j-1), \qquad g(i, j-1) > g(i+1, j-1)$$

and zero otherwise. In other words, $h_V' = h_V$ at (i, j) provided that point (i, j) in g is darker than each of (i, j)'s horizontal neighbors, and the same is also true of (i, j)'s two vertical neighbors. Clearly these conditions are exactly what we mean by saying that a thin dark vertical line passes through (i, j). If the conditions are not met, the output of h_V' is defined to be zero. Note that h_V' no longer has negative values at points adjacent to a dark vertical line.

Exercise 15. Compute the output of h_V' in the vicinity of a step edge and of an isolated point. ∎

Exercise 16. Let g consist of a thin black line (value 1) on a background of salt-and-pepper noise (value 1 has probability p, 0 has probability $1-p$). Compute the probability densities of the values of h_V and h_V': (a) at a point of the black line; (b) at a point of the background. ∎

Examples of the output produced by h_V and h_V' are shown in Fig. 21.

It should be pointed out that although h_V' does not respond to isolated points it does respond to certain spotlike patterns. For example, it responds

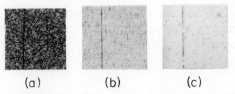

(a)   (b)   (c)

Fig. 21 Vertical line detection by linear and nonlinear filters. (a) Input picture. The noise has normally distributed gray levels with mean 32 and standard deviation 9; the line has constant gray level 50. (b) Result of applying operator h_V (see text). Note that the surviving noise is streaky. (c) Result of applying operator h_V'. The breaks in the line are due to adjacent noise points which cause the operator's detection criteria to be violated.

(with value 1) at the center point of the pattern

$$
\begin{array}{ccccc}
 & & 1 & & \\
 & 1 & 2 & 1 & \\
1 & 2 & 3 & 2 & 1 \\
 & 1 & 2 & 1 & \\
 & & 1 & & \\
\end{array}
$$

Of course, it is not difficult to modify h_V' so that it no longer responds to this pattern; for example, we can ignore responses at points where the nonlinear horizontal detector h_H' also responds (compare Exercise 14). Examples such as this, however, illustrate the nontriviality of designing detectors that respond only to a specified class of patterns.

8.3.4 Curve and Streak Detection

In Section 8.3.3 we described a simple nonlinear filter h_V' designed to detect thin, vertical, dark lines on a light background. We now discuss how to modify this design to handle broken lines, thick lines, lines having other slopes, and lines that contrast with their background in other ways.

To detect light lines on a dark background, we need only reverse the inequalities, and change the signs, in the definition of h_V'. In other words, at each point (i, j) of the picture g, we require that $g(i, j)$ be less than the values of its two horizontal neighbors, and the same at each of (i, j)'s vertical neighbors; and if these conditions are fulfilled, we give h_V' the value

$$
\begin{aligned}
& \tfrac{1}{2}[g(i-1, j+1) + g(i+1, j+1) + g(i-1, j) \\
& \quad + g(i+1, j) + g(i-1, j-1) + g(i+1, j-1)] \\
& \quad - [g(i, j+1) + g(i, j) + g(i, j-1)] \qquad \text{at} \quad (i, j)
\end{aligned}
$$

A more complicated definition is needed if we want to detect, say, a gray line whose background is white on one side and black on the other side. (Such a line might result from digitization of a perfect step edge.) Here we must require that $g(i, j)$ be greater than $g(i-1, j)$ and less than $g(i+1, j)$, and similarly at the points $(i, j\pm 1)$. The directions of the inequalities should be consistent at all three points (i, j) and $(i, j\pm 1)$; otherwise, we would detect the columns of a checker board as "lines," which would normally not be desirable.

Still further modification is needed if we want to detect a line that differs in "texture" from its background, e.g., a line whose points are alternately black and white, on a background of solid gray. We could attempt to detect this line by the fact that it contrasts with the background at each point (even though the sign of the contrast changes from point to point), but this approach would also detect checkerboard columns as "lines." Instead, we can detect the line by the fact that it is "busy" while its background is smooth. For example, we can apply a vertical absolute difference operator to the picture; this will have high value at every point of the line, but zero value elsewhere in the picture. Thus we now have a new picture in which the line is darker than its background, and can be detected using h_V'. The method used here is analogous to our method of detecting "texture edges" by first converting them to average gray level edges, as described in Section 8.2.4.

We next describe how to extend our methods to thick lines ("streaks"), say, having thickness $2t$. Suppose that we average the given picture g, using averaging neighborhoods of radius t; let the average for the neighborhood centered at (i, j) be $g^{(t)}(i, j)$. We can define a thick analog $h_V'^{(t)}$ of our h_V' operator by simply replacing the individual gray levels used in the original definition by adjacent, nonoverlapping averages. For example, we require that $g^{(t)}(i, j)$ be greater than both of $g^{(t)}(i\pm 2t, j)$, and similarly at the two vertically adjacent, nonoverlapping positions $(i, j\pm 2t)$. When these conditions are fulfilled, the value of $h_V'^{(t)}$ is defined analogously to that of h_V', but using $g^{(t)}(u, v)$'s instead of $g(u, v)$'s. Other types of streak contrast are handled similarly. (When we are dealing with thick streaks rather than thin lines, a wide variety of types of "texture contrast" becomes possible; see Section 8.2.4 for examples.) Note that $h_V'^{(t)}$ will detect a streak of width $2t$ most strongly when it is centered on the streak, but it will also detect the streak as we move it sideways to nearby positions, just as was the case with the "coarse" edge detection operators in Section 8.2.4. If we want to locate the streak exactly, we should suppress nonmaxima in the direction across the streak, just as in Section 8.2.4.

A streak detector based on averages over neighborhoods of radius t will detect streaks of width $2t$ strongly; it will also respond to streaks of other widths (between 0 and $6t$), but the more the size differs from $2t$, the weaker its response will be. If we do not know in advance the widths of the streaks that are present in the given picture, we can use streak detectors having a range of sizes, and combine these, or compare them to find a "best" size at each point, just as in Section 8.2.4.

Exercise 17. Show that the maximum response of our streak detector to a streak of width w is proportional to $2tw$ for $0 \leqslant w \leqslant 2t$; to $6t(t-w)$ for $2t \leqslant w \leqslant 6t$; and is zero otherwise. $\blacksquare$

Exercise 18. Design an algorithm for detecting streaks of a given width by finding pairs of step edges (one a step up, the other a step down) that have a given separation. (On the use of this approach to handle a range of widths see Cook and Rosenfeld [9].) $\blacksquare$

We can also extend our approach to broken, e.g., "dotted," lines (or streaks). To this end, we redefine h_V' so that it depends on more than just the three vertically consecutive points (i, j) and $(i, j \pm 1)$. For example, we can examine the five points (i, j), $(i, j \pm 1)$, and $(i, j \pm 2)$, and require that at least three of these points be darker than their two horizontal neighbors. If this condition is satisfied, we give our new h_V' the value

$$\sum_{k=-2}^{2} g(i, j+k) - \tfrac{1}{2} \sum_{k=-2}^{2} [g(i-1, j+k) + g(i+1, j+k)]$$

and value 0 otherwise. The generalizations to handle dotted streaks, and other types of contrast, are analogous.

Finally, we consider operators that detect lines in directions other than the vertical. As long as our operators depend only on small neighborhoods, the number of directions that need be considered is not large. A straight line that passes through the center of a 3×3 neighborhood must intersect that neighborhood in one of the following 12 patterns (the center of the neighborhood has been underlined):

(see Section 9.3.1 on the characterization of digital straight lines). To obtain a complete set of thin line detection operators, we would need an analog of h_V' for each of these patterns. For the five near vertical patterns, the h' operator would be defined by comparing the three "a" points with their horizontal neighbors; for the five near-horizontal patterns, we would compare the points with their vertical neighbors; and for the two diagonal patterns, we could do either type of comparison.

A similar set of operators could be used for streak detection, with the a's representing adjacent nonoverlapping averages rather than individual points. For dotted line detection, a much larger set would be needed, since a straight line can intersect (say) a 5×5 neighborhood in many ways. (*Exercise*: how many?) If linear operators (h rather than h') are used, it is not necessary to use all of the possible operators, since all lines will still be detected, some more strongly than others.

We can use sets of local line detection operators not only to detect straight lines having arbitrary slopes, but also to detect arbitrary smooth curves. (We assume here that a smooth curve never turns as much as 90° within a 3×3 neighborhood; thus we need not consider patterns such as

$$\begin{matrix} a & & a \\ \underline{a}\ , & & \underline{a}a \\ a & & \end{matrix}$$

and their rotations.) Some examples of curve detection, curved streak detection, and dotted curve detection by this method are shown in Figs. 22 and 23.

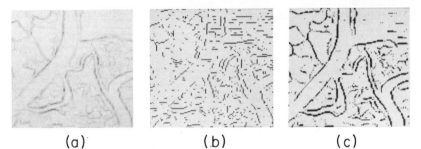

(a)   (b)   (c)

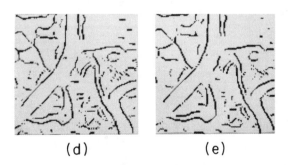

(d)   (e)

Fig. 22   Curve and curved streak detection. (a) Input picture: Output of an edge detection operation applied to part of an ERTS-1 picture. (b) Result of thin curve detection operation (see text), applied to (a). (c) Result of streak detection operation, based on averages over 2×2 neighborhoods (see text), applied to (a). (d)–(e) Results of repeating the same streak detection operation twice more, i.e., applying it again to (c), then again to (d). The noise is now almost all gone.

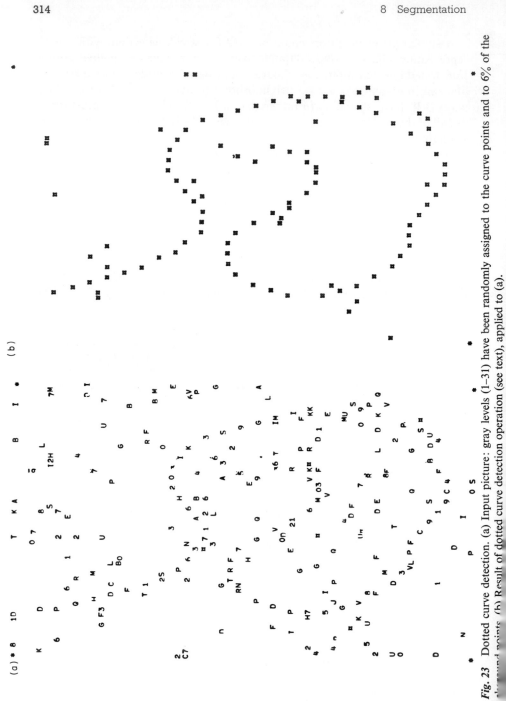

Fig. 23 Dotted curve detection. (a) Input picture: gray levels (1–31) have been randomly assigned to the curve points and to 6% of the background points. (b) Result of dotted curve detection operation (see text), applied to (a).

Methods similar to those used in this and the preceding section can be used to detect types of patterns other than lines and smooth curves. For example, to detect isolated dots in g that contrast with their surroundings, we can use the operator d whose value is $|\nabla^2 g(i, j)|$ when $g(i, j)$ is greater than all four of its neighbors and zero otherwise. (We would not want to use ∇^2 itself as a dot detector, since as pointed out in Section 8.2.1, it responds to edges and lines as well as to dots.) We can generalize d to detect coarse spots by using averages over adjacent, nonoverlapping neighborhoods instead of single points; and we can handle spots that differ texturally from their background using the methods of Section 8.2.4.

Exercise 19. Design operators that detect: (a) ends of curves; (b) sharp angles on curves[†]; (c) branchings or crossings of curves. You need only do this for one orientation. ▮

8.3.5 Search Techniques in Matching

When we match a template against a picture, e.g., by cross correlation, we may expect to find any exact copies of the template that are present in the picture. These copies must be of the same size, however, and have the same orientation, as the template; we will not be able to find a rotated or enlarged version of the template in the picture, since such a version will not, in general, give rise to high values of the cross correlation. Similarly, if a geometrically distorted copy of the template is present, we will have difficulty detecting it by matching unless the distortion is quite small.

If we want to use template matching to recognize a pattern that is subject to rotation, scale change, or other geometrical transformations, a very large number of templates would ordinarily be needed. Another possibility, however, is to use a single template, and search through the space of permissible distortions or transformations of the template, in an attempt to optimize its degree of match with the picture. This "rubber mask" approach may be practical if the space of possibilities is relatively small, and if it is possible to make a good initial guess as to the correct transform of the template that should be used. In some cases, searching may not be necessary, since it may be possible to "normalize" the pictures being matched (see Section 10.1) so as to undo the effects of the transformations; for example, one might be able to transform the pictures' gray scales so that they have some standard (e.g., flat) histogram.

If unlimited amounts of distortion (in gray scale and/or geometry) are allowed, any two pictures can be made to match. Evidently, the "value" of a

---

[†] Another approach to detecting angles on digital curves will be described in Section 9.3.2.

match obtained in this way should depend both on the goodness of the match and on how little distortion was required to achieve it. To illustrate how such a match evaluation measure could be defined, imagine that we have a template composed of subtemplates that are connected by springs. As we search for a match between this template and some part of a picture, we may find many good matches to the various subtemplates, but these partial matches may not be in the right relative positions, so that to achieve them all simultaneously, we may have to stretch or compress the springs. Our goal is to find a combination of partial matches which (a) are as good as possible and (b) require as little tension in the springs as possible. This can be done, in principle, using mathematical programming techniques; but in practice this would usually be computationally costly. A useful shortcut is to first find good matches for the individual subtemplates, and then try to build up a low-tension combination of these matches stepwise, using heuristic search techniques.

Even in the absence of distortions, looking for matches to a template in all possible positions in a picture is a relatively time-consuming process. This process can be shortened if we can develop fast, inexpensive tests for determining positions or regions in which matches are likely to be found, so that it is no longer necessary to look for a match in every possible position. It can be further shortened if, when we are testing for a match in a given position, we can develop methods of rejecting mismatches rapidly, without having to go through the entire point-by-point comparison of the template and picture. These two approaches will be discussed in the following paragraphs.

If the template h is not too highly uncorrelated, there will be fairly good matches to it in the picture g in the vicinity of any ideal match—in other words, the peak(s) in the (normalized) cross correlation C_{gh} will be relatively smooth and broad, rather than spikelike. Hence in this case, we can attempt to find ideal matches by first computing C_{gh} at a coarsely spaced grid of points, and then searching for better matches in the vicinity of those grid points at which C_{gh} is high. In these searches, it may be possible to use hill-climbing techniques, if the correlation peaks are smooth enough.

Another possible approach to reducing the number of match positions that need to be examined is to compute some simple picture property, say a textural property, in the vicinity of every point of g, and to attempt matches first at those points where the value of the property is close to its value for the template h. (Of course, this makes sense only if computing the property at a point is much cheaper than measuring the degree of match between g and h at that point.) This approach should make it possible to find a match sooner, on the average, than if we just tested match positions in some arbitrary order. Incidentally, when we are matching pieces of one picture with another picture (e.g., to determine stereoscopic depth), the sizes of the pieces that should be used also depend on the local texture; when the picture contains fine detail, relatively small pieces can be used.

When we measure the degree of match in a given position, we would like to be able to reject mismatches rapidly, since most positions will be mismatches. For this purpose, it is convenient to use a sum of absolute differences measure of mismatch $\sum |g-h|$. Given a mismatch threshold t, we want to compute $\sum |g-\hat{h}|$ in such a way that, in a mismatch position, the sum can be expected to rise above t as rapidly as possible. One way to do this is to measure $|g-h|$ first for those points of h whose gray levels have high expected absolute differences from the gray level of a randomly selected point of g. The expected contributions of such points to $\sum |g-h|$, when we are not in a match position, are large; hence when we measure $|g-h|$ for these points first, the sum $\sum |g-h|$ should tend to rise more rapidly than if we measured $|g-h|$ for the points of h in some arbitrary order.

8.4 TRACKING

In most of the segmentation methods described up to now, the processing that was done at each point of the picture did not depend on results already obtained at other points. Thus these methods can be regarded as operating on the picture "in parallel," i.e., at all points simultaneously; and they could be implemented very efficiently on a suitable parallel computer. This section deals with segmentation methods in which we do take advantage, in processing a point, of the results at previously processed points. In these inherently sequential methods, the processing that is performed at a point, and the criteria for accepting it as part of an object, can depend on information obtained from earlier processing of other points, and in particular, on the natures and locations of the points already accepted as parts of objects.

Sequential segmentation methods have a potential advantage over parallel methods, with respect to their computational cost on a conventional, sequential computer. In the parallel approach, the same computations must be performed at every point of the picture, since our only basis for accepting or rejecting a given point is the result of its own local computation. If we want our segmentation process to be reliable, these computations may have to be relatively complex. When using the sequential approach, on the other hand, we can often use simple, inexpensive computations to *detect* possible object points. Once some such points have been detected, more complex computations can be used to extend or *track* the object(s). The latter computations need not be performed at every picture point, but only at points that extend objects that have already been detected and are being tracked. The detection computations may have to be performed at every point, but they can be relatively cheap, since they need only detect *some* points of every object while rejecting nonobject points.

Sequential methods are so flexible, and can be defined in so many different ways that we will not attempt to discuss specific techniques in detail as we did in the previous sections. Rather, we will describe general classes of methods, and try to suggest the many variations that are possible.

8.4.1 Raster Tracking

We begin our discussion of tracking methods with a very simple example. Suppose that the objects to be extracted from the given picture are thin, dark, continuous curves whose slopes never differ greatly from 90°. In this case, we can extract the objects by tracking them from row to row of the picture, as we scan the picture row by row in the manner of a TV raster. Specifically, in each row we accept any point whose gray level exceeds some relatively high threshold d; this is our detection criterion. In addition, once a point

```
9     5     8     4 5     3         1         1
   5           6     4 2
      9     1 5         5               1
   6     4         7   2                           1
      3   6 2     9         7                   1       1
   4   6         4   3     5
         5   2 6         6
   6 4     3   7         4 2                   1
              (a)                              (b)
```

```
1     1     1     1 1         1         1
   1           1   1             1           1
      1         1       1         1         1
1       1       1                   1       1
         1         1       1         1       1       1
   1   1       1           1         1       1           1
      1       1       1         1       1       1
   1 1       1       1         1       1       1
           (c)                              (d)
```

Fig. 24 Simple example of raster tracking. (a) Input picture. (b) Result of thresholding (a) at 7. (c) Result of thresholding (a) at 4. (d) Result of tracking with $d = 7$, $t = 4$ (see text).

(h, k) on the kth row has been accepted, we accept any neighbor of (h, k) on the $(k-1)$st row, i.e., we accept any of the points $(h-1, k-1)$, $(h, k-1)$, and $(h+1, k-1)$, provided that the accepted points have gray level above some lower threshold t; this is our tracking criterion.

This trivial example of detection and tracking is illustrated in Fig. 24. Note that if we used the d threshold alone, or the t threshold alone, the curves would not be extracted correctly; but when we combine the two thresholds, in conjunction with row-by-row tracking, we are able to extract the curves. To achieve the same results by a parallel method, we would have to use some sort of line detection operation (see Sections 8.3.3 and 8.3.4) at each point; this would be relatively costly, since in line detection we would have to compare *every* point to its neighbors on the rows above and below it, whereas here we did so only for points already being tracked.

In the following paragraphs we indicate some of the many possible ways that our simple example can be modified and extended. We still consider only raster tracking; extensions to thin curves having arbitrary slopes, and to solid objects, will be discussed in Sections 8.4.2 and 8.4.3.

The detection and tracking acceptance criteria can involve local property values other than gray level. For example, they can be based on local contrast, as measured, for example, by the value of some derivative operator such as the gradient. Note that the gradient defines not only a degree of contrast, but also a direction (of highest rate of change of gray level at the point); in tracking, one could look for successor points along the perpendicular direction, which should be the direction "along" the curve being tracked. The tracking criterion can also depend on a comparison between the current point (h, k) and its candidate neighbors, so as to discriminate against candidates whose gray levels, etc., do not "resemble" that of (h, k), if desired.

In some cases, we may want to accept all candidate points that satisfy the tracking criterion. In other cases, however, it may be preferable only to accept the best one(s). For example, it may turn out that accepting all of them would make a curve thick, which contradicts our *a priori* knowledge that the curves are thin.

The tracking criterion should, in general, apply to some region below each currently accepted point (h, k), rather than just to the point's immediate neighbors on the row below it. Moreover, the criterion can depend on the candidate point's position in this region. For example, it can be more stringent for points that are farther from (h, k), so as to discriminate against large gaps in the curves, or it can be more stringent for points that are less directly below (h, k), so as to discriminate against obliquity.

Similarly, the tracking criterion should depend not just on a single currently accepted point (h, k), but on a set A of already accepted points. This makes it possible, for example, to fit a line or curve to the points in A, and make the criterion more stringent for candidate points that are far from this curve. In this way, we can discriminate against curves that make sharp turns. If desired, we can give greater weight to the more recently accepted points of A than to the "older" points. ("Tracking" based on sets of already accepted points will be further discussed later under the heading of region growing.)

The preceding remarks can be summarized in the following generalized raster tracking algorithm.

(1)   On the first row (or line of the raster) accept all points that meet the detection criterion. Take each such point to be the initial point of a curve C_i that is to be tracked.

(2)   On any current row other than the first row:

(a)   For each curve C_i currently being tracked, apply the appropriate tracking criterion to the points in its acceptance region; adjoin the resulting accepted points to C_i. We recall that this criterion may depend on the distance and direction of these points from the end of C_i or from some curve that extends C_i, as well as on the gray levels, contrasts, etc. of the points. If no new points are accepted into C_i, tracking of C_i has terminated. Note that a curve C_i may branch into two or more curves, in which case we must track them all; or two or more curves may merge into a single curve, in which case we need only track that one from there on.

(b)   In addition, apply the detection criterion to points that do not belong to any acceptance region; if any points meet this criterion, take them to be initial points of new curves C_i.

(3)   When the bottom row is reached, the tracking process is complete.

This algorithm is analogous to an algorithm given in Section 9.1.3 for tracking connected components; but the present algorithm is much less specific, because of the wide variety of detection and tracking criteria that could be used.

8.4.2   Omnidirectional Tracking

Up to now we have restricted ourselves to tracking methods based on a single raster scan. This has the disadvantage that the results depend on the orientation of the raster, and the direction in which it is scanned. For example, if a strong curve gradually becomes weaker (lighter, lower contrast, etc.) as we move from row to row, our tracking scheme may be able to follow it, since the acceptance criteria for a curve that is already being tracked are relatively permissive; but if we were scanning in the opposite direction, so that the curve starts out weak and gets stronger, we might not detect it for quite a while, since our detection criteria are relatively stringent. We can overcome this problem by scanning the picture in both directions, carrying out our tracking procedure for each of the scans independently, and combining the results, but this doubles the computational cost.

Similarly, raster tracking has the disadvantage that it breaks down for curves that are very oblique to the raster lines. For example, if we scan row by row, we cannot track curves that are nearly horizontal, since the crossings of successive rows by such a curve are many columns apart. One way to overcome this problem is to use two perpendicular rasters, e.g., the rows and the columns, so that any curve meets at least one of the rasters at an angle between 45 and 135°. If we carry out detection and tracking independently for the two rasters, and combine the results, we should be able to track every part of the curve using at least one of the rasters. Not only does this approach double the computational cost, but it may also cause us to miss parts of the curve, since when it becomes too oblique to be tracked by one raster, it will be picked up by the other raster only if it meets the more stringent detection criteria.

These directionality problems can be largely avoided if we use omnidirectional curve-following, rather than raster tracking. Given the current point (h, k) on a curve that is being tracked, a curve-following algorithm examines a neighborhood of (h, k), and picks a best candidate for the next point. If the curves being tracked can branch or cross, it may be necessary to pick more than one next point; in that case, all but one of the chosen points are stored for later investigation, and the tracking proceeds with the one remaining point as next point.

The candidates for next point are evaluated on the basis of their satisfying a tracking criterion for acceptance (dark, high contrast, etc.). They need not be immediate neighbors of (h, k), since we want to be able to tolerate small gaps in curves; but the tracking criterion should discriminate against points far from (h, k), since a connected curve would normally be preferable to one with gaps. They need not lie on the side of (h, k) directly opposite the already accepted points, since our curves need not be straight lines; but the criterion should discriminate against points that deviate too much from this direction, since a smooth curve is normally preferable to one that makes sharp turns. All of this is, of course, analogous to the raster tracking case, except that all the neighbors of the current point, not just those on some "next row," are candidates for being the next point.

At any stage, the next point must not be a point that has already been accepted at a previous stage. If there are no candidates that have not already been accepted, the tracking terminates. We can now go back to any points that were stored for later investigation, and resume tracking these, to find other curve branches. In particular, when we first detect a curve point, we track that curve in one direction until it terminates, and then go back to the originally detected point to track the curve in the other direction, if possible. When no points remain to be investigated, we can resume searching the picture for other points that meet the detection criterion; these points must lie on curves that do not cross the curves that we have already tracked.

We can summarize this approach to curve-following in the form of a generalized algorithm:

(1)   Scan the picture systematically, looking for points that meet the detection criterion. When such a point is found, it becomes the "current point."

(2)   Examine the neighborhood of the current point and apply an appropriate tracking criterion. As before, this criterion may depend on the gray level, contrast, etc. of the candidate point, as well as on its distance and direction from the current point, or from some curve that extends the curve branch currently being tracked.

> (a)   If no points as yet unaccepted meet this criterion, tracking of that branch has terminated. In this case, we take the next point on list L (see (c)) as our new current point, and resume tracking. If list L is empty, we go back to step (1).
> (b)   If the unaccepted points meeting the criterion all appear to lie on a single curve, we accept the closest of them as belonging to the curve branch being tracked, take it as the new current point, and go back to step (2).
> (c)   If these points appear to lie on more than one curve, we may conclude that the curve being tracked has branched, or has crossed another curve. In this case we put all but one of the closest ones on list L for later investigation. The remaining closest point is accepted as the new current point, as in case (b), and we go back to step (2).

(3)   When the systematic scan is finished, the algorithm has terminated.

This algorithm is analogous to an algorithm given in Section 9.1.2 for following the borders of connected components; but the present algorithm is much less specific, because of the wide variety of acceptance criteria that could be used.

8.4.3   Region Tracking and Growing

We now consider the problem of "tracking" arbitrary objects or regions, rather than thin curves. One possible approach is to track *runs* of object points from row to row of a raster. On the first row, each run of detected points is taken as the start of an object. On subsequent rows, we accept points that are adjacent to already accepted runs, if they satisfy a tracking acceptance criterion; and we also accept other points if they satisfy the more stringent detection criterion. The accepted points on a new row constitute a new set of runs, and the process repeats. Runs may merge, split, or terminate, as in the thin curve case. The tracking acceptance criterion will depend, in general, on the natures and positions of the candidate points relative to the already accepted runs. On the use of run tracking for shape description see Section 9.4.2.

Another, less direction-dependent approach is to detect and track the edges of the objects. Since edges are curvelike, this can be done using the methods already described. Acceptance criteria for edge points would, of course, be based on local contrast, e.g., on the value of the gradient. Ideally, an edge that is being tracked should never terminate, except by returning to its starting point or running off the picture.

A still more powerful approach is to "grow" the objects, in all directions, starting from points that meet the detection criterion. Let O be a currently accepted piece of an object. We examine all its neighboring points, and incorporate any of them that meet a tracking acceptance criterion. As previously, this criterion can depend both on the natures and positions of the candidate points. For example, acceptance of a point (h, k) may depend on the closeness of (h, k)'s gray level to the average gray level of O; and it may also depend on (h, k)'s position relative to O, and on the size and shape of O itself, since we may want to bias the growing process to favor certain types of object sizes and shapes. When new points are accepted, they are adjoined to O, and the process is repeated with the resulting new O. The growth terminates when no acceptable neighbors exist. A very simple example is shown in Fig. 25.

When tracking or growing arbitrary objects rather than thin curves, the acceptance criteria can be based on texture rather than on gray level or contrast. We can test candidate "cells" (small groups of points), rather than single points, for inclusion in an object, and accept them if their textures resemble that of the object sufficiently closely. For some methods of comparing textures see Section 8.2.4 (regarding texture edge detection). We can also use edge detection to set up barriers to tracking or to growth, i.e., we

```
5 5 8 6      5 5 8 6      5 5 8 6      5 5 8 6
4 8 9 7      4 8 9 7      4 8 9 7      4 8 9 7
2 2 8 3      2 2 8 3      2 2 8 3      2 2 8 3
3 3 3 3      3 3 3 3      3 3 3 3      3 3 3 3
   (a)          (b)          (c)          (d)
```

Fig. 25 Simple example of region growing. The neighbors accepted at each step are those that differ from the average gray level of the current object by less than 2. (a) Input picture, with starting point underlined. (b) Neighbors accepted at first step. (c) Neighbors accepted at second step; no further growth is possible. (d) Result of growth starting from the 6 rather than from the 9.

never accept a neighbor into an object if there is an edge between it and the object.

In any tracking scheme, the results obtained may depend on the starting point. This can lead to objectionable results when we track or grow solid objects, since we do not normally want two different objects to overlap. In Fig. 25, for example, depending on which point we start from, we obtain two overlapping objects. In such cases, we must decide whether to merge the objects, to give one priority over the other, or to establish a growth barrier whenever two growing objects meet.

8.4.4   Partition Building

In region tracking or growing, we start from an initial point and build up a homogeneous region or object by adding neighboring points, or small "cells," one step at a time. In this section we describe another class of region-building methods that start with a partition of the picture into initial regions, and successively improve this partition, by splitting and merging the regions, until a "good" partition into homogeneous regions is obtained. This approach does not really belong to the tracking family of segmentation methods, but we present it here because of its relationship to region growing.

In order to design a partition-building algorithm, we must be given some criterion for the "goodness" of a partition. Typically, this is measured by the closeness with which we can approximate the picture, on each of the regions in the partition, by a function chosen from some given family of functions. (It should also depend on the sizes and shapes of the regions, but we shall be concerned here primarily with the closeness of approximation.) Our goal might be to find a partition into the smallest possible number of regions, such that on each region, the approximation error does not exceed some bound E.

To illustrate this idea, suppose that we want to approximate the picture f as closely as possible, over some region $\mathscr{S}$, by a constant c, and that we use the mean square measure of error

$$E_c = \int\int_{\mathscr{S}} (f-c)^2$$

The value of c that minimizes E_c is found by differentiating E_c with respect to c and equating the result to zero, which gives

$$2 \int\int_{\mathscr{S}} (f-c) = 0$$

so that

$$c = (1/S) \int\int_{\mathscr{S}} f$$

where S is the area of $\mathscr{S}$. In other words, the c that minimizes E_c is just the mean of f, so that the minimum value of E_c is the variance of f. Thus if we use the constant functions to approximate f, and the mean square measure of error, our goal becomes that of finding a partition into as few regions as possible such that, on each region, the variance of f does not exceed E. Of course, other families of functions (e.g., linear rather than constant), and other error criteria, can be used, but in the following discussion we shall use the constant functions and the variance.

A general approach to this partition-finding problem is as follows: We start with an initial partition, and examine each region whose variance exceeds E, with the aim of subdividing it into two or more parts each having variance less than E. When we subdivide, we should also try to shift the boundaries between the parts so as to reduce the variances of the parts; the use of such boundary adjustment can eliminate much unnecessary subdivision. At the same time, we examine pairs of adjacent regions that have variances less than E and whose mean gray levels are close together, to see whether merging them would yield a single region with variance less than E. This process is repeated until every region has variance less than E and no further merging is possible.

The initial partition can be chosen in several different ways. At one extreme, we can start with regions that have zero variance, such as the connected components of constant gray level in the picture, and we can then try to build the desired final partition by merging alone. At the other extreme, we can take the whole picture as a trivial initial partition, and split it (e.g., by successive bisection) until we obtain regions of variance less than E. A reasonable compromise is to start with an intermediate partition and try both merging and splitting, as well as boundary adjustment if possible. In most cases, it would be desirable to have the boundaries coincide with texture edges, especially edges that represent abrupt changes in average gray level (Section 8.2.4).

As indicated earlier, the merging and splitting criteria that are used can also depend on the sizes and shapes of the resulting regions. For example, we may want to ignore small "noise" regions; the criteria for ignoring such a region can depend on whether it is entirely surrounded by a single large region, or is on the boundary between two large regions. As another example, we may be willing to tolerate a higher gray level variance if it results in a simpler shape (see Section 9.4.4 on measures of shape complexity). More generally, the search for an optimum partition should be guided by our model for the class of pictures that we are analyzing; if we know what sorts of objects or regions these pictures should contain, and in what arrangement, we can seek a partition that fits this model. General partitioning criteria such as gray level homogeneity and shape simplicity should only be used in the absence of a more specific model.

8.4.5   Search Techniques in Tracking

The tracking techniques described in Sections 8.4.1–8.4.3 make their acceptance decisions in a specified sequence (row by row of the raster, point by point
along the curve or around the object boundary), and never reverse a decision
once made. An important extension of these techniques is to allow backup,
i.e., if an acceptance decision seems to be leading to a series of poor subsequent acceptances, one can go back and alter the decision.

To illustrate this idea, suppose that we are using raster tracking to find the
best possible thin, dark curve that crosses the picture from top to bottom.
If we pick the darkest candidate point at each successive row, we may get
trapped in a blind alley. As shown in Fig. 26, when backup is allowed, we can
avoid this.

In order to formulate the backup concept more precisely, suppose that we

```
5              5    5
5              5    5
7  4           7         4
3     5        3              5
1        6     1              6
1        6     1              6
         6                    6
      (a)       (b)    (c)
```

Fig. 26   Tracking with backup. (a) Input picture (values not shown are 0's). (b) Result
of picking the darkest candidate point on each row. (c) Result of picking the second-best
choice on the third row.

are given some method of evaluating the curves (or object borders, etc.) that
are being tracked, e.g., using criteria of contrast, smoothness, etc. For any
curve C, let $\varphi(C)$ be the value of C. For example, $\varphi(C)$ might be the average
contrast of C, minus its average curvature. (On the measurement of curvature
for digital curves, see Section 9.3.1.) Given any point (h, k), let $\hat{\varphi}(h, k)$ be
an estimate of the value of the best curve in the picture that passes through
(h, k). For example, we might take $\hat{\varphi}(h, k)$ to be $\max(\varphi(C))$ for all curves C
so far found between the starting point and (h, k).

At any stage of the tracking process, we have a list L of already accepted
points from which the tracking might continue (compare Section 8.4.2).
Initially, L consists of just the starting point. We compute $\hat{\varphi}$ for all the points
on L, and pick the point (h, k) for which $\hat{\varphi}$ is highest. We now examine the
neighbors of (h, k), find those that are candidates for acceptance, and put

them on the list L; this completes one stage of tracking. The point (h, k) need no longer be on L, since its neighborhood has already been examined. We should remember (h, k), however, if we want to avoid reaccepting points that have already been accepted.

If we use this procedure, the point picked for examination of its neighbors at a given stage is not necessarily a neighbor of the point picked at the previous stage. Rather, at each stage, the tracking process backs up, if necessary, to the highest-value point whose neighbors have not yet been examined. Indeed, if we use the $\hat{\varphi}$ just described, this procedure will pick the highest contrast, straightest curve C yet found at the given stage, and extend C by accepting the appropriate neighbors of its endpoint. One could also make the value of a curve depend on its length, so that short curves have higher values than long ones. In practice, this may result in an excessive amount of backup. It may be desirable, in fact, to limit the distance over which backup is allowed. For example, once the point (h, k) has been picked, we can discard from L all points that are more than some given distance away from (h, k).

As a simple example of how this tracking procedure operates, consider Fig. 26. We take the "5" in the top row as the starting point; let its coordinates be $(0, 6)$. Initially, this is the only point available for examination. Its sole candidate neighbor is $(0, 5)$, and we have $\hat{\varphi}(0, 5) = 5 - 0 = 5$ (the average gray level is 5, the curvature is 0). This is now the only available point, but it has two candidate neighbors, $(0, 4)$ and $(1, 4)$. Now

$$\hat{\varphi}(0, 4) = \tfrac{17}{3} - 0 = \tfrac{17}{3}$$

while

$$\hat{\varphi}(1, 4) = \frac{14}{3} - \frac{1}{3}\left(\frac{\pi}{4}\right)$$

since there is a 45° turn on the best curve leading to $(1, 4)$; hence we pick $(0, 4)$ as the point to investigate next. Its neighbor $(0, 3)$ has $\hat{\varphi}(0, 3) = 5$, and this in turn has neighbor $(0, 2)$ with $\hat{\varphi}(0, 2) = 21/5$. Since $\hat{\varphi}(0, 2) < \hat{\varphi}(1, 4)$, the tracking process now backs up to $(1, 4)$ and examines its neighbors, thus discovering the stronger branch of the curve.

The $\hat{\varphi}$ evaluation function just described is only one of many devices that could be used to control the order in which the tracking process examines points. In general, the tracking problem can be formulated as one of searching for an optimum curve between the top and bottom rows of the picture (say), where optimality might be defined in terms of maximizing a function such as φ. In principle, mathematical programming techniques can be used to find optimum curves, although this approach tends to be computationally costly. Alternatively, heuristic search techniques can be used; if it is properly defined, such a technique is guaranteed eventually to find an optimum path. Ideally, a heuristic technique would not only evaluate a given point on the basis of how it can be reached from the starting point, but would also attempt to look ahead and estimate how the goal (a curve that crosses the picture) can be reached from the given point.

Search methods are an improvement over classical tracking techniques, since they allow look-ahead and backup, rather than making irrevocable decisions at each step. These methods are, however, still not as flexible as one might wish. The evaluation functions used in heuristic search are static; they are not normally modified in the course of the search. In the example previously given, the value of $\hat{\phi}$ at (h, k) depends on the values of paths from the starting point to (h, k), but not on anything that may have been discovered elsewhere in the picture. Thus there is no way of allowing this tracking procedure to learn from its mistakes, i.e., to modify itself when failures occur, in a manner that depends on the nature of these failures.

A more flexible approach to tracking would allow its constituent procedures to modify themselves in accordance with information gained during the course of the search. In such an approach, it is also easier to incorporate any available information about the class of pictures that are being analyzed, e.g., as regards the shapes that the curves are expected to have, and relationships that are expected to hold among them.

8.5 BIBLIOGRAPHICAL NOTES

Segmentation of a picture by thresholding its gray levels is a very old technique; it has been used extensively for character recognition, among many other applications. Edge detection by thresholding the output of an edge-sensitive operation and shape or pattern detection by thresholding the output of a matching operation are also very old ideas; references on edge detection and on matching will be given below. For early examples of textured region extraction by locally averaging and then thresholding, see Narasimhan and Fornango [29] and Rosenfeld et al. [36].

The p-tile method of threshold selection was suggested by Doyle [10]; on the more general idea of choosing a threshold so that the resulting f_t has some desired property, see Bartz [3]. The mode method is given by Prewitt and Mendelsohn [33]. On its application to texture discrimination, based on modes of local property values, see Tomita et al. [42] and Tsuji and Tomita [43]. Segmentation based on clusters of cooccurring gray levels is treated by Rosenfeld et al. [37].

Minimum-error thresholding of a mixture of two normal gray level probability densities is analyzed by Chow and Kaneko [7], who apply the method to finding local thresholds which can then be interpolated to other parts of the picture. The use of two thresholds, with points that satisfy the less stringent criterion accepted only if they lie near other accepted points, is described in Narasimhan and Fornango [29]; see also reference [31] in Section 6.6. On the use of significant local maxima for detecting edges, lines, spots, etc., see Wolfe [46], Kasvand [20], Rosenfeld and Thurston [39], and Rosenfeld *et al.* [40].

The use of derivative operators, particularly the gradient, for edge detection dates back to Kovasznay and Joseph (see references [15, 16] in Section 6.6). Many authors have proposed digital approximations to these operators; Roberts [34] is one example. Edge detection by fitting a function to the local gray levels is discussed by Prewitt [32]. The approach based on best-fitting ideal step edges is described by Hueckel [18, 19].

On the probability densities of gradient and Laplacian values in noisy pictures see Cofer and Tou [8]. Minimum-error line detection, based on measurements made at sets of points, is treated in Griffith [13]. On minimum-error edge location see Beckenbach and Desilets [4]; the related problem of detecting run ends in the presence of noise is treated by Kubba [22].

Detection of abrupt changes in average gray level, or average local property value, is discussed by Rosenfeld [35], Rosenfeld *et al.* [38], Rosenfeld and Thurston [39], and Rosenfeld *et al.* [40]. On the use of weighted averages see Macleod [23]. The "planning" approach is due to Kelly [21]. Averages of the second derivative were used by Herskovits and Binford. Fisher-like distances, and "distances" between histograms, are discussed by Holmes *et al.* [16], Muerle and Allen [27], and Muerle [26].

Picture-matching techniques have a long history; as indicated at the beginning of Section 8.3, they were used extensively in early work on optical character recognition, as well as on automatic target recognition, automatic change detection, map-matching navigation, and automatic stereophotogrammetry. For the minimum-error motivation of cross correlation see Harris [15]; for an early reference on the two-valued case see Horowitz [17]. On matched filtering see, e.g., Turin, [44]; on the importance of derivative filtering in the case of a correlated scene see Arcese *et al.* [1].

Line and curve detection, as well as spot detection, line end detection, etc., are discussed by Rosenfeld *et al.* [38], Rosenfeld and Thurston [39], and Rosenfeld *et al.* [40]. "Rubber masks" are discussed by Widrow [45]; the spring-loaded subtemplate model is described in Fischler and Elschlager [11]. On methods of finding good match positions, see Barnea and Silverman [2] and Nagel and Rosenfeld [28].

Raster tracking is an old technique that is used extensively in bubble chamber picture processing, among other applications. Two-dimensional edge and curve following was developed nearly 20 years ago in conjunction with work on automatic target recognition; for some early published references, applying it to character recognition, see Greanias *et al.* [12] and Bradshaw [5]. On raster tracking of regions see Grimsdale *et al.* [14].

Region growing is described by Muerle and Allen [27]. A mathematical discussion of partition building is given by Pavlidis [31]; some good examples of scene analysis by partition building are Brice and Fennema [6] and Yaki-movsky and Feldman [47]. A mathematical programming approach to curve tracking was used by Montanari [25]; a heuristic search approach is taken by Martelli [24]. A good recent example of knowledge-based line finding is Shirai [41].

REFERENCES

1. A. Arcese, P. H. Mengert, and E. W. Trombini, Image detection through bipolar correlation, *IEEE Trans. Informat. Theory* **IT-16**, 1970, 534–541.
2. D. I. Barnea and H. F. Silverman, A class of algorithms for fast digital image registration, *IEEE Trans. Comput.* **C-21**, 1972, 179–186.
3. M. R. Bartz, Optimizing a video preprocessor for OCR, *Proc. Int. Joint Conf. Artificial Intelligence, 1st* 1969, 79–90.
4. E. S. Beckenbach and D. T. Desilets, The computerization of high speed cineangio-cardiographic left ventricular volume determination, *in* "Pattern Recognition Studies" (H. L. Kasnitz and G. C. Cheng, eds.), pp. 173–192. S.P.I.E., Redondo Beach, California, 1969.
5. J. A. Bradshaw, Letter recognition using a captive scan, *IEEE Trans. Electron. Comput.* **EC-12**, 1963, 26.
6. C. R. Brice and C. L. Fennema, Scene analysis using regions, *Artificial Intelligence* **1**, 1970, 205–226.
7. C. K. Chow and T. Kaneko, Automatic boundary detection of the left ventricle from cineangiograms, *Comput. Biomed. Res.* **5**, 1972, 388–410.
8. R. H. Cofer, Jr. and J. T. Tou, Some approaches toward scene extraction, *in* "Pattern Recognition Studies" (H. L. Kasnitz and G. C. Cheng, eds.), pp. 63–79. S.P.I.E., Redondo Beach, California, 1969.
9. C. M. Cook and A. Rosenfeld, Size detectors, *Proc. IEEE* **58**, 1970, 1956–1957.
10. W. Doyle, Operations useful for similarity-invariant pattern recognition, *J. ACM* **9**, 1962, 259–267.
11. M. A. Fischler and R. A. Elschlager, The representation and matching of pictorial structures, *IEEE Trans. Comput.* **C-22**, 1973, 67–92.
12. E. C. Greanias, P. F. Meagher, R. J. Norman, and P. Essinger, The recognition of handwritten numerals by contour analysis, *IBM J. Res. Develop.* **7**, 1963, 14–21.
13. A. K. Griffith, Mathematical models for automatic line detection, *J. ACM* **20**, 1973, 62–80.
14. R. L. Grimsdale, F. H. Sumner, C. J. Tunis, and T. Kilburn, A system for the automatic recognition of patterns, *Proc. IEE* **106B**, 1959, 210–221.
15. J. L. Harris, Resolving power and decision theory, *J. Opt. Soc. Amer.* **54**, 1964, 606–611.
16. W. S. Holmes, H. R. Leland, and G. E. Richmond, Design of a photo interpretation automaton, *Proc. Eastern Joint Comput. Conf.* 1962, 27–35.
17. M. Horowitz, Efficient use of a picture correlator, *J. Opt. Soc. Amer.* **47**, 1957, 327.
18. M. H. Hueckel, An operator which locates edges in digital pictures, *J. ACM* **18**, 1971, 113–125.

19. M. H. Hueckel, A local visual operator which recognizes edges and lines, *J. ACM* **20**, 1973, 634–647.

20. T. Kasvand, Some thoughts and experiments on pattern recognition, *in* "Automatic Interpretation and Classification of Images" (A. Grasselli, ed.), pp. 391–398. Academic Press, New York, 1969.

21. M. D. Kelly, Edge detection in pictures by computer using planning, *in* "Machine Intelligence" (B. Meltzer and D. Michie, eds.), Vol. 6, pp. 379–409. University Press, Edinburgh, 1971.

22. M. H. Kubba, Automatic picture detail detection in the presence of random noise, *Proc. IEEE* **51**, 1963, 1518–1523 (See also the comments by A. Sekey, *ibid.* **53**, 1965, 75–76).

23. I. D. G. Macleod, On finding structure in pictures, *in* "Picture Language Machines" (S. Kaneff, ed.), pp. 231–256. Academic Press, London, 1970.

24. A. Martelli, Edge detection using heuristic search methods, *Comput. Graph. Image Proc.* **1**, 1972, 169–182.

25. U. Montanari, On the optimum detection of curves in noisy pictures, *Comm. ACM* **14**, 1971, 335–345.

26. J. L. Muerle, Some thoughts on texture discrimination by computer, *in* "Picture Processing and Psychopictorics" (B. S. Lipkin and A. Rosenfeld, eds.), pp. 371–379. Academic Press, New York, 1970.

27. J. L. Muerle and D. C. Allen, Experimental evaluation of techniques for automatic segmentation of objects in a complex scene, *in* "Pictorial Pattern Recognition" (G. C. Cheng *et al.*, eds.), pp. 3–13. Thompson, Washington, D. C., 1968.

28. R. N. Nagel and A. Rosenfeld, Ordered search techniques in template matching, *Proc. IEEE* **60**, 1972, 242–244.

29. R. Narasimhan and J. P. Fornango, Some further experiments in the parallel processing of pictures, *IEEE Trans. Electron. Comput.* **EC-12**, 1963, 748–750.

30. T. Pavlidis, Piecewise approximation of functions of two variables through regions with variable boundaries, *Proc. ACM Nat. Conf.* 1972, 652–662.

31. T. Pavlidis, Segmentation of pictures and maps through functional approximation, *Comput. Graph. Image Proc.* **1**, 1972, 360–372.

32. J. M. S. Prewitt, Object enhancement and extraction, *in* "Picture Processing and Psychopictorics" (B. S. Lipkin and A. Rosenfeld, eds.), pp. 75–149. Academic Press, New York, 1970.

33. J. M. S. Prewitt and M. L. Mendelsohn, The analysis of cell images, *Ann. N. Y. Acad. Sci.* **128**, 1966, 1035–1053.

34. L. G. Roberts, Machine perception of three-dimensional solids, *in* "Optical and Electrooptical Information Processing" (J. T. Tippett *et al.*, eds.), pp. 159–197. M I T Press, Cambridge, Massachusetts, 1965.

35. A. Rosenfeld, A nonlinear edge detection technique, *Proc. IEEE* **58**, 1970, 814–816 (See also the discussion *ibid.* **59**, 1971, 285–287; **60**, 1972, 344).

36. A. Rosenfeld, C. Fried, and J. N. Orton, Automatic cloud interpretation, *Photogrammetr. Eng.* **31**, 1965, 991–1002.

37. A. Rosenfeld, H. K. Huang, and V. B. Schneider, An application of cluster detection to text and picture processing, *IEEE Trans. Informat. Theory* **IT-15**, 1969, 672–681.

38. A. Rosenfeld, Y. H. Lee, and R. B. Thomas, Edge and curve detection for texture discrimination, *in* "Picture Processing and Psychopictorics" (B. S. Lipkin and A. Rosenfeld, eds.), pp. 381–393. Academic Press, New York, 1970.

39. A. Rosenfeld and M. Thurston, Edge and curve detection for visual scene analysis, *IEEE Trans. Comput.* **C-20**, 1971, 562–569.

40. A. Rosenfeld, M. Thurston, and Y. H. Lee, Edge and curve detection: further experiments, *IEEE Trans. Comput.* **C-21**, 1972, 677–715.
41. Y. Shirai, A context sensitive line finder for recognition of polyhedra, *Artificial Intelligence* **4**, 1973, 95–119.
42. F. Tomita, M. Yachida, and S. Tsuji, Detection of homogeneous regions by structural analysis, *Proc. Int. Joint Conf. Artificial Intelligence, 3rd* 1973, 564–571.
43. S. Tsuji and F. Tomita, A structural analyzer for a class of textures, *Comput. Graph. Image Proc.* **2**, 1973, 216–231.
44. G. L. Turin, An introduction to matched filters, *IRE Trans. Informat. Theory* **IT-6**, 1960, 311–329.
45. B. Widrow, The "rubber-mask" technique, *Pattern Recognit.* **5**, 1973, 175–211.
46. R. N. Wolfe, A dynamic thresholding technique for quantization of scanned images, *Proc. Symp. Automatic Pattern Recognit.* 1969, 142–162.
47. Y. Yakimovsky and J. A. Feldman, A semantics-based decision theory region analyzer, *Proc. Int. Joint Conf. Artificial Intelligence, 3rd* 1973, 580–588.
48. L. A. Zadeh, Fuzzy sets, *Informat. Contr.* **8**, 1965, 338–353.

Chapter 9

Geometry

In Chapter 8 we described a variety of methods for segmenting a picture into parts, e.g., into object(s) and background; in general, these parts can be arbitrary subsets of the picture. In this chapter we discuss geometrical properties of picture subsets; these are properties that depend only on which points of the picture belong to the given subset, and not on the gray levels of these points.[†] Examples of such properties are connectedness, area, elongatedness, convexity, and straightness. We also treat mathematical relationships between picture subsets—containment, adjacency, distance, etc. In addition, we discuss how to use these concepts to derive new subsets from given ones, e.g., intersections, borders, "skeletons," hulls, etc.

The properties discussed in this chapter are, in general, very sensitive to noise. For example, the number of connected components of a given picture subset S increases drastically if S is noisy, since every isolated point of S is a component; the perimeter of S may increase greatly if S has a ragged border; the skeleton of S is very sensitive to the presence of small holes in S; and so on. Thus before we measure such properties, we should try to be sure that we have extracted S from the picture correctly. We can sometimes suppress undesirable types of noise from S by applying noise-cleaning operations to it, e.g., shrinking S slightly and reexpanding the result should remove small

---

[†] Picture properties that do depend on the gray levels, such as moments, textural properties, etc., will be discussed in Chapter 10.

isolated parts of S, and should smooth out roughness in S's border (see Sections 6.5.1 and 9.2.4). One way of assessing the noisiness of S might be to see how sensitive its property values are to small amounts of such noise cleaning.

Let S be any subset of a picture. As mentioned at the beginning of Chapter 8, we will often find it convenient to represent S by an "overlay"—a two-valued picture having value 1 at the points of S and 0 elsewhere. We shall denote this overlay by φ_S. It corresponds to the mathematical concept of the "characteristic function" of the set S.

In the following paragraphs, we review some elementary concepts from set theory.

The *complement* of S, denoted by $\bar{S}$, is the set of points that are not in S. Evidently, we have $\varphi_{\bar{S}} = 1 - \varphi_S$; in terms of Boolean algebra, this is the logical negation of φ_S. More generally, for any picture subset T, the *difference* between T and S, denoted by $T - S$, is the set consisting of points that are in T but not in S.

For any picture subsets S and T, the *intersection* $S \cap T$ is the set of points that are in both S and T. Similarly, the *union* $S \cup T$ is the set of points that are in either S or T (or both). Readily, $\varphi_{S \cap T}$ is the logical AND of φ_S and φ_T, while $\varphi_{S \cup T}$ is their logical OR. These concepts can be generalized to unions and intersections of arbitrary collections of picture subsets.

Exercise 1. Prove that for all R, S, T, we have

(a) $\overline{S \cap T} = \bar{S} \cup \bar{T}$; $\overline{S \cup T} = \bar{S} \cap \bar{T}$; $T - S = T \cap \bar{S}$;

(b) $R \cup (S \cap T) = (R \cup S) \cap (R \cup T)$;
     $R \cap (S \cup T) = (R \cap S) \cup (R \cap T)$;

(c) $R \cap (S - T) = (R \cap S) - (R \cap T)$. ∎

Exercise 2. Define the *symmetric difference* $S \triangle T$ of S and T to be $(S - T) \cup (T - S)$. Show that $S \cup T = (S \cap T) \cup (S \triangle T)$. Prove also that $\varphi_{S \triangle T}$ is the logical "exclusive or" of φ_S and φ_T; in other words, $S \triangle T$ is the set of points that are in either S or T, but not both. Show also that $\varphi_{S \triangle T} = \varphi_S + \varphi_T$ (modulo 2), while $\varphi_{S \cap T} = \varphi_S \varphi_T$. ∎

We say that S *contains* T, or T is contained in S, if every point of T is a point of S; that S *meets* T, if some point of T is a point of S; and that S is *disjoint* from T, if no point of T is a point of S. For example, let s and t be thresholds, with $s \leqslant t$; if S is the set of points whose gray levels $\geqslant s$, and T is the set whose levels $\geqslant t$, then S contains T.

Exercise 3. Prove that:

(a) If S contains T, then $\bar{T}$ contains $\bar{S}$, and conversely.

(b) If S contains T, then $\bar{S}$ is disjoint from T, and conversely. ∎

The set of picture subsets $\{S_1, ..., S_k\}$ is a *partition* of the picture if the union of the S_i's is the whole picture, and any two of the S_i's are disjoint. For example, if the possible gray levels are $z_1, ..., z_k$, and S_i is the set of picture points that have gray level z_i, $1 \leqslant i \leqslant k$, then $\{S_1, ..., S_k\}$ is a partition. If S is any picture subset, $\{S, \bar{S}\}$ is a partition.

9.1 ADJACENCY AND CONNECTEDNESS

In this section we discuss *topological* properties and relationships of digital picture subsets, based on the notions of adjacency and connectedness.[†] As we shall see, these notions are somewhat complex because, in a square grid array, there are two different types of adjacency (horizontal/vertical and diagonal), and it turns out that both of them must be taken into account when we study connectedness. Matters are somewhat simpler in a hexagonal array; see Exercise 4.

Section 9.1.1 defines the basic concepts of adjacency, borders, connectedness, components, and surroundedness. Section 9.1.2 describes border-following algorithms, and discusses their extension to edge and curve following. Section 9.1.3 deals with methods of labeling and counting components.

9.1.1 Connectedness and Borders

Let (i, j) be a point of the given picture. Then (i, j) has four[‡] horizontal and vertical neighbors, namely the points

$$(i-1, j), \quad (i, j-1), \quad (i, j+1), \quad (i+1, j)$$

These points are called the *4-neighbors* of (i, j), and are said to be *4-adjacent* to (i, j). In addition, (i, j) has four diagonal neighbors, namely

$$(i-1, j-1), \quad (i-1, j+1), \quad (i+1, j-1), \quad (i+1, j+1)$$

Both these and the 4-neighbors are called *8-neighbors* of (i, j). More generally, if S and T are picture subsets, we say that S is 4- or 8-adjacent to T if some point of S is 4- or 8-adjacent to some point of T.

[†] There is an unfortunate tendency in the picture processing literature to use the term "topological" for all types of geometrical properties. Properly speaking, a property or relationship is topological only if it is preserved when an arbitrary "rubber-sheet" distortion is applied to the picture. Adjacency and connectedness are topological; area, elongatedness, convexity, straightness, etc. are not.

[‡] If (i, j) is on the border of the picture, some of these neighbors will be outside the picture.

Exercise 4.   Define the *6-neighbors* of (i, j) as the 4-neighbors together with

$$(i-1, j-1) \quad \text{and} \quad (i-1, j+1) \qquad \text{if } i \text{ is odd}$$

$$(i+1, j-1) \quad \text{and} \quad (i+1, j+1) \qquad \text{if } i \text{ is even}$$

These are just the neighbors of (i, j) in a "hexagonal" array constructed from a square array by shifting the odd-numbered rows half a unit to the right; see Section 1.2.2. Develop hexagonal versions of all the results in this section. ∎

A *path* from (i, j) to (h, k) is a sequence of distinct points

$$(i, j) = (i_0, j_0), (i_1, j_1), \ldots, (i_n, j_n) = (h, k)$$

such that (i_m, j_m) is adjacent to (i_{m-1}, j_{m-1}), $1 \leqslant m \leqslant n$. (*Note that there are two versions of this and the following definitions, depending on whether "adjacent" means "4-adjacent" or "8-adjacent".*) For example, if we denote the points of the path by 1's, then $\begin{smallmatrix}1&1&1\\&&1\end{smallmatrix}$ is a 4-path (if its points are suitably ordered), while $\begin{smallmatrix}&1&1\\1\end{smallmatrix}$ is an 8-path but not a 4-path. Here n is called the *length* of the path.

If $p = (i, j)$ and $q = (h, k)$ are points of a picture subset S, we say that p is *connected* to q (in S) if there is a path from p to q consisting entirely of point of S. For any point (i, j) of S, the set of points of S that are connected to (i, j) is called a connected *component* of S. If S has only one component, it is called "connected." For example, $\begin{smallmatrix}1&\\&1\end{smallmatrix}$ is 8-connected but not 4-connected.

Exercise 5.   Prove that "is connected to" is an equivalence relation, i.e., that for all points u, v, w in S:

(1) u is connected to u;
(2)   if u is connected to v, then v is connected to u;
(3)   if u is connected to v, and v is connected·to w, then u is connected to w.

Corollary 1 to Exercise 5.   Any two points of a component are connected to each other.

Corollary 2 to Exercise 5.   Distinct components are disjoint. ∎

Exercise 6.   Prove that every 8-component is a union of 4-components. ∎

Exercise 7.   Prove that if S contains T, every component of T is contained in a unique component of S. As an application of this, suppose that we are given a set of picture subsets $S_1, \ldots, S_k$ such that S_i contains S_{i+1}, $1 \leqslant i < k$, e.g., S_i might be the set of points having gray level at least z_i, where $z_1, \ldots, z_k$ are the allowable gray levels. Let $\mathscr{G}$ be the graph whose nodes are the components of the S_i's, and where nodes A and B are joined by an arc if, for some j, A and B are components of S_j and S_{j+1}, respectively, and A contains B. Prove that $\mathscr{G}$ is a tree. ∎

Let $\bar{S}$ be the complement of S. Since points outside the picture are not in S, it follows that all points of $\bar{S}$ that are connected (in $\bar{S}$) to points on the border of the picture belong to the same component of $\bar{S}$. We call this component the *background* of S. All other components of $\bar{S}$ (if any) are called *holes* in S. If S is connected and has no holes, it is called *simply connected*.

In giving examples of these last concepts we have to decide whether to use 4- or 8-connectedness for $\bar{S}$ as well as for S. For example, consider the S whose points are

$$
\begin{array}{ccc}
& 1 & \\
1 & & 1 \\
& 1 &
\end{array}
$$

This has either one or four components, depending on whether we use 8- or 4-connectedness for S; and it has one or no holes, depending on whether we use 4- or 8-connectedness for $\bar{S}$.

It turns out that there are many advantages in using opposite types of connectedness for S and for $\bar{S}$: If we use the 4-definitions for S, then we use the 8-definitions for $\bar{S}$, and vice versa. (Intuitively: in the pattern

$$
\begin{array}{cc}
1 & 0 \\
0 & 1
\end{array}
$$

if the diagonally adjacent 1's are connected, then the diagonally adjacent 0's cannot be, and vice versa.) This makes it easier to define some of the algorithms that are developed later in this section; e.g., see Exercises 9, 10, and 14.

Let S and T be disjoint picture subsets. We say that T *surrounds* S, or that S is *inside* T, if any path from any point of S to the border of the picture must meet T. (There are two definitions here, depending on whether we mean 4-path or 8-path). For any S, the background component B of $\bar{S}$ evidently surrounds S. Any other component D of $\bar{S}$ is a hole in S, and is surrounded by S (if not, there would be a path in $\bar{S}$ from D to the picture border, so that D would be the same as B).

Exercise 8. Prove that "surrounds" is a strict partial order relation, i.e., that for R, S, T all disjoint, we have:

(a) If T surrounds S, then S does not surround T;
(b) If T surrounds S, and S surrounds R, then T surrounds R. ∎

Exercise 9. It can be shown that if S is any picture subset, C is a component of S, and D is a component of $\bar{S}$ that is adjacent to C, then either C surrounds D or D surrounds C. This is not true if we use 4-connectedness for both S and $\bar{S}$. For example, if S is

1

1   1

1

the central point is a 4-component of $\bar{S}$, and any one of the 1's is a 4-component of S, but neither of these components surrounds the other. It is certainly not true for other types of partitions of a picture; for example, connected components of constant gray level, in a picture that has more than two gray levels, can be adjacent without surrounding one another, e.g.,

0 1

0 2

Prove (using Exercise 8a) that exactly one such D surrounds C.   ∎

Exercise 10.   Given any partition $\{S_1, ..., S_k\}$ of a picture, we can define its *adjacency graph* as the graph whose nodes are the S_i's, and in which nodes S_i and S_j are joined by an arc if S_i is adjacent to S_j. Let S be any picture subset, and let $C_1, ..., C_k$ be the components of S and of $\bar{S}$ (where we use opposite kinds of connectedness for S and for $\bar{S}$; it does not matter what kind of adjacency we use, since if a component of S and a component of $\bar{S}$ are 8-adjacent, they are also 4-adjacent). Prove that $\{C_1, ..., C_k\}$ is a partition of the picture, and (using Exercise 9) that its adjacency graph is a tree.   ∎

Given S and a point p not in S, one often wants to know whether or not S surrounds p. The simplest way to determine this, for an arbitrary S, is to mark all the points of B (see Section 9.1.3 on component labeling). Now, given p, we know immediately that it is inside S if and only if it is unmarked. Many other methods, which will not be described here, have been developed to solve this problem in special cases, e.g., when S is a polygon. If S is a thin closed curve, we can move (say) to the right from p, and count the number of times we cross S (but not when we merely touch S without crossing it) until we reach the picture border; if this number is odd, p is inside S, and if even, outside.

By the *border* of a picture subset S we mean the set of points of S that have neighbors in $\bar{S}$. (As before, there are two versions of this definition, one using 4-neighbors, the other using 8-neighbors.) We shall denote the border of S by S'. The difference set $S - S'$ is called the *interior* of S. By our convention that points outside a picture are not in S, any point of S on the border of the picture is a border point of S.

Exercise 11.   Prove that $(S')' = S'$.   ∎

Given the overlay φ_S representing S, it is straightforward to construct $\varphi_{S'}$; we simply scan φ_S systematically, and whenever a 1 is found, examine its neighbors. If any neighbor is a 0, we change the 1 to a 2. When the scan is complete, points of S' have all becomes 2's. If φ_S contains more 1's than 0's, it is more economical to examine the neighbors of 0's rather than the neighbors of 1's; in this case, if any neighbor is a 1, we change *it* to a 2.

On a parallel array-processing computer φ_S can be constructed from φ_S very efficiently as follows:

(1)   Construct $\varphi_{\bar{S}}$, the logical negation of φ_S.

(2)   Compute the logical OR of $\varphi_{\bar{S}}(i-\alpha, j-\beta)$, i.e., $\varphi_{\bar{S}}$ shifted by (α, β), for shifts corresponding to the positions of a point's neighbors (i.e., $(0,1)$, $(0,-1)$, $(1,0)$, $(-1,0)$, in the 4-neighbor case; and also $(\pm 1, \pm 1)$, in the 8-neighbor case). Call the result $\varphi_{\bar{S}}^{\cup}$.

(3)   The logical AND of φ_S and $\varphi_{\bar{S}}^{\cup}$ is the desired $\varphi_{S'}$.

As a simple example, let φ_S be

$$
\begin{array}{ccccc}
0 & 0 & 0 & 0 & 0 \\
0 & 1 & 1 & 1 & 0 \\
0 & 1 & 1 & 1 & 0 \\
0 & 1 & 1 & 1 & 0 \\
0 & 0 & 0 & 0 & 0
\end{array}
$$

Thus $\varphi_{\bar{S}}$ is

$$
\begin{array}{ccccc}
1 & 1 & 1 & 1 & 1 \\
1 & 0 & 0 & 0 & 1 \\
1 & 0 & 0 & 0 & 1 \\
1 & 0 & 0 & 0 & 1 \\
1 & 1 & 1 & 1 & 1
\end{array}
$$

The four shifted versions of $\varphi_{\bar{S}}$ are

$$
\begin{array}{ccccc|ccccc|ccccc}
1 & 0 & 0 & 0 & 1 & 0 & 0 & 0 & 0 & 0 & 0 & 1 & 1 & 1 & 1 \\
1 & 0 & 0 & 0 & 1 & 1 & 1 & 1 & 1 & 1 & 0 & 1 & 0 & 0 & 0 \\
1 & 0 & 0 & 0 & 1 & 1 & 0 & 0 & 0 & 1 & 0 & 1 & 0 & 0 & 0 \\
1 & 1 & 1 & 1 & 1 & 1 & 0 & 0 & 0 & 1 & 0 & 1 & 0 & 0 & 0 \\
0 & 0 & 0 & 0 & 0 & 1 & 0 & 0 & 0 & 1 & 0 & 1 & 1 & 1 & 1
\end{array}
$$

and

$$
\begin{array}{ccccc}
1 & 1 & 1 & 1 & 0 \\
0 & 0 & 0 & 1 & 0 \\
0 & 0 & 0 & 1 & 0 \\
0 & 0 & 0 & 1 & 0 \\
1 & 1 & 1 & 1 & 0 \\
\end{array}
$$

so that their OR $\varphi_S{}^\cup$ is

$$
\begin{array}{ccccc}
1 & 1 & 1 & 1 & 1 \\
1 & 1 & 1 & 1 & 1 \\
1 & 1 & 0 & 1 & 1 \\
1 & 1 & 1 & 1 & 1 \\
1 & 1 & 1 & 1 & 1 \\
\end{array}
$$

The AND of this with φ_S is thus

$$
\begin{array}{ccccc}
0 & 0 & 0 & 0 & 0 \\
0 & 1 & 1 & 1 & 0 \\
0 & 1 & 0 & 1 & 0 \\
0 & 1 & 1 & 1 & 0 \\
0 & 0 & 0 & 0 & 0 \\
\end{array}
$$

which is just $\varphi_{S'}$.

Exercise 12. We call the point (i, j) of S *isolated* if it has no neighbors in S. Describe sequential and parallel algorithms for finding the ioslated points of a given S. ∎

It is not possible, in general, to construct S when only S' is given. For example, if S' is

$$
\begin{array}{ccc}
1 & 1 & 1 \\
1 &   & 1 \\
1 & 1 & 1 \\
\end{array}
$$

we have no way of telling whether S is

$$
\begin{array}{ccccccc}
1 & 1 & 1 &   & 1 & 1 & 1 \\
1 &   & 1 & \text{or} & 1 & 1 & 1 \\
1 & 1 & 1 &   & 1 & 1 & 1 \\
\end{array}
$$

If, however, we know which neighbors of the points of S' are not in S, it becomes easy to "fill in" S; see Section 9.1.2.

In the next section we describe how to "follow" the border(s) of a given S. This makes it easy to construct chain code representations of the borders, if desired. It can also be less costly computationally than the simple scanning method of border finding, if there are very few border points.

9.1.2   Border Following

Let C be a component of S and D a component of $\bar{S}$. By the D-border of C we mean the set of points of C that are adjacent to points of D. There are several possible versions of this definition, depending on whether C and D are 4- or 8-components, and on whether we use 4- or 8-adjacency. In what follows we shall assume that C is an 8-component and D a 4-component, and shall use 4-adjacency; see Exercise 14 regarding other cases.

As a simple example, if C is

$$
\begin{array}{cccc}
1 & 1 & 1 & 1 \\
1 & & 1 & 1 \\
1 & 1 & 1 & 1
\end{array}
$$

and D is the background of C, then the D-border of C is

$$
\begin{array}{cccc}
1 & 1 & 1 & 1 \\
1 & & & 1 \\
1 & 1 & 1 & 1
\end{array}
$$

while if D is the hole in C, the D-border is

$$
\begin{array}{ccc}
& 1 & \\
1 & & 1 \\
& 1 &
\end{array}
$$

(or

$$
\begin{array}{ccc}
1 & 1 & 1 \\
1 & & 1 \\
1 & 1 & 1
\end{array}
$$

if we use 8-adjacency). Note that two different D-borders can have points in common; they can even be identical, e.g., if C is

$$
\begin{array}{ccc}
1 & 1 & 1 \\
1 & & 1 \\
1 & 1 & 1
\end{array}
$$

Exercise 13. On how many different D-borders can a given point of C be? ▌

We now describe an algorithm for finding all the points of the D-border of C, given an initial pair of adjacent points c, d with c in C, d in D. We first check that C is not just an isolated point; if C has just the one point c, this point is C's D-border, and there is nothing to do. Otherwise, our algorithm, which we shall call BF, proceeds as follows:

(1)   Change the values of c and d to 3 and 2, respectively.

(2)   Let the 8-neighbors of c in (say) clockwise order, starting with d and ending with the first occurrence of 1, 3, or 4, be $e_1, ..., e_k$.

   (a)   If $c = 3$, $e_k = 4$, and $e_h = 2$ for some $h < k$, change the 3 to 4, the 2 to 0, and stop.

   (b)   Otherwise, change the value of c to 4 (if it was 1). Take e_k as the new c and e_{k-1} as the new d, and return to step (2).

When BF stops, the 4's are exactly the points of the D-border of C.

The following is an example of the operation of BF. Let C be

$$
\begin{array}{cc}
\overset{*}{\underline{1}} & 1 \\
1 & 1 \\
 & 1
\end{array}
$$

where the initial c is underlined, and the position of the initial d is starred.

In Table 1 we show the situation at successive entries into step (2) of BF, with the current c underlined; e_1 starred; $e_2, ..., e_{k-1}$ marked $u, v, w, ...$; and e_k primed. At the last entry we have $c = 3$, $e_3 = 2$, $e_5 = 4$, so that we can change the 3 to 4, the 2 to 0, and stop.

As another example, consider the pattern

$$
\begin{array}{cc}
 & 1 \\
* & \underline{1} \\
 & 1
\end{array}
$$

Here the successive stages are shown in Table 2. BF does not stop at the third entry, even though $c = 3$, since the 2 is not one of the e's. It does not

9.1 Adjacency and Connectedness   343

Entry into step (2)	Pattern			Entry into step (2)	Pattern		
1	2*	u		5	2		
	3	1'			3	4	
	1	1			1'	4	
		1			u	*	4
2	2	*	u	6	2		
	3	4	v		w 3'	4	
	1	1'	w		v 4	4	
		1			u *	4	
3	2			7	u 2	w	
	3	4			* 3	4'	
	1	4	*		4	4	
		1'					4
4	2						
	3	4					
	1 4'	*	u				
	z 4	v					
	y x	w					

stop at the fourth entry, even though $e_6 = 2$, since $c \neq 3$. It does, however, stop at the fifth entry, since $c = 3$, $e_1 = 2$.

Exercise 14. Define a modification of algorithm BF for the case where C is 4-connected, D is 8-connected, and we still use 4-adjacency. (*Hint*: At step (2), let e_k be the first 4-neighbor that has value 1, 3, or 4. If e_{k-1} does not have one of these values, take e_k as the new c and e_{k-1} as the new d. If it does have one of these values, take e_{k-1} as the new c and e_{k-2} as the new d.) What happens if we use other combinations of connectedness and adjacency? ▌

Exercise 15. Define a modification of algorithm BF in which we start with two consecutive points of the D-border of C, and stop when we again find these two points consecutively in the same order. ▌

Exercise 16. Let C be a 4-connected 8-component (i.e., S is 4-connected, and no point of $S - C$ is 8-adjacent to C). Let b_1, b_2 be a pair of 4-adjacent

TABLE 2

Entry into step (2)	Pattern	Entry into step (2)	Pattern
1	$u\quad v\quad 1'$ $2^*\ \underline{3}$ 1	4	$\qquad 4$ $2\ \ 3'$ $x\ \ \underline{4}\ \ *$ $w\ \ v\ \ u$
2	$\quad u\quad v\quad w$ $\quad *\ \ \underline{4}\ \ x$ $2\ \ 3'\ z\ \ y$ 1	5	$u\quad v\quad 4$ $2^*\ \underline{3}$ 4
3	$\qquad 4$ $2\ \ \underline{3}\ \ *$ $1'\ u$		

points of C and D, respectively, and define b_k inductively as follows: If b_{k-1} is in S, turn left; if in $\bar S$, turn right. (More precisely: let θ_i be the direction from b_{i-1} to b_i. If b_{k-1} is in S, then $\theta_k = \theta_{k-1} + 90°$; if b_{k-1} is in $\bar S$, then $\theta_k = \theta_{k-1} - 90°$.) Then the 1's in the sequence $b_1, b_2, \ldots$ include all points of the D-border of C. Show that if C is not 4-connected, this is not true. (*Hint*: Let C be $_1{}^1$.) ∎

Exercise 17. Let (c_i, d_i) be a pair of adjacent points of C, D, respectively, where C is 4-connected, D is 8-connected, and we use 4-adjacency. Let e_i, f_i be the pair of points that form a 2×2 square with c_i, d_i and that are in front of us when c_i is on our left, with e_i adjacent to c_i and f_i to d_i. If $e_i = 0$, take $c_{i+1} = c_i$, $d_{i+1} = e_i$; if $e_i = 1$ and $f_i = 0$, take $c_{i+1} = e_i$, $d_{i+1} = f_i$; if $e_i = f_i = 1$, take $c_{i+1} = f_i$, $d_{i+1} = d_i$. This procedure follows the "cracks" between C and D along the D-border of C, by "keeping its left hand on the wall" of C. Define analogous procedures for other types of connectivity. ∎

Using the BF algorithm, we can define a detection and tracking scheme for finding all border points of a given picture subset S:

(1)   Scan the picture in a TV raster until a 1 or 2 is found immediately to the right of a 0, or at the left end of a row. This point (i, j) must be a border point of S on a border (say the D-border) that has not yet been followed.

(2)   Use BF to follow the D-border and mark it by

(a)   changing 1's or 2's to 3's if they have points of D as left hand neighbors;

(b)   changing 1's to 2's if they do not have points of D as left hand neighbors;

(c)   3's are left unchanged.

(3)   When BF has finished, we are back at (i, j), and can resume the scan. When the scan is finished, every border point of S has been marked 2 or 3.

If we did not use two distinct marks (2 and 3), we would have to check all neighbors of a 1 at step (1), rather than just their left-hand neighbors, or else we might miss some border points. For example, suppose S is

$$\begin{array}{ccc} 1 & 1 & 1 \\[4pt] 1 & \underline{1} & 1 \\[4pt] 1 & & 1 \\[4pt] 1 & 1 & 1 \end{array}$$

so that $\bar{S}$ consists of the background region B and a one-point hole H. We first hit S at its upper left corner; if we follow its B-border and mark it all with 3's, this marks all 1's except the underlined 1 (which is the only point of the H-border that is not also on the B-border), and since this 1 has no 0 on its left, it will never be found. On the other hand, if we use both 2's and 3's, then after following the B-border, we have

$$\begin{array}{ccc} 3 & 2 & 2 \\[4pt] 3 & 1 & 2 \\[4pt] 3 & & 2^* \\[4pt] 3 & 2 & 2 \end{array}$$

so that the scan will find the H-border when it reaches the starred 2. It is easily seen that if S has no holes, one mark is sufficient.

Successive border points (i.e., successive c's) found by the BF algorithm are always 8-neighbors. The sequence of these points can be represented by a chain code (see Section 1.2.3) consisting of the directions between successive points of the path (multiples of 45°), together with the coordinates of an initial c.

Conversely, suppose that we are given the chain code of the D-border of C, together with the coordinates of a point c_1 of this border and of a neighbor d_1 of c_1 that lies in D. Thus d_1 and c_2 are neighbors of c_1. Mark the neighbors of c_1 in clockwise order from d_1 to (but not including) c_2 as 0's, and mark c_1 as 1. Take the neighbor just preceding c_2 as d_2, and repeat the process with

c_2, d_2, c_3 replacing c_1, d_1, c_2. (On subsequent steps, some of the points may already be marked.) When all the c's defined by the chain code have been processed, the points marked 1 are just the D-border of C, and the points marked 0 are just the C-border of D.

If all the borders of a given S have been marked in this way, it becomes easy to "fill in" the rest of S. In fact, any horizontal run of unmarked points with 1's at its beginning and end must belong to S, and any point of S must belong to such a run; those runs can be turned into 1's in a single raster scan of the picture. On a parallel array processor, S can be filled in by "propagating" the 1's into the unmarked points, i.e., repeatedly turning unmarked neighbors of 1's into 1's; the number of propagation steps required is at most proportional to half the diameter of the picture.

Our algorithms for finding and following borders can be applied, with suitable modifications, to an unsegmented picture f, provided that it is easy to tell which points of f belong to the subset S of interest and which do not. For example, suppose that S consists of the points whose gray levels belong to a given set Z. Then we can find border points of S by looking for points of f whose gray levels are in Z, and which have neighbors whose gray levels are not in Z. Similarly, in the BF algorithm, we start with a pair of points (c, d) such that $f(c)$ is in Z and $f(d)$ is not, and we let e_k be the first 8-neighbor of c (in clockwise order, starting with d) whose gray level is in Z. Of course, the marks that we use in these algorithms (e.g., 2, 3 and 4, in BF) must be chosen so that they cannot be confused with gray level values.

It is much less trivial to devise algorithms for following edges, e.g., points where the gradient has a high value, on a picture f. One might, for example, try to follow an edge by moving in the direction perpendicular to the gradient at each point. It is difficult, however, to say how well such an approach would work, because edges, unlike borders of objects, need not form closed, connected curves.[†] It is better to use edge following algorithms that incorporate tracking capabilities, as discussed in Section 8.4.

Another nontrivial problem is to define algorithms for following arcs and

---

[†] If an edge is defined by the presence of a gray level difference of 1 (or greater) the edges do form connected, closed curves; indeed, the edges then surround the connected components of points having constant gray level, so that we have a special case of the situation considered in the preceding paragraph. On the other hand, if a gray level difference greater than 1 is needed in order for an edge to be present, then edges can consist of isolated arcs. As a trivial example, in the picture

$$
\begin{array}{cccc}
1 & 1 & 1 & 1 \\
1 & 2 & 0 & 1 \\
1 & 1 & 1 & 1
\end{array}
$$

there is an isolated edge segment between the 2 and the 0.

curves. (When BF follows a nonclosed curve, each point of the curve is visited twice; it would be useful to devise a curve-following algorithm that only had to visit each point once.) If the curve is thin, and never touches or crosses itself, so that every curve point (except the endpoints) has exactly two neighbors on the curve, it is easy to define such an algorithm; but in the real world, these conditions are rarely satisfied.

In the case of a thick curve, one possible approach is to follow just one side of the curve, and stop when the ends are reached (assuming we can recognize the ends); but this could give widely differing results depending on which side is followed, if the curve makes sharp bends. A more symmetrical approach is to follow both sides of a thick curve simultaneously, using two BF's; to keep one BF from getting too far ahead of the other, we can keep track of the distance between their current positions, and if this distance begins to exceed the thickness of the curve, we can stop one of them until the other one catches up. The average of the two BF positions is then a sort of "skeleton" of the curve. (See Section 9.4.5 on other ways of obtaining skeletons of elongated objects.)

For curves that can cross themselves (or cross other curves), one can design a curve-following algorithm that will take the straightest of the possible continuations whenever it comes to a crossing. For example, at each curve point, one can look first for a continuation in the direction exactly opposite the previous point; then in the directions $\pm 45°$ away from this; and so on. When curve crossings are "thick," this approach becomes complicated; it is better to give the curve-following algorithm tracking capabilities, as in Section 8.4.

9.1.3   Component Labeling and Counting

One often wants to treat each component of a given picture subset S as a separate object. In the first part of this section we discuss methods of assigning a distinct "label" to each component of a given S; in other words, for any component C, we want all points of C to have the same value, and no point not in C to have that value.

A simple procedure for labeling the components of S involves searching and "propagating." We scan S until a 1 is found, and change its value to that of the first label not yet used, say v. We then repeatedly (in parallel, if desired) "propagate" v's into 1's, i.e., we change 1's to v's if they have v's as neighbors. When no further changes are possible, all the 1's connected to the initial v have evidently become v's. We now resume the scan; if another 1 is found, it cannot belong to the v component, so that we can give it a new label and repeat the process. This procedure, while simple, may be the very time-consuming, since each "propagation" process, even if performed in parallel, may require a number of repetitions of the order of the picture area.

Another approach to component labeling is to use the border finding technique described in Section 9.1.2, but modified to mark outer borders (i.e., borders of components of S where they are adjacent to the components of $\bar{S}$ that surround them; see Exercise 9) distinctively, and to use a different label for the outer border of each component. Once this has been done, the labels of the outer borders can be "propagated" into their components, if desired; if performed in parallel, this requires a number of repetitions which can at most equal the picture radius. The procedure for labeling the outer borders (which will not be described here in detail; it is somewhat more complex than that in Section 9.1.2) is time consuming, however, since it involves a scan and many uses of BF, each requiring a number of steps that can grow with the picture area.

For most purposes, the following technique, which does labeling by tracking runs of 1's rather than borders, is the best choice. On the first row of the picture, we give each run a distinct label. On the second (and succeeding) rows, we examine the runs of 1's and compare their positions with the runs on the previous row. If the run ρ is adjacent to no runs on the previous row, we give ρ a new label. If ρ is adjacent to just one run on the previous row, we give ρ the label of that run. If ρ is adjacent to two or more runs on the previous row, we give ρ (say) the lowest-valued of their labels, but we also make a note of the fact that these labels all belong to the same component. When the picture has all been scanned in this way, we can sort out our notes and determine the classes of equivalent labels. If desired, we can then rescan the picture and replace each label by, for instance, the lowest valued equivalent label.

Many of the methods described in this section can be used to *count* the connected components of a given picture subset S. In particular, the component labeling methods can all be used for this purpose, since the number of (inequivalent) labels needed is just the number of components.

Exercise 18.   Generalize all of the preceding results to define algorithms for labeling and counting the connected components of constant gray level in a picture.   ∎

The topology-preserving shrinking operations described in Section 9.2.5 can also be used to count components, since each component eventually becomes an isolated 1 when these operations are applied, and it is easy to keep count of isolated 1's.

If the components of S are known to be all simply connected, most of the algorithms described in this section can be simplified. (One should, however, be wary of making this assumption in practice, since if a picture is noisy, subsets derived by segmenting it are also likely to be noisy.) More generally, there exist simple methods for computing the number of components of S minus the number of holes in S (provided we use opposite types of connectedness for S and $\bar{S}$). If all components are simply connected, this is the same as the number of components. This quantity is called the *genus*, or Euler number, of S.

The following is a very simple run-tracking method of computing the genus: On the first row, count the runs of 1's. On the second and succeeding rows, add one to the count for each run that does not overlap any run on the preceding row; and subtract k from the count for each run that overlaps $k+1$ runs on the preceding row (so that if $k = 0$, the count does not change). When the scan is complete, the final count is the genus.

Two other methods of computing the genus are as follows:

(1)   Let V be the number of 1's in S, E the number of $\begin{smallmatrix}1\\1\end{smallmatrix}$'s and $\begin{smallmatrix}1&1\end{smallmatrix}$'s, D the number of $\begin{smallmatrix}1&\\&1\end{smallmatrix}$'s and $\begin{smallmatrix}&1\\1&\end{smallmatrix}$'s, T, the number of $\begin{smallmatrix}1&1\\1&\end{smallmatrix}$'s, $\begin{smallmatrix}1&1\\&1\end{smallmatrix}$'s, $\begin{smallmatrix}1&\\1&1\end{smallmatrix}$'s, and $\begin{smallmatrix}&1\\1&1\end{smallmatrix}$'s, and F, the number of $\begin{smallmatrix}1&1\\1&1\end{smallmatrix}$'s. Then

(a)   if we use 4-connectedness for S and 8-connectedness for $\bar{S}$, the genus of S is

$$V - E + F$$

(b)   if we use 8-connectedness for S and 4-connectedness for $\bar{S}$, the genus of S is

$$V - E - D + T - F$$

(2)   Let Q_V be the number of 2×2 point squares in the picture that contain just one 1, Q_T the number that contain just three 1's, and Q_D the number that contain just two diagonally adjacent 1's. Then the genus of S is

$$\tfrac{1}{4}(Q_V - Q_T \pm 2Q_D)$$

where we use the plus sign if we have 4-connectedness for S and 8-connectedness for $\bar{S}$, and the minus sign in the reverse case. Note that if we regard S as composed of small squares, this formula can be interpreted as the number of convex corners of S minus its number of concave corners. Indeed, each Q_V pattern has a convex corner at its center point; each Q_T pattern has a concave corner; and each Q_D pattern has two convex corners if diagonally adjacent 1's are not connected, but two concave corners if they are connected.

9.2   SIZE AND DISTANCE

In this section we discuss geometrical properties of digital picture subsets based on concepts of size and distance. In Section 9.2.1 we discuss the area and perimeter of a subset, and in Section 9.2.2 deal with the concept of distance between subsets. In Section 9.2.3, distance concepts are applied to define the "skeleton" of a subset S as a set of points whose distances from $\bar{S}$ are local maxima. In Sections 9.2.4 and 9.2.5 we describe "propagation" and shrinking operations on subsets, and some of their applications.

9.2.1   Area and Perimeter

The *area* of a subset S of a digital picture is just the number of points in S. If the points of S are easily recognizable (in particular, if we are explicitly given the overlay φ_S that has 1's at the points of S and 0's elsewhere), we can compute the area of S by simply scanning the picture and counting these points. More generally, given a partition $\{S_1, \ldots, S_k\}$ of the picture, we can count the numbers of points in all of the parts in a single scan of the picture. We simply set up k counters, one for each set in the partition; each point reached by the scan is identified, say as belonging to S_i, and we add 1 to the ith counter.

We can also compute the areas of all the connected components of a given subset S, or of the sets in a given partition (e.g., the connected components of constant gray level)[†] in a single scan of the picture. To do this, we proceed as follows: On the first row, count the number of 1's in each run of 1's, store each of these numbers in a counter, and create a pointer from each run to its counter. On succeeding rows,

(1)   If a run ρ is adjacent to no run on the previous row, create a pointer from ρ to a new counter, and store the length of ρ in that counter.

(2)   If ρ is adjacent to exactly one run σ on the previous row, add the length of ρ into σ's counter, and transfer the pointer from σ to each such ρ.

(3)   If ρ is adjacent to two or more runs $\sigma_1, \ldots, \sigma_n$ on the previous row, add the contents of the $\sigma_2, \ldots, \sigma_n$ counters into the σ_1 counter, add the length of ρ into it, and transfer the pointer from σ_1 to ρ.

The area of a picture subset S, as well as the areas of various subsets that can be derived from S (e.g., its connected components, the set of points at a given distance from it, its convex hull or its "shadows", and so forth) are useful descriptive properties of S. Area-based criteria can also be used to derive new subsets from a given S, or a given partition. To give just one example, we can delete all connected components whose areas are below a given threshold—or, less trivially, all such components that satisfy some adjacency criterion (e.g., they are adjacent to only one other component).

It is somewhat less trivial to define the *perimeter* of a digital picture subset S. Some possible definitions are:

(a)   The sum of the lengths of the "cracks" separating points of S from points of $\bar{S}$—in other words, the number of pairs of points (p, q) with p in S and q in $\bar{S}$ (see Exercise 17).

[†] It has been found empirically (S. Nishikawa, R. J. Massa, and J. C. Mott-Smith, Area properties of television pictures, *IEEE Trans. Informat. Theory* **IT-11**, 1965, 348–352) that the number of components of constant gray level that have area N is roughly proportional to $1/N^4$.

(b) The number of steps taken by a border-following algorithm (e.g., BF of Section 9.1.2) in following all the borders of S.

(b') The same, but with diagonal steps counting $\sqrt{2}$ each, while horizontal and vertical steps count only 1 each.

(c) The number of border points of S, i.e., the area of the border S' of S. To illustrate these definitions, consider the following three S's:

```
                                          1

                                     1

     1  1       1  1  1  1       1

     1  1                        1
```

The perimeters of these S's, for definitions (a)–(c), are as follows:

(a)	8	10	16
(b)	4	6	6
(b')	4	6	$6\sqrt{2} \doteq 8\frac{1}{2}$
(c)	4	4	4

Exercise 19. Using these four definitions, compute the perimeters of

```
        1              1  1  1

     1  1  1    and    1        1

        1              1  1  1
```

In methods (b) and (c) what happens if we use 8-adjacency rather than 4-adjacency in defining the border? ▌

We conclude this section by briefly discussing how the area and perimeter of a subset of a digital picture depend on the fineness with which the picture was sampled when it was digitized. Let f be a real (nondigital) picture, and let A be a subset of f, say the set of points at which the gray level exceeds some threshold t. If we digitize f using a square grid array of sample points, spaced (say) ε apart, then the digital picture subset $\hat{A}_\varepsilon$ corresponding to A is just the set of grid points that lie inside A. Let the number of these grid points be $n(\varepsilon)$; thus in units of ε^2, the area of $\hat{A}_\varepsilon$ is $n(\varepsilon)$. As ε goes to zero, the area $\varepsilon^2 n(\varepsilon)$ of $\hat{A}_\varepsilon$ converges to the (real) area of A (assuming, of course, that A has a well defined area, in the sense of measure theory). Thus the digital area of $\hat{A}_\varepsilon$, as defined at the beginning of this section, is a good approximation to the real area of A, provided that we use ε^2 as the unit of area.

The situation with perimeter is quite different; here the perimeter of $\hat{A}_\varepsilon$ will usually grow exponentially as ε goes to zero. This is because the edges of objects become very ragged when they are examined microscopically. Thus the number of border points in $\hat{A}_\varepsilon$ grows rapidly as ε becomes small; and similarly for the other measures of digital perimeter described above.[†] It is important to choose ε appropriately when we digitize a picture, if we expect to measure the perimeters of objects in the resulting digital picture. If ε is too small, our digital perimeter measurements may be swamped by "noise" resulting from edge raggedness.

9.2.2 Distance

Let d be a function that takes pairs of points into nonnegative numbers. We call d a *metric*, or a *distance* function, if for all points p, q, r we have

(1) $d(p, q) = 0$ if and only if $p = q$,
(2) $d(q, p) = d(p, q)$,
(3) $d(p, r) \leqslant d(p, q) + d(q, r)$.

Example 1 (*Euclidean distance*)

$$d_e((i, j), (h, k)) = \sqrt{(i-h)^2 + (j-k)^2}$$

For this function, the points having distance $\leqslant t$ from (i, j) are just those points contained in a circle of radius t centered at (i, j). The distance between the opposite corners of an $n \times n$ picture is $n\sqrt{2}$. ■

Example 2 (*City block distance*)

$$d_4((i, j), (h, k)) = |i-h| + |j-k|$$

Here the points having distance $\leqslant t$ from (i, j) form a diamond (i.e., a square with sides inclined at $\pm 45°$) centered at (i, j) and with side length $t\sqrt{2}$, e.g., the points of distance $\leqslant 3$ are

```
                3
             3  2  3
          3  2  1  2  3
       3  2  1  0  1  2  3
          3  2  1  2  3
             3  2  3
                3
```

[†] The same is true when we measure the perimeter of A by approximating A with an inscribed polygon whose sides have length ε. See J. Perkal, On the ε-length, *Bull. Acad. Sci. Polon.* **4**, 1956, 399–403; B. Mandelbrot, How long is the coast of Britain?, *Science* **156**, 1967, 636–638.

where the 0 indicates the position of (i, j), and the points have values equal to their distances. In particular, the points at distance 1 from (i, j) are just the 4-neighbors of (i, j). The distance between the opposite corners of an $n \times n$ picture is $2n$. ■

Example 3   (*Chessboard distance*)

$$d_8((i, j), (h, k)) = \max(|i-h|, |j-k|)$$

Here the points at distance $\leqslant t$ from (i, j) form a square (with sides horizontal and vertical) centered at (i, j) and with side length $2t$, e.g., the points of distance $\leqslant 3$ from the 0 are

$$
\begin{array}{ccccccc}
3 & 3 & 3 & 3 & 3 & 3 & 3 \\
3 & 2 & 2 & 2 & 2 & 2 & 3 \\
3 & 2 & 1 & 1 & 1 & 2 & 3 \\
3 & 2 & 1 & 0 & 1 & 2 & 3 \\
3 & 2 & 1 & 1 & 1 & 2 & 3 \\
3 & 2 & 2 & 2 & 2 & 2 & 3 \\
3 & 3 & 3 & 3 & 3 & 3 & 3
\end{array}
$$

In particular, the points at distance 1 from (i, j) are just its 8-neighbors, and the distance between the opposite corners of an $n \times n$ picture is n. ■

Exercise 20.   Prove that $[d_e]$, the greatest integer not exceeding d_e, is not a metric, and similarly for $\{d_e\}$, the integer closest to d_e. ■

Exercise 21.   Prove that the sum or max of two metrics is a metric, but that the min of two metrics need not be a metric. ■

Exercise 22.   Prove that $d_4((i, j), (h, k))$ is the length of the shortest path between (i, j) and (h, k), where a path consists of horizontal and vertical steps of unit length. Similarly, prove that $d_8((i, j), (h, k))$ is the smallest number of horizontal, vertical, or diagonal steps required to get from (i, j) to (h, k). ■

Distances between points of a picture subset S are often useful as descriptive properties of S. Indeed, if S and T are digital picture subsets having the same number of points, and there exists a one to one correspondence h between the points of S and T such that $d_e(p, q) = d_e(h(p), h(q))$ for all points p, q of S, then S and T must be congruent, i.e., they can differ only by a translation, a rotation by a multiple of $90°$, and/or a reflection in a horizontal or vertical line. (When S and T have just three points, this is the familiar theorem that if two triangles have the same side lengths, they are congruent.)

Exercise 23.   Show by example that for the metrics d_4 and d_8, equality of corresponding interpoint distances does not guarantee congruence. ▮

Incidentally, if we only know that S and T have the same *set* of interpoint distances, but we are not given a distance-preserving one-to-one correspondence between them, they need not be congruent. This can be seen from the example (due to Minsky and Papert [22]) of

$$
\begin{array}{ccc}
1 \quad 1 \quad 1 & & 1 \quad 1 \quad 1 \\
1 & \text{and} & 1 \qquad 1 \\
1 & &
\end{array}
$$

for both of which the set of distances between pairs of points is 1 from four pairs, 2 from two pairs, $\sqrt{2}$ from two pairs, and $\sqrt{5}$ from two pairs.

One can also use the notion of distance to define new picture subsets from a given subset S. For example, we can construct the set S_t of points of S that are at distance t from $\bar{S}$. (By the distance $d(p, A)$ from a point p to a set A we mean the smallest of the distances between p and all the points of A.) For $t = 1$, S_t is just the border of S. For other values of t, we obtain "rings" of points that "parallel" the border of S. The d_4- and d_8-distances to $\bar{S}$ are shown in Fig. 1 for two S's, an upright rectangle and a diamond; note that these distances are the same in the first case but different in the second.

We shall now describe some simple algorithms which, for any given picture subset S, compute the distances from all the points of S to $\bar{S}$. These algorithms work for integer-valued metrics such as d_4 and d_8 which have the following important property:

(4)   For all p, q such that $d(p, q) \geqslant 2$, there exists an s different from p and q such that $d(p, q) = d(p, s) + d(s, q)$.

A metric with property (4) will be called *regular*.[†]

Exercise 24.   Prove that d is regular if and only if for all p, q, where $p \neq q$, there exists an s such that

$$d(p, s) = 1 \qquad \text{and} \qquad d(p, q) = 1 + d(s, q) \qquad ▮$$

The following "parallel" algorithm, which we denote by PD, computes the distances to $\bar{S}$ from all points of S, given the "overlay" φ_S which has 1's at the points of S and 0's elsewhere. Let $f^{(0)}(i, j)$ be the value of φ_S at the point

---

[†] Regularity is a necessary and sufficient condition for an integer-valued metric to be the distance on a graph; see F. Harary, "Graph Theory," p. 24, Exercise 2.8. Addison-Wesley, Reading, Mass., 1969.

```
    1 1 1 1 1 1 1 1                              1
    1 2 2 2 2 2 2 1                            1 2 1
    1 2 3 3 3 3 2 1                          1 2 3 2 1
    1 2 3 4 4 3 2 1                        1 2 3 4 3 2 1
    1 2 3 3 3 3 2 1                    1 2 3 4 5 4 3 2 1
    1 2 2 2 2 2 2 1                      1 2 3 4 3 2 1
    1 1 1 1 1 1 1 1                        1 2 3 2 1
          (a)                               1 2 1
                                              1
                                             (b)


            1                                 1
          1 1 1                             1 2 1
        1 1 2 1 1                         1 2 2 2 1
      1 1 2 2 2 1 1                     1 2 2 3 2 2 1
  1 1 2 3 2 2 1 1               1 2 2 3 4 3 2 2 1
      1 1 2 2 2 1 1                     1 2 2 3 2 2 1
        1 1 2 1 1                         1 2 2 2 1
          1 1 1                             1 2 1
            1                                 1
          (c)                               (d)
```

Fig. 1 Distances to $\bar{S}$ for a rectangle and a diamond. (a) $d_4 = d_8$ for the rectangle. (b) d_4 for the diamond. (c) d_8 for the diamond. (d) "Octagonal" distances for the diamond.

(i, j), and for $m = 1, 2, 3, \ldots$, define $f^{(m)}(i, j)$ inductively by

$$f^{(m)}(i, j) = f^{(0)}(i, j) + \min_{d((u, v), (i, j)) \leqslant 1} f^{(m-1)}(u, v)$$

Then if d is a regular metric, we have $f^{(r)}(i, j) = d((i, j), \bar{S})$ for all $r \geqslant d((i, j), \bar{S})$. In particular, if we take r at least equal to the greatest distance between any two points of the picture, then this must hold for every (i, j).

To see how algorithm PD operates, note first that 0's in φ_S (i.e., points of $\bar{S}$) always remain 0's, since if $f^{(0)}(i, j) = 0$ we have

$$\min_{d((u, v), (i, j)) \leqslant 1} f^{(m)}(u, v) = f^{(m)}(i, j) = 0 \qquad \text{for all } m$$

Similarly, 1's at distance 1 from $\bar{S}$ remain 1's, since for such 1's too, the min is 0. On the first iteration of PD, however, all 1's at distance >1 from $\bar{S}$ become 2's, since for such 1's, the min is 1. On the second iteration, all 2's at distance >2 from $\bar{S}$ become 3's, since the min for such 2's is 2. Similarly, on the third iteration, all 3's at distance >3 from $\bar{S}$ become 4's; and so on.

The PD algorithm can be implemented efficiently on a parallel array-processing machine, since at each point, the value at the mth iteration can be computed from the values of neighboring points at the $(m-1)$th iteration. This would not be particularly efficient on a conventional sequential computer, since each point must be processed at each iteration. We can, however, also define a simple sequential algorithm that computes all distances to $\bar{S}$ in only two scans of the picture. For city block distance, this algorithm is as follows: Let

$$g(i,j) = \begin{cases} 0 & \text{if } f(i,j) = 0 \\ \min(g(i-1,j)+1, g(i,j+1)+1) & \text{if } f(i,j) = 1 \end{cases}$$

$$h(i,j) = \min(g(i,j), h(i+1,j)+1, h(i,j-1)+1)$$

We assume here that there are only 0's on the border of the picture. Thus for all (i,j), we can compute $g(i,j)$ in a single TV raster scan of φ_S, and we can then compute all the $h(i,j)$'s in a single *reverse* TV raster scan of the array of $g(i,j)$'s. It can be shown that, for all (i,j), $h(i,j)$ is the distance from (i,j) to $\bar{S}$. As an example,

$$\text{if } f \text{ is} \quad \begin{matrix} 0 & 0 & 0 & 0 & 0 & 0 \\ 0 & 0 & 1 & 1 & 1 & 0 \\ 0 & 1 & 1 & 1 & 0 & 0 \\ 0 & 1 & 1 & 1 & 1 & 0 \\ 0 & 0 & 0 & 0 & 0 & 0 \end{matrix}$$

$$\text{then } g \text{ is} \quad \begin{matrix} 0 & 0 & 0 & 0 & 0 & 0 \\ 0 & 0 & 1 & 1 & 1 & 0 \\ 0 & 1 & 2 & 2 & 0 & 0 \\ 0 & 1 & 2 & 3 & 1 & 0 \\ 0 & 0 & 0 & 0 & 0 & 0 \end{matrix} \quad \text{and } h \text{ is} \quad \begin{matrix} 0 & 0 & 0 & 0 & 0 & 0 \\ 0 & 0 & 1 & 1 & 1 & 0 \\ 0 & 1 & 2 & 1 & 0 & 0 \\ 0 & 1 & 1 & 1 & 1 & 0 \\ 0 & 0 & 0 & 0 & 0 & 0 \end{matrix}$$

The metrics d_4 and d_8 are quite non-Euclidean; as we have seen, the set of points at distance t from a given point is a diamond or a square, rather than a circle, when we use these metrics. We can compute a distance in which "circles" are octagons by using d_4 and d_8 at alternate iterations in algorithm PD. For example, the points of "octagonal" distance $\leqslant 3$ from a single point are

$$
\begin{array}{ccccccc}
 & & 3 & 3 & 3 & & \\
 & 3 & 2 & 2 & 2 & 3 & \\
3 & 2 & 2 & 1 & 2 & 2 & 3 \\
3 & 2 & 1 & 0 & 1 & 2 & 3 \\
3 & 2 & 2 & 1 & 2 & 2 & 3 \\
 & 3 & 2 & 2 & 2 & 3 & \\
 & & 3 & 3 & 3 & & \\
\end{array}
$$

For a diamond-shaped S, the "octagonal" distances to $\bar{S}$ are shown in Fig. 1d. It can be shown that octagons are the most circular possible shapes that can be obtained by iteration of isotropic local operations such as those used in the PD algorithm.

Algorithms similar to PD can be devised to count the number of shortest paths between every (i, j) and $\bar{S}$, the number of points of $\bar{S}$ at a given distance from every (i, j), and the like. We can also *construct* shortest paths from every (i, j) to $\bar{S}$ by proceeding as follows: In algorithm PD, when m reaches $d((i, j), \bar{S})$, $f^{(m)}(i, j)$ stops increasing, and at that stage, any neighbor of (i, j) whose value is minimal must lie on a shortest path from (i, j) to $\bar{S}$. Thus if we create a pointer at (i, j) that points to any such neighbor, then when algorithm PD is finished, we can follow a chain of these pointers from any point, along a shortest path, to $\bar{S}$.

It is also of interest to define the "gray-weighted" distance between two points p and q in a (multivalued) picture; this is the smallest possible sum of gray levels of the points along a path from p to q. Note that a path which yields the smallest sum is not necessarily a shortest path from p to q. If we apply algorithm PD to an arbitrary picture (not necessarily two-valued), it can be shown that the value at any point (i, j) eventually becomes equal to the gray-weighted distance between (i, j) and the set of points that have value 0. Note that if the picture points can have only the two values 0 and 1, the gray-weighted distance to the set of 0's is the same as the ordinary unweighted distance.

9.2.3   Skeletons

The set of points of S whose distances from $\bar{S}$ are locally maximum (i.e., no neighboring point has greater distance from $\bar{S}$) constitutes a sort of *skeleton* of S. If we know the points of this skeleton, and their associated distance values, we can reconstruct S exactly. Thus the skeleton of S provides a concise way of defining and representing S, just as the border of S does. In this section we discuss methods of defining the skeleton of an arbitrary picture subset S, as well as some applications of this concept. A somewhat different kind of "skeleton," obtained by "thinning" an elongated object, will be discussed in Section 9.4.5. The skeleton discussed in this section is sometimes called the "medial axis" or "symmetric axis" of S.

Let the "disk" $D_t(i, j)$ be the set of points that are at distance $\leqslant t$ from (i, j). As we saw earlier, for the d_4 metric, $D_t(i, j)$ is a diamond; for the d_8 metric, it is a square; and so on. Given the picture subset S, let $d((i, j), \bar{S})$ be the distance from (i, j) to $\bar{S}$; this is zero if (i, j) is in $\bar{S}$, and positive otherwise. We first make two simple observations about these concepts:

(1)   For all $t < d((i, j), \bar{S})$, the disk $D_t(i, j)$ is contained in S and contains (i, j).

Proof: A point in $D_t(i, j)$ has distance $t < d((i, j), \bar{S})$ from (i, j); hence no such point can be in $\bar{S}$, since the smallest distance from (i, j) to any point of $\bar{S}$ is $d((i, j), \bar{S})$.   ∎

It follows that the union of all of these disks, for all (i, j) in S, is equal to S, since each of the disks is contained in S, and each point (i, j) of S is in at least one of the disks.

(2)   If (u, v) is a neighbor of (i, j), then for any t, $D_t(i, j)$ is contained in $D_{t+1}(u, v)$.

Proof: Any point (h, k) at distance $\leqslant t$ from (i, j) is at distance $\leqslant t+1$ from (u, v), since if ρ is any path from (i, j) to (h, k), and we adjoin the point (u, v) to the beginning of ρ, we obtain a path from (u, v) to (h, k) which is just 1 longer than ρ.   ∎

Let S^* be the set of points of S whose distances from $\bar{S}$ are local maxima; in other words, the point (i, j) of S is in S^* if and only if

$$d((i, j), \bar{S}) \geqslant d((u, v), \bar{S})$$

for all neighbors (u, v) of (i, j). We call S^* the *skeleton* of S. Using (1) and (2), we can now prove that S is the union of the "disks"

$$D_{d((i, j), \bar{S}) - 1}(i, j)$$

for all (i, j) in S^*. Indeed, by (1), S is the union of all disks $D_t(i, j)$ for (i, j) in S and $t < d((i, j), \bar{S})$; and we can certainly discard any disk $D_t(i, j)$ such that $t < d((i, j), \bar{S}) - 1$, since this disk is contained in $D_{d((i, j), \bar{S}) - 1}(i, j)$. Moreover, if (i, j) is not in S^*, we must have $d((i, j), \bar{S}) < d((u, v), \bar{S})$ for some neighbor (u, v) of (i, j). Then, however, $d((u, v), \bar{S}) - 1$ is at least 1 greater than $d((i, j), \bar{S}) - 1$, so that by (2), the disk centered at (i, j) is contained in the one centered at (u, v), and so can be discarded.

We have thus shown that if we are given the set S^* and the distance from each point of S^* to $\bar{S}$, then we can construct the original set S, since S is the union of the disks $D_{d((i,j),\bar{S})-1}(i,j)$ for all (i,j) in S^*. The skeletons of a rectangle and a diamond, for both the d_4 and d_8 metrics, are shown in Fig. 2; each skeleton point is labeled with its associated distance value. These skeletons were obtained from the arrays of distances shown in Fig. 1 for the same rectangle and diamond. The skeletons consist of the points of the arrays whose distance values are 4-neighbor and 8-neighbor local maxima, respectively. The values shown here are 1 less than the values in the distance arrays, in accordance with the preceding discussion. One can also define a skeleton for the "octagonal" metric of Section 9.2.2, by taking 8-neighbor maxima for points whose distances to $\bar{S}$ are odd, and 4-neighbor maxima for points whose distances are even; this choice guarantees that S will still be the union of the maximal octagonal "disks" centered at the skeleton points.

It will be noted from the examples in Fig. 2 that skeletons, as defined here,

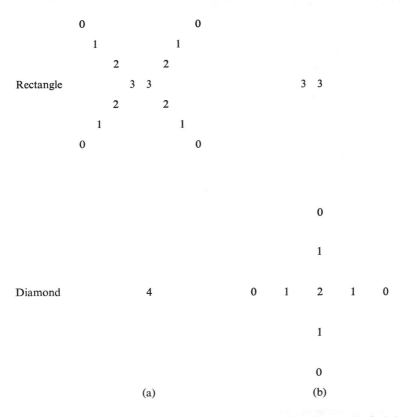

Fig. 2 Skeletons of a rectangle and a diamond. (a) d_4 skeletons. (b) d_8 skeletons.

are usually not connected.[†] Indeed, two points of a d_4-skeleton can never be 4-adjacent unless they have the same distance value, since otherwise one of them would not be a local maximum; and similarly, two 8-adjacent points of a d_8-skeleton must have the same distance value. We can make the skeleton connected by discarding fewer points when we construct it. For example, in the d_8-case, we might keep the point (i, j) if

(a) it is a 4-neighbor of a skeleton point;
(b) at most one of its 8-neighbors has a larger distance value.

(A rule of this type will tend to keep points at which there are sharp corners in the contours S_t of equal distance from $\bar{S}$.) It can be verified that this would yield an 8-connected d_8-skeleton for the diamond. Of course, when we discard fewer points, the skeleton that we obtain is a less economical representation of S.

The skeleton S^* of an arbitrary S is usually somewhat less economical, as a representation of S, than is the border S' of S. This is because the points of S' always lie on connected, closed curves, and so can be represented very compactly by chain codes (together with the coordinates of one starting point; a neighboring point outside S is also needed, to determine which side of the border S is on). On the other hand S^* is often not connected, so that there is no simple way to represent its points by chain codes. Moreover, besides the points of S^* themselves, we also need their distances to $\bar{S}$ if we want to reconstruct S from S^*.

In spite of its comparative lack of economy, the skeleton representation of S has advantages over the border representation for some purposes. For example, suppose that we will often have to determine whether a given point p lies in S or in $\bar{S}$. If S is represented by its border S', this is very hard to do without actually constructing φ_S from S'. When we are given the skeleton S^* of S, on the other hand, we can tell whether p is in S or in $\bar{S}$ by computing the distance $d(p, q)$ between p and each point q of S^*; p is in S if and only if, for some q, this distance is less than $d(q, \bar{S})$.

The skeleton representation can also be used to create simplified versions of a given S. In particular, suppose that we delete from S^* points whose distances to $\bar{S}$ are low. The picture subset $\hat{S}$ defined by this modified skeleton will be contained in S, but small components of S (if any) will be missing, and small irregularities in the shape of S will be smoothed out. As a simple example, if S is

	1	1	1	
	1	1	1	1
1	1	1	1	

[†] The thinning operations of Section 9.4.5, on the other hand, are designed to yield connected "skeletons."

then its d_8-skeleton is

$$\begin{array}{cc} 1 & 0 \end{array}$$

$$0$$

If we delete the 0's from this skeleton, the resulting $\hat{S}$ is

$$\begin{array}{ccc} 1 & 1 & 1 \\ 1 & 1 & 1 \\ 1 & 1 & 1 \end{array}$$

The example just given shows that the skeleton of S does not change greatly if points are added to S, or deleted from it, near $\bar{S}$; this will primarily affect the low-distance parts of S^*. On the other hand, S^* may change drastically if points far from $\bar{S}$ are deleted from S^*. For example, the d_8-skeletons of

$$\begin{array}{ccccc} 1 & 1 & 1 & 1 & 1 \\ 1 & 1 & 1 & 1 & 1 \\ 1 & 1 & 1 & 1 & 1 \\ 1 & 1 & 1 & 1 & 1 \\ 1 & 1 & 1 & 1 & 1 \end{array} \quad \text{and} \quad \begin{array}{ccccc} 1 & 1 & 1 & 1 & 1 \\ 1 & 1 & 1 & 1 & 1 \\ 1 & 1 & & 1 & 1 \\ 1 & 1 & 1 & 1 & 1 \\ 1 & 1 & 1 & 1 & 1 \end{array}$$

are

$$2 \quad \text{and} \quad \begin{array}{ccccc} 0 & 0 & 0 & 0 & 0 \\ 0 & 0 & 0 & 0 & 0 \\ 0 & 0 & & 0 & 0! \\ 0 & 0 & 0 & 0 & 0 \\ 0 & 0 & 0 & 0 & 0 \end{array}$$

This suggests that it may be very desirable to suppress such noise from a given S before we construct S^*.

"Skeletons" analogous to S^* can also be defined for multivalued digital pictures. To see how this can be done, note that in the two-valued case, a point of S is a skeleton point if and only if it does not lie on a shortest path from any other point of S to $\bar{S}$. (Indeed, if p were on a shortest path from q to $\bar{S}$, the point just preceding p on this path would be a neighbor of p and would have higher distance to $\bar{S}$ than p.) In the multivalued case, we can define a minimum-cost path from p to $\bar{S}$ as a path having the smallest possible sum of gray levels (see the end of Section 9.2.2). Based on this concept, we can call p a "skeleton point" if it does not lie on a minimum-cost path from any other point q to $\bar{S}$. Unfortunately, the original picture is not, in general, reconstructible from this type of skeleton.

Another approach to representing a picture as a union of "disks" (specifically: squares) is to subdivide the picture successively, say, into quarters. If, at any stage, one of the parts has constant gray level, it is not subdivided further; for parts not having constant gray level, the subdivision process continues until the level of single picture elements is reached. (We may assume here that the picture is a square array of size $2^n \times 2^n$, for some n, so that it is repeatedly divisible into quarters.) The results of this subdivision process can be represented by a tree whose nodes correspond to the parts; the root node corresponds to the entire picture, and each node is either terminal (if the corresponding part was not further subdivided) or has four successors. The terminal nodes are labeled with the (constant) gray levels of their corresponding picture parts. This "hierarchical array" method of representing a picture may be advantageous over other methods if the picture contains relatively few, large regions of constant gray level, particularly if these regions' locations are simple fractions of the dimensions of the picture.

9.2.4 Propagation and Shrinking

For any picture subset S, the set of points of the picture that are within a given distance t of S can be obtained by "propagating" S into $\bar{S}$: we start with φ_S, and change 0's into 1's if they have any 1's as neighbors; this is repeated t times. The type of neighbors used must correspond to the distance function; we use 4-neighbors for d_4, 8-neighbors for d_8, and we alternate them for "octagonal" distance. Similarly, the set of points that are not within a given distance t of $\bar{S}$ can be obtained by shrinking S, i.e., starting with φ_S, changing 1's into 0's if they have any 0's as neighbors, and repeating this t times. Note that shrinking S is the same as expanding ("propagating") $\bar{S}$. Both propagation and shrinking can be done very efficiently on a parallel machine.

A variety of useful results can be obtained by combining propagation and shrinking operations. Several of these will be described in the following paragraphs. For brevity, we first introduce the following notation: $S^{(t)}$ denotes the set of points that are within distance t of S, where $t \geqslant 0$ (note that $S^{(0)} = S$), i.e., $S^{(t)}$ is the result of propagating S t times. Similarly, $S^{(-t)}$ denotes the result of shrinking S t times. To represent a combination of propagation and shrinking, we can combine these notations; e.g., $(S^{(m)})^{(-n)}$ denotes the result of propagating S m times, and then shrinking the result n times.

We can compute the distances from all points of S to $\bar{S}$ by repeatedly shrinking and adding. In fact, it is easily verified that $f^{(m)}(i,j)$ in the distance-computing algorithm PD is equal to

$$\varphi_S + \varphi_{S(-1)} + \varphi_{S(-2)} + \cdots + \varphi_{S(-m)}$$

It is important to realize that the propagation and shrinking operations do not commute with one another; $(S^{(m)})^{(-n)}$ is not necessarily the same as $(S^{(-n)})^{(m)}$, and neither of them is the same as $S^{(m-n)}$. For example, if S consists of a single 1, then $S^{(1)}$ is

$$\begin{array}{ccc} 1 & 1 & 1 \\[6pt] 1 & 1 & 1 \\[6pt] 1 & 1 & 1 \end{array}$$

(if we use 8-neighbor propagation), so that $(S^{(1)})^{(-1)}$ is the same as S. On the other hand, $S^{(-1)}$ is empty (S vanishes when we shrink it once), hence so is $(S^{(-1)})^{(1)}$.

Exercise 25.   Prove that $(S^{(k)})^{(-k)}$ contains S, and $(S^{(-k)})^{(k)}$ is contained in S.   ∎

We can use propagation and shrinking to obtain the skeleton S^* of a given S. In fact, consider $(S^{(-t-1)})^{(1)}$ and $S^{(-t)}$; readily, the first of these is contained in the second. Let S_t^* be the difference set $S^{(-t)} - (S^{(-t-1)})^{(1)}$. Any point p of S_t^* is at distance exactly t from $\bar{S}$, since if p were farther than t from $\bar{S}$, it would be in $(S^{(-t-1)})$, hence in $(S^{(-t-1)})^{(1)}$. Moreover, p has no neighbors whose distances from $\bar{S}$ are greater than t, since any such neighbor would be in $S^{(-t-1)}$, so that p would be in $(S^{(-t-1)})^{(1)}$. Hence the points of S_t^* are just the skeleton points that are at distance t from $\bar{S}$.

Propagation and shrinking can also be used to determine isolated components and clusters of components of a given S, since these concepts depend on the distances between the components. In fact, suppose that we propagate S k times, then shrink the result k times, and examine the connected components of what remains. Note that a single, isolated point of S, under these operations, expands into a "disk" of radius k, which then shrinks back to a point, thus yielding a one-point component. Conversely, if a small component is obtained, it must have arisen from a small, isolated part of S. On the other hand, if a component is obtained whose area is large (relative to k^2), it must have arisen from the "fusion" of a cluster of parts of S, as illustrated in Fig. 3. The larger k, the greater the degree of isolatedness, and the sparser the degree of clustering that is detected by this method.

If we do the shrinking and propagation in the opposite order, i.e., we first shrink S k times and then reexpand the result, we obtain a simplified version of S from which small components or pieces have been deleted, as illustrated

```
              1                 1                  1
      1                                                        1

                                          1
                  1         1     1
          1
      1                 1           1                        1
                  1                       1
                        1
      1                           1                  1
                  1         1         1

                                          1
              1
      1                     1                  1                  1
                        1
```

(a)

```
          1 1 1       1 1 1         1 1 1
    1 1 1 1 1 1       1 1 1         1 1 1       1 1 1
    1 1 1 1 1 1       1 1 1         1 1 1       1 1 1
    1 1 1                           1 1 1       1 1 1
                  1 1 1 1 1 1 1 1     1 1 1
          1 1 1   1 1 1 1 1 1 1 1     1 1 1
  1 1 1 1 1 1     1 1 1 1 1 1 1 1           1 1 1
  1 1 1 1 1 1     1 1 1 1 1 1 1 1 1 1       1 1 1
  1 1 1           1 1 1 1 1 1 1 1 1 1       1 1 1
          1 1 1   1 1 1 1 1 1 1 1 1 1   1 1 1
          1 1 1   1 1 1 1 1 1 1 1 1 1   1 1 1
          1 1 1   1 1 1 1 1 1 1 1 1 1   1 1 1
                  1 1 1 1 1 1 1 1 1 1 1 1
          1 1 1                         1 1 1
  1 1 1 1 1 1 1 1 1               1 1 1 1 1 1 1 1 1
  1 1 1 1 1 1 1 1 1     1 1 1 1 1 1         1 1 1
  1 1 1           1 1 1   1 1 1 1 1 1         1 1 1
                      1 1 1
```

(b)

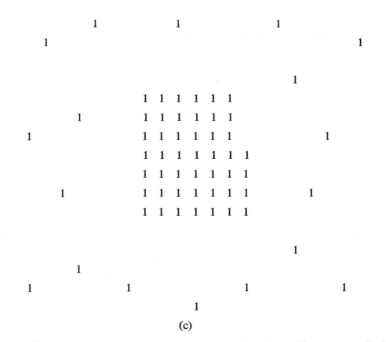

(c)

Fig. 3   Cluster detection by expanding and reshrinking. (a) Original picture. (b) Result of 8-neighbor expansion of (a). (c) Result of 8-neighbor shrinking of (b).

```
1            1
1  1  1
1  1  1  1                         1              1  1  1
1  1  1                     1  1                  1  1  1
      1  1  1  1                                  1  1  1
1
                (a)                        (b)           (c)
```

Fig. 4   Simplification by shrinking and reexpanding. (a) Original picture. (b) Result of 8-neighbor shrinking of (a). (c) Result of 8-neighbor expansion of (b).

in Fig. 4. Note that thin pieces of S are also deleted; this may not be desirable. We can avoid this, if we wish, by using the shrinking and propagation process to detect elongated parts of S, and not allowing such parts to be deleted. Methods of detecting elongatedness will be described in Section 9.4.5.

We conclude this section by describing how propagation can be used to fill in regions, given a set of points on their edges, where the points may not form connected, closed region borders. Suppose, for example, that we have

```
                    L
          L       L  D                L          L  L  L
   L             L  D              L  L  L  L  L  D  D
 L  D       D                   L  L  D  D  D  D  D  D  L
                       D  L        L  D           D      D  L  L
 L  D                           L  L  D  D                 D  L
                       D  L        L  L  D  D  D  D  D  L  L
    L  D  D  D                  L  L  D  D  D  D  D  L
       L  L                        L  L  L  L  L  L
                                         L  L

          (a)                              (b)
```

Fig. 5 Region filling by propagation from edge points. (a) Detected edge points, represented by pairs of symbols (D = dark, L = light) just on opposite sides of the edge. (b) Result of 4-neighbor expansion of the D's and L's into the previously unmarked points. When an unmarked point had both L's and D's as neighbors, it was given the L label. It is clear that as the expansion continues, the D's will fill up the hole, and the L's will fill the rest of the picture.

detected edges between a dark region and a light region at the points shown in Fig. 5a; here we have represented each detected edge point by a pair of symbols (D, L), with the D just on the dark side of the edge, and the L just on its light side. We can now fill in the regions by propagating the D's and L's into the unmarked points, as shown in Fig. 5b.

This technique can create reasonable regions from discontinuous borders, as seen in Fig. 5. It is, however, sensitive to false alarms in the edge detection process; a single spurious edge point can create a large region that should not really be present. This problem can be alleviated by examining the regions in the original picture that correspond to the regions of L's and D's on the "map" that results from the propagation process. If a region of L's is too dark, the L's can be changed to D's, and vice versa. The "map" can also be improved by using region-merging techniques, as described in Section 8.4.4.

9.2.5 Topology-Preserving Shrinking

The shrinking operation of Section 9.2.4 can disconnect a picture subset S, or even erase it completely. In this section we describe shrinking operations that preserve connectedness. As mentioned in Section 9.1.3, these operations can be used to count the connected components of S, since they shrink such components to isolated 1's, which are easy to detect and count.

The point p of S is called a *simple* point if the set of 8-neighbors of p that lie in S has exactly one component that is adjacent to p. For example, p is 4-simple if its 8-neighborhood is

$$
\begin{array}{ccc}
0 & 1 & 1 \\
0 & p & 0 \\
1 & 0 & 0
\end{array}
$$

since only one 4-component of 1's is 4-adjacent to p; but p is not 4-simple if its 8-neighborhood is

$$
\begin{array}{ccccccc}
0 & 1 & 1 & & 0 & 1 & 0 \\
0 & p & 0 & \text{or} & 0 & p & 1 \\
0 & 1 & 0 & & 0 & 0 & 0
\end{array}
$$

On the other hand, p is 8-simple in the third case, but not in the first two cases.

It is easily seen that deleting a simple point p from S does not change the connectedness relationships in either S or $\bar{S}$; specifically, $S-\{p\}$ has the same components as S, except that one of them now lacks the point p, and $\bar{S} \cup \{p\}$ has the same components as $\bar{S}$, except that p is now in one of them. (Here if p is 4-simple, we must use 4-connectedness for S and 8-connectedness for $\bar{S}$, and vice versa if p is 8-simple; in what follows we take p to be 4-simple.) Adjacencies between pairs of components also continue to hold after p is deleted. Thus deletion of simple points (one at a time!) is a topology-preserving operation. Moreover, if we perform this operation repeatedly, it can be shown that all simply connected components of S shrink to isolated points. Thus we can use this operation to count these simply connected components.

The usefulness of this shrinking operation would be greatly enhanced if we could delete all simple points simultaneously, rather than one at a time; the operation could then be performed very efficiently on a parallel computer. Unfortunately, we cannot do this indiscriminately; for example, in

$$
\begin{array}{cccccc}
1 & 1 & 1 & \cdots & 1 & 1 \\
1 & 1 & 1 & \cdots & 1 & 1
\end{array}
$$

every point is simple, so that this component would vanish under a single parallel application of simple-point deletion.

The situation can be improved if we use operations that are direction dependent. Let Φ be the operation that deletes the point p from S if

(a) p has just one neighbor in S (so that p is simple) or
(b) p is simple, and its lower and right-hand (or left-hand) neighbors are in $\bar{S}$.

```
      1
    1 1                            1
    1 1 1              1 1              1              1
      1 1 1              1 1            1
      (a)                (b)          (c)            (d)
```

Fig. 6   Example of shrinking operation Φ. (a) Original simply connected object. (b)–(d) Successive shrinking steps.

If we apply Φ in parallel, it can be shown that any component C of S that has more than two points remains connected, and that repeated application of Φ shrinks simply connected C's to either pairs of points (which then vanish) or single points. We can easily modify Φ to test for pairs of points, and delete just one of them; the modified Φ shrinks all simply connected C's to single points. An example of how Φ works is shown in Fig. 6.

It is of interest to note that unless we modify Φ to test for pairs of points, we cannot guarantee that C's will never vanish. Indeed, let Φ be any operation such that $\Phi(0) = 0$, and $\Phi(1)$ depends only on the values of the 1's eight neighbors. Suppose that when Φ is applied in parallel to a two-valued picture, repeatedly, every simply connected component of 1's shrinks to a single point. By considering the objects

$$
\begin{array}{ccc}
1 & 1 & 1 \\
1 & & 1 \\
1 & & 1
\end{array}
\quad \text{and} \quad
\begin{array}{ccc}
1 & & 1 \\
1 & & 1 \\
1 & 1 & 1
\end{array}
$$

we can see that for the neighborhoods

$$
\begin{array}{ccc}
0 & 1 & 0 \\
0 & 1 & 0 \\
0 & 0 & 0
\end{array}
\quad \text{and} \quad
\begin{array}{ccc}
0 & 0 & 0 \\
0 & 1 & 0 \\
0 & 1 & 0
\end{array}
$$

Φ must change the center 1 to 0, since otherwise these objects could not shrink to points. Hence Φ causes two-point objects to vanish.

The operation Φ is useful when all components of S are simply connected, but it is of limited usefulness in more general cases. We shall now describe a more powerful operation that shrinks *all* components of S to single points (which then vanish).

Let Ψ be the operation which takes 1's into 0's if their right-hand and lower neighbors are both 0's; takes 0's into 1's if their right-hand, lower, and lower-right diagonal neighbors are all 1's; and leaves all other 0's and 1's unchanged. In other words, given the points $\begin{smallmatrix} a & b \\ c & d \end{smallmatrix}$, then:

For $a = 1$, we have $\Psi(a) = 0$ if and only if $b = c = 0$;
for $a = 0$, we have $\Psi(a) = 1$ if and only if $b = c = d = 1$.

Exercise 26. Prove that $\Psi(a) = 1$ if and only if $a+b = 2$, $a+c = 2$, or $b+c+d = 3$. ▌

For any subset A of the given picture, let A_1 be the set of 1's that are either in A, or immediately above and to the left of A, after Ψ is applied, and let A_0 be the set of such 0's. It can be shown that if C is a component of 1's that has more than one point, then C_1 is a component of 1's; and similarly if D is a component of 0's with more than one point, D_0 is a component of 0's. (Here we must use 4-connectedness for C and 8-connectedness for D.) Moreover, if C is 4-adjacent to D, then C_1 is 4-adjacent to D_0. In this sense, Ψ is a topology-preserving operation, provided no component of S or $\bar{S}$ consists of a single isolated point.

Exercise 27. If we use 8-connectedness for C and 4-connectedness for D, an analogous operation $\bar{\Psi}$ is defined by:

For $a = 1$, we have $\bar{\Psi}(a) = 0$ if and only if $b = c = d = 0$;
for $a = 0$ we have $\bar{\Psi}(a) = 1$ if and only if $b = c = 1$.

Prove that $\bar{\Psi}(a) = 1$ if and only if $a+d = 2$ or $a+b+c \geqslant 2$. ▌

When Ψ is applied repeatedly, it can be shown that any component C of S shrinks to the single point (i_c, j_c), where i_c is the smallest i-coordinate, and j_c the smallest j-coordinate of any point of C; this isolated point then vanishes. The number of steps required for this to happen is just the largest city block distance of any point of C from (i_c, j_c). The same is true for any component D of $\bar{S}$. A component shrinks to a point even if it originally has holes; this is possible because the holes themselves shrink to isolated points, and disappear, long before the component does. An example of the operation of Ψ is shown in Fig. 7.

```
1  1  1  1      1  1  1  1      1  1  1  1      1  1  1      1  1        1

1        1      1     1  1      1  1  1         1  1         1

1  1  1  1      1  1  1         1  1            1

   (a)             (b)            (c)            (d)        (e)       (f)
```

Fig. 7 Example of shrinking operation Ψ. (a) Original object. (b)–(f) Successive shrinking steps.

9.3 ARCS AND CURVES

This section deals with geometrical properties of arcs and curves in digital pictures, and with their application to the segmentation of curves and the analysis of their shapes. Shape analysis for arbitrary digital picture subsets will be treated in Section 9.4. Curves are treated first because some of the material in Section 9.4 makes use of them, e.g., concepts such as cross sections, convexity, etc. are defined in terms of straight lines.

Section 9.3.1 discusses the concepts of slope, straightness, and curvature for digital arcs and curves. Section 9.3.2 deals with the segmentation of curves, while Section 9.3.3 considers the use of various types of transformations to aid in the analysis of curves.

A digital picture subset S is arclike or curvelike if most of the points of S have exactly two neighbors in S. We must allow some points to be *end points*, having just one neighbor in S (compare with Exercise 12), and some to be *branch points* having more than two neighbors in S; but such exceptional points must not be too common, or else our intuitive notions of "arc" and "curve" will be violated. In the example given below, end points have been labeled E, branch points B, and ordinary arc points A:

```
     E  A
           A                              E
                B            A  A
                B   B   A
                B
     E  A  B  B  B  B  A  A  E                                    .
           B
           E
```

If S is arclike, we can evidently break it up into a small set of simple arcs (having two end points and no branch points) or simple closed curves (having no end or branch points). Each of these can then be represented by a chain code, by starting from one of the ends (or, for a closed curve, from any point) and recording the sequence of moves from neighbor until the other end (or the starting point) is reached. For example, we can break up the S just shown into three arcs

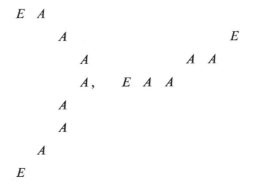

and

$$E \quad A \quad A \quad A \quad A \quad A \quad A \quad A \quad E$$

which have chain codes (see Section 1.2.3)

$$07765655, \qquad 00101, \qquad \text{and} \qquad 00000000$$

We will assume in most of this section that S is a simple arc or simple closed curve. Thus S consists of a sequence of points $p_0, \ldots, p_n$, where p_i is a neighbor of p_{i-1}, $1 \leqslant i \leqslant n$ (and in the closed curve case, p_n is a neighbor of p_0). Note that there are two possible definitions here, depending on whether "neighbor" means "4-neighbor" or "8-neighbor"; compare with Section 9.1.1.

9.3.1   Arc Length, Slope, and Curvature

In the real plane, an arc S can be defined by parametric equations of the form $x = f(t)$, $y = g(t)$, where t can take on values in some interval, say $0 \leqslant t \leqslant 1$. The *slope* of S at t_1 [i.e., at the point $(f(t_1), g(t_1))$] is $dy/dx = (dg/dt)/(df/dt)]_{t=t_1}$. The *arc length* of S between t_1 and t_2 is

$$\int \sqrt{1 + (dy/dx)^2}\, dx = \int_{t_1}^{t_2} \sqrt{(df/dt)^2 + (dg/dt)^2}\, dt$$

The *curvature* of S at t_1 is the derivative of its slope as a function of arc length

$$\frac{d^2y}{dx^2} \bigg/ \left(1 + \left(\frac{dy}{dx} \right)^2 \right)^{3/2} = \left[\frac{df}{dt}\frac{d^2g}{dt^2} - \frac{dg}{dt}\frac{d^2f}{dt^2} \right] \bigg/ \left(\left(\frac{df}{dt} \right)^2 + \left(\frac{dg}{dt} \right)^2 \right)^{3/2} \Bigg]_{t=t_1}$$

For a digital arc $S = p_0, \ldots, p_n$, we can define the arc length between p_i and p_j by counting the steps required to move from p_i to p_j along the arc. Here horizontal or vertical steps should count 1 each, and diagonal steps $\sqrt{2}$ each (compare the definition of perimeter in Section 9.2.1). For example, the arc lengths of the three arcs shown above are $3+5\sqrt{2}$, $3+2\sqrt{2}$, and 8, respectively. This definition is the discrete analog of the continuous definition given previously; the integral becomes a sum, and the derivatives become differences, only one of which is 1 for a horizontal or vertical step, while both are 1 for a diagonal step.

Slope and curvature are somewhat less trivial to define for a digital arc. At any point of the arc, the directions of the neighboring points (i.e., the arc's chain code values) define "slopes," and the difference between these directions defines a "curvature." These definitions, however, are of only limited value, since they are too much influenced by the discreteness of the digitization—in fact, they are always multiples of 45°.

A more flexible approach to defining digital slope and curvature is to allow the use of an arbitrary amount of smoothing. For the digital arc $S = p_0, \ldots, p_n$, we can define the left and right k-slopes of S at p_i as the directions from p_i to p_{i-k} and from p_i to p_{i+k}, respectively, where $k \geqslant 1$. Similarly, we define the k-curvature of S at p_i as the difference between the left and right k-slopes.

The k-slope (and k-curvature) can take on angular values whose tangents are rational numbers with denominator $\leqslant k$; thus the larger k, the more nearly continuous is the range of possible values. Of course, the k-slopes are not both defined if the point p_i is within k of an end of the arc. For a closed curve, they are defined at any point, provided that we add or subtract k modulo the length of the curve, i.e., we use $i+k-n$ if $i+k > n$, and $i-k+n$ if $i-k < 0$. The optimum choice of k depends on the particular application.

As an example of the advantages of using k-slopes, consider the arc

$$\cdot \cdot$$
$$A \quad A \quad A$$
$$A \quad A \quad A$$
$$A \quad A \quad A$$
$$\cdot \cdot$$

Here the right 1-slope is $0°$ at some points and $45°$ at others. The right 3-slope, on the other hand, is $\tan^{-1}(\tfrac{1}{3})$ at every point; this constant value reflects the fact that the given arc is the digitization of a straight line. The right k-slopes for $k > 3$ are not all constant, but for large k, their values are close to $\tan^{-1}(\tfrac{1}{3})$.

Exercise 28.   Compute the k-curvatures of the arc shown in the preceding paragraph; show that they are zero when k is a multiple of 3, and approach zero for any large k.   ▮

As pointed out at the beginning of Section 9.3, a digital arc can be represented by its chain code, i.e., by the set of moves from neighbor to neighbor that we make in going from one end of the arc to the other. This is analogous to the fact that a real arc is determined, except for its position, by specifying its slope as a function of arc length. Similarly, a real arc is determined, except for position and orientation, by specifying its curvature as a function of arc length; in the digital case, this corresponds to using a "difference chain code" in which we specify the angles between the successive directions from neighbor to neighbor. For example, the difference chain codes of the three arcs shown at the beginning of Section 9.3, expressed in multiples of $45°$, are $-1, 0, -1, -1, 1, -1, 0; 0, 1, -1, 1$; and $0, 0, 0, 0, 0, 0, 0$, respectively. These methods of specifying a curve, using its slope or curvature as a function of arc length, are called *intrinsic equations* for the curve.

Useful information about a digital arc can be obtained if we know its

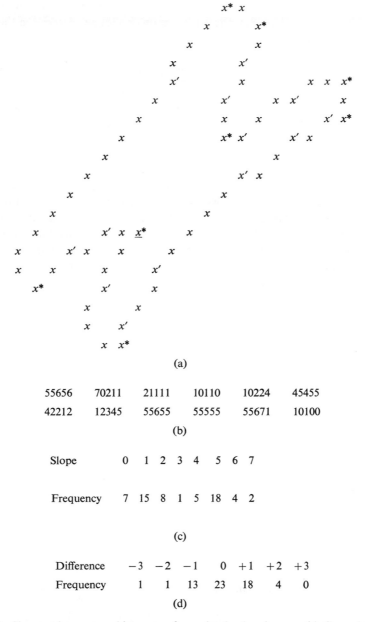

(a)

55656   70211   21111   10110   10224   45455
42212   12345   55655   55555   55671   10100

(b)

Slope   0 1 2 3 4 5 6 7

Frequency   7 15 8 1 5 18 4 2

(c)

Difference   −3 −2 −1   0 +1 +2 +3
Frequency   1   1 13   23 18   4   0

(d)

Fig. 8 Slope and curvature histograms for a simple closed curve. (a) Curve (starting point underlined); asterisks show curvature maxima, primes show inflections (curvature zero-crossings). (b) Chain code of (a). (c) 1-slope histogram of (a) (d) 1-curvature histogram of (a), modulo 8.

histogram of k-slopes or curvatures, i.e., how often each value of slope or curvature occurs. Of course, these histograms no longer determine the shape of the arc; for example,

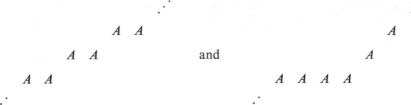

have the same histogram of right 1-slopes, but have different shapes. Useful descriptions of the arc can, however, be derived from such histograms. For example, a large peak on the k-slope histogram indicates that the arc has a strong directional bias. Similarly, a preponderance of low k-curvatures indicates a relatively straight, smooth curve, while a large number of high k-curvatures corresponds to a "busy" curve. The 1-slope and 1-curvature histograms for a simple digital closed curve are shown in Fig. 8. The slope histogram is sometimes called the "directionality spectrum."

It is important for many purposes to be able to characterize digital *straight lines*—in other words, to determine conditions under which a digital arc can be the digitization of a real straight line segment. For example,

```
                        A   A   A
                    A   A   A
                A   A   A
```

and

```
                                A   A   A
                            A   A
                        A   A   A
                    A   A   A
                A   A
```

are digital straight line segments, but

```
            A                           A
            A       and             A
        A   A                   A   A   A
```

and

$$A \quad A$$
$$A \quad A$$
$$A \quad A \quad A \quad A$$

are not.

Let the given arc S be represented by a chain code in which each slope is a multiple of 45°. Then the following are necessary conditions for S to be a digital straight line segment:

(a)   At most two slopes occur in the chain, and these slopes differ by 45°. This is violated by

$$A$$
$$A$$
$$A \quad A$$

(b)   At least one of the slopes occurs only in runs of length 1. This is violated by

$$A$$
$$A$$
$$A \quad A \quad A$$

(c)   The other slope occurs only in runs of at most two lengths (except possibly at the ends of the arc, where shorter runs could occur), and these lengths differ by 1. This is violated by

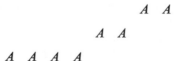

These conditions are not sufficient; in fact, it turns out that the runs in (c) themselves occur in runs—one of them only in runs of length 1, and the other in runs of at most two lengths that differ by 1; and so on. Basically, condition (c) and its refinements ensure that the runs of length 1 are spaced as evenly as possible in the chain. For most purposes, one would not care whether a digital arc satisfied these conditions exactly; it would be sufficient to know that the arc is approximately a straight line, e.g., that its smoothed slope is approximately constant, or that it is a sufficiently close fit to a real straight line.

9.3.2 Segmentation of Curves

As indicated at the beginning of Section 9.3, a digital picture subset that is composed of arcs can be segmented into its individual arcs (or closed curves) by breaking it at its branch points. In this section, we discuss methods for segmenting simple arcs or curves into parts. Many of these methods are analogous to the methods of picture segmentation described in Chapter 8.

Parts of a curve where the k-slope (or curvature) lies in a given range can be extracted by simple "thresholding," i.e., suppressing points at which the slope lies outside the given range. This elementary type of segmentation might be of use in separating a drawing, or an alphanumeric character, into directional "strokes," or into straight versus curved parts.

A more interesting use of curvature for segmenting a curve is to detect *curvature maxima,* or *"angles,"* on the curve. This is analogous to detecting edges in a picture, since angles are points where there is an abrupt change in (average) slope, just as edges are abrupt changes in (average) gray level. Many of the edge detection methods described in Section 8.2 can be adapted to angle detection. For example, as in Section 8.2.4, we can take average slopes over adjacent, nonoverlapping segments of a curve, subtract them (this is essentially how we defined k-curvature in Section 9.3.1), and suppress nonmaxima, to determine good "angles" on the curve. The angles found in this way for the shape in Fig. 8a are marked by asterisks.

Another useful class of curve segmentation points is its *points of inflection.* For a curve in the real plane, these points are the zero crossings of the curvature, and separate the curve into convex and concave parts ("peninsulas" and "bays"). Zero crossings of the smoothed curvature are marked by primes in Fig. 8a.

If the curve S being segmented is the border of a solid object T, a very simple technique can sometimes be used to detect angles. Given a point p on S, we count the number k of points of T that lie in some given neighborhood of p. If k is about half the area m of the neighborhood, then S must be relatively straight at p. If k is much larger or much smaller than $m/2$, on the other hand, T must have a sharp concave or convex angle, respectively, at p. Here again, we can suppress nonmaxima or nonminima to locate the angle sharply.

Local shape features on an arc, such as angles, bays, etc., can sometimes be detected by constructing chords. For example, given the point p_i of the curve, we can draw a chord from p_{i-k} to p_{i+k}, say, and measure how far p_i lies from this chord. If this distance is a large fraction of k, then p_i must be at or near a sharp angle; a point where this distance is a maximum would be the best choice for locating the angle. On the other hand, if p_i is very close to this chord, then either the curve is relatively straight between p_{i-k} and p_{i+k}, or else p_i lies near a point of inflection.

Another way of using chords to detect shape features on a curve is to measure the chord/arc ratio. If the chord from p_{i-k} to p_{i+k} is almost as long as the arc between these points, the curve must be relatively straight there. The chord/arc ratio can, in fact, be used as a measure of straightness. On the other hand, if the chord is much shorter than the arc, there must be a "spur" or "bulb" between the points. The size of the features that can be detected in this way depends, of course, on the choice of k.

Curves can also be segmented by using matching techniques to find places where given sequences of slopes (or curvatures) are present. For example, let $s_1, \ldots, s_n$ be the successive slopes (i.e., the chain code) of the curve S, and let $t_1, \ldots, t_m$ be similarly defined for the "template" arc T. We can measure the degree of match between the T slope sequence and a part of the S sequence, beginning at the $(k+1)$st point of S, by computing an expression of the form

$$\sum_{i=1}^{m} h(t_i, s_{i+k})$$

where h is some measure of absolute angular difference. This is analogous to measuring the match between two pictures by cross-correlating them. As in picture matching (Section 8.3.5), search techniques can be very useful when we are looking for matches between arcs or curves.

Still another approach to curve segmentation is to use piecewise approximation techniques analogous to the "partition building" methods described in Section 8.4.4. Here the curve is initially broken into pieces in some arbitrary way, and real curves of some standard form, e.g., straight line segments, are fitted to the pieces. If the fit is bad for a given piece, it is subdivided; if the fits to two adjacent pieces match well, they are merged. As a simple example of this approach, we can pick an arbitrary set of points on the curve and join them by lines to form an inscribed polygon. If the curve does not lie close enough to some sides of this polygon, we can subdivide that piece of the curve further, e.g., pick the point farthest from s as a new vertex. The curvature maxima of a curve are generally good places to put the vertices of an approximating polygon, since the curve is turning relatively rapidly at such a maximum, so that it cannot be fit very well there by a single straight line segment.

9.3.3 Transformations of Curves

It is sometimes easier to analyze or recognize the shape of a curve if the curve is represented in a particular way, e.g., in a transform domain or in a special coordinate system. In this section we describe several such approaches to curve representation.

The intrinsic equation of a closed curve, which specifies the slope of the curve as a function of arc length, is a periodic function with period L (the total arc length). This function can therefore be expanded in a Fourier series on the interval $[0, L]$, and we can take the coefficients of this series, or combinations of these coefficients, as descriptors of the curve's shape. We can also derive useful approximations to the curve by truncating the series. If desired, the coefficients in some other orthonormal expansion, e.g., Walsh coefficients, can be used as descriptors in place of the Fourier coefficients.

Such Fourier descriptors of a curve are independent of the curve's position, since they are derived from an intrinsic equation. We can also make them independent of orientation, by taking the power spectrum of the Fourier series (i.e., using the sums of the squares of the cosine and sine Fourier coefficients as descriptors), since this is independent of the phase of the original periodic function. Scale independence can also be achieved by scaling the Fourier series so that its constant term has some specified value, say 1.

This method can be applied to digital curves if we regard them as having piecewise constant slope (i.e., each link in the chain code corresponds to a straight line segment whose length is 1 or $\sqrt{2}$, and whose slope is a multiple of 45°). For such polygonal curves, the Fourier coefficients can be expressed as discrete sums. Of course, if we truncate the Fourier series, the curve reconstructed from the truncated series will no longer be a digital curve, but will have to be redigitized.

The Fourier expansion method can also be applied to other equations for a closed curve, provided that the curve is represented by a periodic function. For example, suppose that the curve has a single-valued equation $r = h(\theta)$ in polar coordinates, i.e., that each ray emanating from some origin hits the curve exactly once. Since h is periodic with period 2π, we can expand it in a Fourier series as above. As another example, let $x = f(t)$, $y = g(t)$ be the parametric equations of a curve, where $0 \leqslant t \leqslant 1$; then f and g are periodic with period 1, and can be expanded in a Fourier series.

Another approach to curve transformation involves a change of coordinates in the space domain. Suppose, for example, that we want to determine whether a given (digital) curve is (approximately) a straight line segment. Let p be a point on the curve, and let us choose coordinates so that the origin is at p. If the curve is really a straight line segment, its equation should now be of the form $y = mx$. Thus if we take the ratio y/x (where y and x are measured relative to p) for all points on the curve, and find that this ratio is approximately constant, we can conclude that the curve is approximately a line segment. Note that this method works even if the curve is a dotted curve, i.e., we can use it to determine whether a disconnected set of points lies approximately on a line.

As a less trivial example, suppose that we want to know whether the given curve is approximately an arc of a circle. Let p be a point on the curve, and let us choose coordinates so that the origin is at p and the x-axis is the tangent to the curve at p. If the curve is a circle, say of radius r, its equation in this coordinate system is $x^2+(y-r)^2 = r^2$. This is equivalent to $(x^2+y^2)/y = 2r$. Thus if we compute the quantity $(x^2+y^2)/y$ (with x and y measured relative to p) for all points on the curve, and find that this is approximately constant, we can conclude that the curve is approximately circular. Note that this method depends on our being able to determine a tangent to the curve at p; this is nontrivial for a digital curve.

We conclude this section by describing an interesting type of transformation, taking points into straight lines, that can be used to determine whether a given set of points lies on a line. For each given point (a_i, b_i), we construct the line $y = a_i x + b_i$. It is easily verified that if the points $(a_1, b_1), (a_2, b_2), \ldots$ are collinear, then the lines $y = a_1 x + b_1, y = a_2 x + b_2, \ldots$ all pass through a common point. (If the points lie on a line that is nearly parallel to the y-axis, the corresponding lines become nearly parallel; we can use lines of the form $y = b_i x + a_i$ to handle this case, if desired.[†]) Suppose then that each time we find a point (a, b) in the given picture, we add 1's to the values of all points along the line $y = ax + b$ in a new picture. When this has been done for a large number of collinear (a, b)'s, we can expect to find a point in the new picture (or perhaps a small neighborhood of a point, due to roundoff errors) that has a high value. The presence of many collinear points in the original picture can thus be detected by thresholding the new picture. In fact, the position of the point in the new picture determines the slope and intercept of the line in the original picture.

Exercise 29. Verify that collinear points (a_i, b_i) give rise to concurrent lines $y = a_i x + b_i$, and determine the coordinates of the common point on these lines in terms of the slope and intercept of the line through the (a_i, b_i)'s. ∎

9.4 DIRECTIONALITY AND SHAPE

Descriptions of pictures often refer to the shapes of objects or regions in the pictures. Unfortunately, shape is a very difficult concept to describe; it is not a single property, but a large family of properties. Studies of human judgments of shape similarity have shown that shape comparision is a highly multidimensional process. In the latter part of this section we discuss a few specific shape properties—extent, convexity, dispersedness, elongatedness—which can be treated quantitatively.

[†] More generally, given any two-parameter representation of the lines in the plane, we can define an analogous transformation by letting these parameters correspond to the coordinates of a point. Similar methods can be used for other two-parameter families of curves.

```
                                        B  B  B
                    A                   B  B  B
                                        B  B  B
                          (a)

                    A

                                        B  B  B
                                        B  B  B
                                        B  B  B
                          (b)

B  B  B  B  B  B  B  B  B  B  B  B  B  B  B  B  B
                                        B  B  B
                    A                   B  B  B
                                        B  B  B
                          (c)

      B  B  B  B  B  B  B  B  B  B  B  B  B  B
      B  B  B                              B
      B  B  B                 A            B
      B  B  B
                          (d)

                    A  A  A  A  A  A  A  A
                    A
                    A              B  B  B
                    A              B  B  B
                    A              B  B  B
                          (e)

                    A  A  A  A  A  A  A  A  A  A  A
                    A
                    A              B  B  B
                    A              B  B  B
                    A              B  B  B
                          (f)
```

Fig. 9   The difficulty of defining "to the left of." In which of these cases is object A to the left of object B?

In describing a picture, we often want to refer to the relative positions of objects in the picture, and to make statements like "*A* is to the left of *B*," "*C* is above *D*," or "*Y* is between *X* and *Z*." Even these simple-seeming relationships, unfortunately, are quite nontrivial to define. For example, consider Fig. 9. Clearly *A* is to the left of *B* in (a); but in the remaining parts of the figure, the situation becomes less and less clearcut.

It is not difficult to define a reasonable concept of "to the left of" for objects *A* and *B* that are single points. For example, we can say that point *A* is to the left of point *B* if the line from *A* to *B* has slope in the range $\pm 45°$; or, more realistically, we can treat "to the left of" as a quantitative, or "fuzzy" relation (see reference [48] in Section 8.5), and say that the strength of this relation is 1 when the line *AB* has slope 0, and decreases monotonically as the slope departs from 0.

For extended objects, on the other hand, defining "to the left of" becomes much more complicated, as Fig. 9 shows. Do we require that every point of *A* be to the left of *every* point of *B*? This would exclude all of the cases in Fig. 9 except (a) and (b); but excluding case (e) seems unreasonable. Do we require only that every point of *A* be to the left of *some* point of *B*? This would allow cases (a)–(d), and exclude (e) and (f); but allowing (d) seems unreasonable. Do we require that the centroid of *D* be to the left of the centroid of *B*? This allows all cases except (d); but allowing (f) seems dubious.

It has been suggested that "to the left of" must be defined in a complex fashion, e.g.,

(1) the centroid of *A* is to the left of the leftmost point of *B*, and
(2) the rightmost point of *A* is to the left of the rightmost point of *B*.

As applied to Fig. 9, this rule excludes cases (c), (d), and (f). Even a rule of this type will, however, not be adequate in all cases, since the concept "to the left of" is context sensitive; it can be influenced by the presence of other objects, by the meanings of the objects, etc. The problem of defining relations of relative position in a picture still lacks a satisfactory solution.

The first part of this section deals with geometrical properties and relationships which are direction-dependent. In Sections 9.4.1 and 9.4.2, directional projections and cross sections of a picture are defined, and their use in segmenting and analyzing picture subsets is discussed. (Reconstruction of pictures from projections is not treated here; on the related problem of estimating point spread functions from line spread functions see Section 7.1.2.)

The rest of the section, as previously mentioned, treats various basic shape properties of picture subsets, including convexity and elongatedness. A region all of whose cross sections are connected is convex; properties and segmentation techniques related to convexity are considered in Section 9.4.3. Measures of the complexity of a picture subset (dispersedness, asymmetry, etc.) are introduced in Section 9.4.4, which also describes methods of segmenting a picture by merging regions so as to achieve shape simplification. The property of elongatedness is treated in Section 9.4.5, where we also describe methods of thinning an elongated subset into a set of arcs and curves.

9.4.1 Projections

Let f be an $m \times n$ digital picture. The vector of column sums

$$\left(\sum_{j=1}^{n} f(1,j), \sum_{j=1}^{n} f(2,j), \ldots, \sum_{j=1}^{n} f(m,j) \right)$$

is called the *x-projection* of f, and the vector of row sums

$$\left(\sum_{i=1}^{m} f(i,1), \sum_{i=1}^{m} f(i,2), \ldots, \sum_{i=1}^{m} f(i,n) \right)$$

is called the *y-projection* of f. For example, if f is the 3×5 picture

$$
\begin{array}{ccccc}
0 & 1 & 1 & 2 & 2 \\
1 & 1 & 2 & 3 & 2 \\
1 & 3 & 2 & 2 & 3
\end{array}
$$

then its *x*-projection is (2, 5, 5, 7, 7), and its *y*-projection is (11, 9, 6).

More generally, we can define the projection of a picture f on any line, say a line having slope θ, by summing (or in the nondigital case, integrating) the gray levels of f along the family of lines perpendicular to θ. For a digital picture, these lines can be determined as in Section 9.3.1. Note that in this case, the sums do not all have the same number of terms, since some of the lines will be long lines that cross the picture near its center, while others will be short lines that cross it near its corners. As an example, the projection of the 3×5 picture shown above, on a line having slope 45°, is computed by summing the gray levels along all lines having slope 135°; the vector of sums is (1, 4, 3, 5, 7, 4, 2).

Still more generally we can define a "projection" of f by summing its gray levels along some family of curves, e.g., along all the lines through some point, along all the circles centered at some point, etc. Note that in the digital case, nearby curves in such a family may turn out to have many points in common, e.g., digital lines through a point whose slopes are almost the same will have all of their points in common until they get sufficiently far away from the point. Thus in the digital case the sums taken along a family of curves may be quite redundant.

We now show how projections can be valuable in detecting and locating objects in a picture. Suppose, for example, that the picture contains a relatively large, relatively compact object that is darker (i.e., has higher average gray level) than its background. Then, as shown in Fig. 10a, the x- and y-projections of the picture have plateaus at the approximate x- and y-positions of the objects, respectively. Thus examination of these projections gives a good indication that there is such an object in the picture, and where the object is located.

As another example, suppose that the picture contains a thin dark vertical line on a noisy background, as shown in Fig. 10b. Then the x-projection of the picture will have a peak at the position of the line, if the noise level is not too high. Note that the line can be very discontinuous, as long as there is a significantly higher average gray level along it than there is along parallel lines that traverse only the background. More generally, "projection" along a family of curves can be used to detect the fact that, in the given picture, some curve of that family has high average gray level compared to its background.

The methods of curve recognition by coordinate conversion described in Section 9.3.3 can be regarded as based on this approach. For example, when we compute y/x for each point of a curve through the origin, and conclude that the curve is a straight line if the values we obtain are nearly constant, we are essentially projecting along all lines through the origin [the point (x_0, y_0) on the curve is counted as belonging to the line $y = (y_0/x_0)x$], and detecting a straight line if one of these projections detects a high count or sum (i.e., many points have the same y/x value).

Projections in various directions also provide properties that can be useful in the identification of objects, particularly if the objects have known orientations. For example, consider the set of numerals

```
1  1  1          1          1  1  1

1     1          1                1

1     1          1                1        etc.

1  1  1          1          1  1  1
```

(where the blanks are 0's). These have x-projections

$$(4, 2, 4), \quad (4), \quad (2, 3, 3)$$

and y-projections

$$(3, 2, 2, 3), \quad (1, 1, 1, 1), \quad (3, 1, 1, 3)$$

which are quite distinctive.

Given a picture subset S, we can define its *extent* in a given direction θ as the length of its projection on a line in that direction. Here the length of a

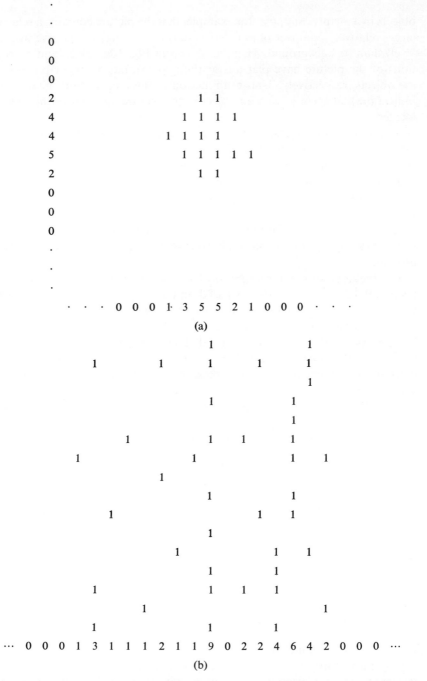

Fig. 10 Detection of objects by analysis of projections. (a) Picture containing a dark blob; the x- and y-projections show plateaus that indicate its position. (b) Picture containing dark vertical lines; the x-projection shows peaks at their positions.

projection is the distance between its farthest apart nonzero values. Thus the extent of S is the distance between a pair of parallel lines perpendicular to θ that just bracket S. For example, the vertical extent, or *height*, of S is the distance between the uppermost and lowermost rows of the picture that contain points of S. Similarly, the horizontal extent, or *width*, of S is the distance between the leftmost and rightmost columns that contain points of S. The *diameter* of S is defined as its greatest extent in any direction.

Exercise 30. Prove that the diameter of S is the greatest distance between any two points of S. ∎

Let S_x be the projection of S on the x-axis, and let x_L, x_R be the leftmost and rightmost x's for which S_x has nonzero values; thus $x_R - x_L$ is the width of S. The points of S whose x-coordinates are x_L or x_R must be border points (if they were interior points, x_L and x_R could not be rightmost or leftmost). These points are the leftmost and rightmost points of S; we can refer to them as the horizontal *extremum points*[†] of S. In general, let S_θ be the projection of S on a line having slope θ; then we can analogously define the extremum points in direction θ. Extremum points are sometimes useful places to segment the border of S.

9.4.2 Cross Sections

For any (digital) picture f and any straight line L, the sequence of gray levels of f at the points of L is called the *cross section* of f along L. (More generally, we can speak of the cross section of f along any curve C, not necessarily a straight line.) The projection of f on a line having slope θ is just the sequence of sums of the cross sections of f along the family of lines whose slopes are $\theta + (\pi/2)$. For example, the cross sections of the picture

$$0 \quad 1 \quad 1 \quad 2 \quad 2$$
$$1 \quad 1 \quad 2 \quad 3 \quad 2$$
$$1 \quad 3 \quad 2 \quad 2 \quad 3$$

along the set of lines having slope $135°$ are 1; $3, 1$; $2, 1, 0$; $2, 2, 1$; $3, 3, 1$; $2, 2$; and 2.

When a picture contains objects whose orientations are known, cross sections in suitable directions can be very useful for detecting, locating, and

---

[†] Note that these are "global" extremum points, i.e., no point of S is further to the left or right. It is easier to detect *local* extremum points, i.e., points that are farther to the left or right than any of their *neighboring* points of S (and similarly for directions other than horizontal). Note that these too must be border points.

identifying these objects. For example, if a picture contains a line having slope θ, one of its cross sections in direction θ will detect this line. As another example, consider the numerals composed of 1's and 0's that were shown in Section 9.4.1, where 1's are black and 0's white. If we look at horizontal cross sections, we can distinguish the numeral 1 from the other numerals because it meets each row in a single black point; we can distinguish the numeral 0 because it meets the top and bottom rows in runs of three black points, and the intermediate rows in black–white–black patterns; and so on.

Given a subset S of the picture f, a line L across f will in general contain runs of points of S, alternating with runs of points of $\bar{S}$. If S has known orientation, the patterns of runs obtained in this way from various lines, or the numbers of lines that yield specified patterns of runs, provide useful descriptive properties of S, as just seen. Even if the orientation of S is unknown, some of the basic geometrical properties of S can be estimated by studying cross sections along a family of random lines. For example, it can be shown that the expected number of times that a random line meets S is proportional to the perimeter of the convex hull of S; while the expected total length of intersection of a random line with S (i.e., the sum of the lengths of the runs in which the line meets S) is proportional to the area of S. These results, which belong to the field of *integral geometry*, depend on the way the concept of a random line is defined, and are strictly valid only for continuous pictures. In principle, they could form the basis for a class of Monte Carlo methods of estimating the values of picture properties.

The patterns of runs in which lines meet S can also be used to define useful segmentations of the picture. For example, consider the set of points p such that the half-line through p in direction θ meets S. Any such point p would be in S's *shadow* if we illuminated the picture from direction $\theta + \pi$. Some of the shadows defined in this way by a simple S are shown in Fig. 11. More generally, one can consider points p such that a line through p in some direction meets S at least a specified number of times, or has at least a specified

```
0  0  0  0  0  0  0                                    B  B  B  B  B

0  1  1  1  1  1  0

0  1  0  0  0  1  0          A  A  A                   B  B  B

0  1  1  1  1  1  0

0  1  0  0  0  1  0          A  A  A

0  1  0  0  0  1  0          A  A  A

0  0  0  0  0  0  0       A  A  A  A  A
        (a)                    (b)                        (c)
```

Fig. 11 Segmentation using shadows. (a) Original picture. (b)–(c) Parts of background of (a) that are in shadow from above and below, respectively.

total length of intersection with S, or meets S in a run of at least a certain length, and so on. Those definitions single out parts of the picture that stand in various simple spatial relationships to parts of S.

Comparison of successive cross sections in a given direction also provides a useful method of further segmenting a given picture subset S. (This method is best used when S has known orientation, since its results are sensitive to the direction of the cross sections.) Suppose, for concreteness, that we examine the cross sections of S along the successive rows of the picture. As previously mentioned, each row contains runs of points of S separated by runs of points of $\bar{S}$. We recall from Sections 9.1.3 and 9.2.1 that tracking the runs of S points from row to row can be used to label the connected components of S, measure their areas, etc. We can also use the tracking process to break the components into pieces having simple shapes. For example, whenever a new run is found (not adjacent to runs on the previous row), we can regard this as the beginning of a new piece of S; and similarly, when a run splits into two or more runs, we can regard them as starting new pieces. On the other hand, when a run has nothing adjacent to it on the following row, or when two or more runs merge into a single run, we can regard the corresponding piece(s) of S as having come to an end. If desired, we can also start and end pieces of S whenever there is a drastic change in the lengths or positions of the runs being tracked, e.g., when we detect successions of runs such as

```
1  1  1  1  1              1  1  1  1  1  1  1  1

1  1  1  1  1              1  1  1  1  1  1  1  1
                    or
   1  1  1  1  1  1               1  1

   1  1  1  1  1  1               1  1
```

we can regard a piece of S as having ended and a new one as having begun. An example of segmentation based on this method is shown in Fig. 12. Of course, this method can be used with cross sections in any desired direction, not necessarily horizontal. The pieces found in this way can be characterized by the successions of lengths and positions of their constituent runs, e.g., as straight or slanted bars

```
   1  1  1          1  1  1

   1  1  1             1  1  1

   1  1  1    or          1  1  1

   1  1  1                   1  1  1

   1  1  1                      1  1  1
```

```
        1  1  1                            A  A  A
        1  1  1                            A  A  A
        1  1  1                            A  A  A
        1                                     B
   1  1  1  1  1  1  1               C  C  C  C  C  C  C
   1     1  1  1     1               D     E  E  E     F
   1     1  1  1     1               D     E  E  E     F
   1     1  1  1     1               D     E  E  E     F
1  1     1  1  1     1  1            G  G     E  E  E     H  H
         1  1  1                              E  E  E
         1  1  1                              E  E  E
         1  1  1                              E  E  E
         1     1                              I     J
         1     1                              I     J
         1     1                              I     J
      1  1     1  1                        K  K     L  L
            (a)                                  (b)
```

Fig. 12 Segmentation by run tracking. (a) Original picture. (b) Parts found by run tracking.

or as wedges

```
   1  1  1  1                1  1  1  1
      1  1  1                   1  1  1
                 or
         1  1                      1  1

         1                            1
```

9.4.3 Convexity

Let S be a picture subset, and p a point of S. Suppose that every straight line through p meets S exactly once, i.e., that the picture cross section along any line through p contains exactly one run of points of S. We then say that S is *star-shaped* from p. This means that one can see any point of the border of S by standing at p and looking in the proper direction. Thus if we draw a ray from p in any direction θ, it meets the border of S exactly once. The distance $r = f(\theta)$ from p to the border, as a function of θ, evidently determines S

completely. The equation $r = f(\theta)$ provides a convenient way to represent S; moreover, since $r(\theta)$ is periodic with period 2π, we can use its Fourier series expansion to define other convenient descriptors of S, as in Section 9.3.3.

Exercise 31. Prove that if S is star-shaped from some point, S must be simply connected. ∎

Exercise 32. Prove that the following statements are equivalent:

(a) S is star-shaped from p.
(b) For any point q of S, the straight line segment from p to q lies entirely in S. ∎

It is not hard to see that the following properties of a picture subset S are all equivalent (compare with Exercise 32):

(a) S is star-shaped from every point p of S.
(b) For any two points p, q of S, the straight line segment from p to q lies entirely in S.
(c) For any two points p, q of S, the midpoint of the straight line segment from p to q lies in S.
(d) Every straight line meets S at most once.

If S has these properties, it is called *convex*. Thus a convex S has no "concavities," which would be places where a line segment between two points of S would pass outside S. It is also clear that a convex S is connected and can have no holes (see Exercise 31), and that an arc is convex if and only if it is a straight line segment.

These observations apply to subsets S of a continuous picture. We can continue to use these definitions for digital pictures, since they depend only on the concept of a straight line (see Section 9.3.1). Property (c) must be slightly modified, since the midpoint of two digital points (i, j) and (h, k) is $((i+h)/2, (j+k)/2)$, which is not necessarily a digital point. We can, however, modify property (c) to require that if (i, j) and (h, k) lie in S, then some digital point obtained from its midpoint by increasing or decreasing each coordinate by $\frac{1}{2}$ (if necessary) also lies in S.

Any digital S *could* be the digitization of a concave real object R, since R could have tiny concavities which are missed by the digitization process. On the other hand, not all digital S's can be the digitizations of convex real objects; if S has a major concavity, large in comparison with the spacing of the sampling array, there is no way that S could be the digitization of a convex R. If S *is* the digitization of a convex R, it can be shown that S must have the midpoint property described in the previous paragraph. The converse, unfortunately, is not true: S can have the midpoint property and yet not be the digitization of any convex R.

Any intersection of convex sets is convex (in the continuous case), since if p and q are in the intersection, they are in each of the sets; hence the line segment joining them lies in each of the sets, and so in the intersection. Thus, for any set S, there is a smallest convex set containing S, namely the intersection of *all* the convex sets that contain S. This set is called the *convex hull* of S; let us denote it by S_H. Clearly S is convex if and only if $S = S_H$. Constructing S_H is one way to detect the concavities of S, since any point of S_H that is not in S must belong to a concavity in S.

It is easily shown that if S is connected, then S_H is the union of all the line segments whose endpoints lie in S. We can use this observation to define an algorithm for constructing S_H in the digital case; this algorithm is particularly suitable for implementation on a parallel array-processing computer. Let $(S)_\theta$ denote the result of "smearing" S in direction θ, i.e., it is the union of all possible shifts of S, by amounts $0, 1, 2, \ldots$, in direction θ. (In fact, we need not use shifts greater than the diameter of S.) Thus $(S)_\theta \cap (S)_{\theta+\pi}$ is the set of points that have points of S on both sides of them in direction θ. (Compare the discussion of "shadows" in Section 9.4.2.) In other words, any point in $(S)_\theta \cap (S)_{\theta+\pi}$ is on some line segment, of slope θ, both of whose ends are in S; and conversely, any point on such a line segment is in $(S)_\theta \cap (S)_{\theta+\pi}$. Thus the union of the sets $(S)_\theta \cap (S)_{\theta+\pi}$ for all θ is just S_H. Note that we cannot get away with using just a few directions θ in this construction; for example,

$$1$$

$$1 \quad 1$$

$$1 \quad 1 \quad 1 \quad 1 \quad 1 \quad 1$$

contains every line segment between a pair of its points whose slope is a multiple of 45°, but it is not convex.

Another construction for S_H, more suitable for a conventional sequential computer, is as follows: Let p_1 be the leftmost of the topmost points of S, and let L_0 be the horizontal line through p_1. Rotate L_0 counterclockwise about p_1 until it hits S; call the resulting rotated line L_1, and let p_2 be the point of S farthest from p_1 along L_1. Repeat the process with p_2 replacing p_1 and L_1 replacing L_0, i.e., rotate L_1 counterclockwise about p_2 until it hits S, let L_2 be this rotated line, and let p_3 be the point of S farthest from p_2 along L_2. It is not hard to show that this process, after going completely around S counterclockwise, eventually yields $p_n = p_1$ and $L_n = L_1$ for some n. If S is connected, it can be shown that the polygon whose vertices are $p_1, p_2, \ldots, p_{n-1}, p_n = p_1$, together with its interior, is just S_H.

Exercise 33.   Prove the assertions in the preceding paragraph. Also prove the following: let p, q be points on the outer border of S such that the chord

pq lies outside S, and let $(S)_{pq}$ be the part of $\bar{S}$ surrounded by S and by pq. Prove that if $(S)_{pq}$ is maximal (i.e., is not contained in any other $(S)_{p'q'}$), then p and q must be on one of the lines L_i. ∎

An arbitrary S can be decomposed into convex pieces in various ways. For example, if S is a polygon, we can break it into convex pieces (e.g., triangles) by drawing some of its diagonals, or by extending some of its sides until they intersect other sides. As another example, suppose that S is a union of convex pieces some of which overlap; then if we want to cut apart S into these pieces, we should begin the cuts at the deepest points of concavities, and try to cut so as to join pairs of the concavities.

9.4.4 Complexity

Human judgments of the complexity of shapes depend on a number of factors. (We will assume here that a "shape" is a simply connected picture subset S.) In general, the more curvature maxima ("angles") S's border has, the more complex S is judged to be. The judged complexity also depends, however, on the variability of the magnitudes of these angles; an equiangular polygon would be judged to be less complex than one whose angles are all different.

Intuitively, the judged complexity of a shape should depend on the amount of information that must be given in order to specify the shape. Thus a shape that has fewer angles, or whose angles are of fewer sizes, is simpler than a shape whose angles are more numerous or more variable. For the same reason, *symmetry* is a significant factor in complexity judgments; it takes at least twice as much information to specify an asymmetrical shape as to specify a symmetrical one.

Shape complexity appears to play an important role in the way certain ambiguous pictures are perceived, as discussed in Section 3.5. For example, Fig. 3.18a is easy to see as a perspective drawing of a three-dimensional wireframe cube, while Fig. 3.18b is easier to see as a two-dimensional polygon. This can be explained on the grounds that Fig. 3.18a, when seen in two dimensions, has considerable variability in the lengths of its lines and the sizes of its angles; thus a three-dimensional interpretation of this drawing is "simpler." Unfortunately, straightforward formulations of the concept of simplicity–complexity have not been very successful in predicting how people will judge ambiguous pictures.

A significant complexity factor not yet mentioned here is the "dispersedness" of a shape. Let us consider shapes S that have a given area A. Intuitively, if the perimeter of P of S is small, S must be relatively "compact"; while as P gets large, S becomes more and more "dispersed."

In the real plane, a theorem known as the "isoperimetric inequality" states that $P^2/A \geqslant 4\pi$, with equality holding if and only if S is a circle. This suggests that one can take P^2/A as a measure of dispersedness; a circle is the most compact possible shape, since it has the smallest possible perimeter for a given area. Note that if two S's are geometrically similar, i.e., they have the same shape but different sizes, then they have the same P^2/A, since perimeter increases linearly with size, while area increases with the square of the size.

For digital pictures, one can similarly use P^2/A as a dispersedness measure, with A and P computed as in Section 9.2.1. Unfortunately, it turns out that the digital P^2/A is not a minimum for digitized circles; depending on how P is measured, P^2/A is smaller for certain octagons than it is for circles. Thus in the digital case, P^2/A is not as precise a measure of dispersedness as it is in the real case.

Shape complexity measures such as dispersedness are useful in formulating region-merging criteria for building good partitions of a picture, as in Section 8.4.4. The cost of a merge should reflect a tradeoff between the increase in gray level variability and the decrease in shape complexity. If the gray levels of the regions are kept fixed, the cost of this merge should depend only on the shape complexity factor. Let the two regions S_1 and S_2 have perimeters P_1, P_2 and areas A_1, A_2, respectively, and let Q be the length of their common border, i.e., the number of pairs of adjacent points p_1, p_2 with p_1 in S_1 and p_2 in S_2. We shall assume here that P_i is measured by the number of pairs of adjacent pairs of points (p, q) with p in S_i and q not in S_i; thus $Q \leqslant \min(P_1, P_2)$. For the merged region, the P^2/A measure is $(P_1 + P_2 - 2Q)^2/(A_1 + A_2)$. This is evidently greatest when $Q = 0$, i.e., when the regions have no common border; and it is least when one of the regions surrounds the other, since Q is then as large as possible. Thus if we use the P^2/A of the merged region to measure the cost of a merge, we will be more likely to merge two regions the larger their common border is relative to their perimeters.

A merge criterion which takes both local gray level differences and shape complexity into account can be formulated by letting Q_t be the length of the "weak" part of the common border of S_1 and S_2, i.e., the number of pairs of adjacent points p_1, p_2 with p_1 in S_1, p_2 in S_2, and where the gray levels of p_1 and p_2 differ by less than some threshold t. We can then measure the cost of merging S_1 and S_2 by the smallness of Q_t relative to $\min(P_1, P_2)$.

A rather different application of shape criteria to region-merging involves the interpretation of perspective line drawings as three-dimensional bodies. Consider, for example, the three types of branch points shown in Fig. 13. The "Y" type, in which the three angles are all less than 180°, looks like a trihedral angle viewed from above its vertex; thus the three regions that meet at a Y could be the faces of such an angle, i.e., could belong to a single body.

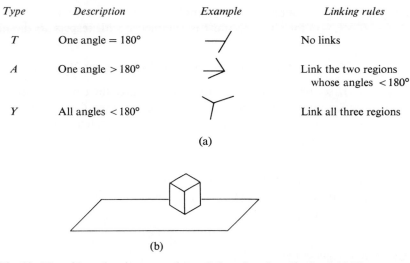

Type	Description	Example	Linking rules
T	One angle = 180°		No links
A	One angle > 180°		Link the two regions whose angles < 180°
Y	All angles < 180°		Link all three regions

(a)

(b)

Fig. 13 Use of branch point properties to link regions into "bodies." (a) Three types of branch points and their linking rules. (b) If these rules are applied to the simple drawing shown here, the faces of the cube become doubly linked to one another, but no other links are formed.

On the other hand, an "*A*" branch point, which has one angle greater than 180°, looks like a trihedral angle viewed from the side; here the two regions bounded by the angles that are less than 180° could be faces of the trihedral, thus belonging to a single body, whereas the third region appears not to belong to the same body. Finally, a "*T*" branch point, where one angle is exactly 180°, looks like a place where one body disappears behind another, and there is no good evidence that any of the three regions belong to the same body.

Given a line drawing, suppose that we "link" regions that meet at branch points in accordance with the rules just suggested: all three regions at a *Y* are linked, the two regions at an *A* whose angles are less than 180° are linked, and nothing is linked at a *T*. For the drawing shown in Fig. 13b, these rules create two links between each pair of faces of the cube, and no links between any other regions in the drawing. Thus, using these rules, we would have grounds for merging the three cube faces into a single body. Note that these rules can be regarded as based on shape simplicity, since they interpret the branch points as views of simple three-dimensional configurations; compare this with the discussion of Fig. 3.18 earlier in this section.

We conclude this section with some remarks about the measurement of symmetry for arbitrary pictures. A picture *f* is called *symmetric* with respect to a point or line if it remains unchanged when it is reflected in the given point or line. For example, f is symmetric with respect to the origin if $f(x, y) = f(-x, -y)$ for all x, y; and f is symmetric with respect to the line $y = x \tan \theta$ if

$$f(x, y) = f(x \cos 2\theta + y \sin 2\theta, x \sin 2\theta - y \cos 2\theta)$$

for all x, y. If we let f' denote the reflected picture, then we can use any measure (see Section 8.3.1) of the degree of match between f and f' as a symmetry measure, and any measure of the degree of mismatch as an asymmetry measure. On the use of moments as asymmetry measures, see the end of Section 10.2.2.

Exercise 34. Compute the asymmetry of a disk of radius r about a line at distance $d < r$ from its center. (You may assume that $f = 1$ inside the disk and $f = 0$ outside.) ▌

9.4.5 Elongatedness and Thinning

As in Section 9.2.4, let $S^{(t)}$ denote the result of propagating (or expanding) S t times, and $S^{(-t)}$ the result of shrinking S t times. Suppose that $S^{(-t)}$ is empty, i.e., S vanishes completely when we shrink it t times; thus every point of S is within distance t of the complement $\bar{S}$. We can thus say that the "width" of S is at most $2t$. Let A be the area of S. If A is large compared to t^2, say $A \geq 10t^2$, then we can reasonably call S *elongated*, since its "length" ($= \text{area/width}, \geq 10t^2/2t = 5t$) is at least $2\frac{1}{2}$ times its "width."

These remarks suggest that we can define the *elongatedness* of a picture subset S as A/w^2, where A is the area of S and w its width, i.e., the smallest number of shrinking steps required for S to disappear completely.[†] (Note that w is quite unrelated to the "width," in the sense of horizontal extent, which we defined in Section 9.4.1.) It should be pointed out that elongatedness is not the same as dispersedness, as measured by P^2/A (Section 9.4.4); a disk with a jagged border may have a high P^2/A, but it is not signicantly elongated, even though it may vanish a little sooner than a smooth disk when we shrink it.

If we know that S is everywhere elongated, then A/w^2 provides an overall measure of its elongatedness. Alternatively, we can analyze elongatedness by counting the number of points of S that are at a given distance d from $\bar{S}$. If this number decreases only slightly as d increases, and then drops abruptly to zero, S must be highly elongated; whereas if it decreases smoothly to zero, S is nonelongated. These measures are, however, of little value if S is partly

---

[†] This measure of elongatedness is unreliable when w is small. For example, ¦ ¦ and ¹ ¹ ¹ ¹ both have $A = 4$ and $w = 1$, but only the second of them would be called elongated.

elongated and partly not, e.g.,

```
1   1   1
1   1   1   1   1   1   1   1
1   1   1
```

We shall now show how a shrinking and reexpanding technique can be used to define and detect *elongated parts* of S.

Consider $(S^{(-k)})^{(k)}$, which is the result of shrinking S k times and then expanding the result k times. It is easy to show (Exercise 25) that $(S^{(-k)})^{(k)}$ is contained in S. Let C be a connected component of the difference set $S - (S^{(-k)})^{(k)}$. Clearly every point of C must have been within distance k of $\bar{S}$, i.e., the width of C is at most $2k$. Thus if C has sufficiently large area relative to k^2, it must be an elongated part of S. An example of this definition is given in Figs. 14a–d. If we perform shrinking and reexpansion on a given S for

```
    1  1  1  1                         1  1  1  1  1  1
 1  1  1  1  1  1  1  1  1  1          1  1  1  1  1  1  1
 1  1  1  1  1  1  1  1  1  1          1        1  1        1  1
    1  1  1  1                         1  1  1  1  1  1  1
       1                               1  1  1  1  1  1  1
```
                                                    (a)
```
    1  1
    1  1
```
                                                    (b)
```
    1  1  1  1
    1  1  1  1
    1  1  1  1
    1  1  1  1
```
                                                    (c)
```
                                           1  1  1  1  1  1  1
 1              1  1  1  1  1               1  1  1  1  1  1  1
 1              1  1  1  1  1               1        1  1        1  1
                                           1  1  1  1  1  1  1
       1                                   1  1  1  1  1  1  1
```
                                                    (d)

Fig. 14 Shrinking and reexpansion for elongated part detection. (a) Original picture. (b) Result of shrinking (a). (c) Result of expanding (b). (d) Points in (a) but not in (c). Note that the 10-point component is elongated, but the 33-point component is not.

```
  1  1  1  1  1  1                    1  1  1  1  1  1  1  1  1
1 1  1  1  1  1  1  1  1  1  1  1      1  1  1  1  1  1  1  1  1
1 1  1  1  1  1  1  1  1  1  1  1      1  1  1  1  1  1  1  1  1
1 1  1  1  1  1  1  1  1  1  1  1      1  1  1  1  1  1  1  1  1
1 1  1  1  1  1  1  1  1  1  1  1      1  1  1  1  1  1  1  1  1
    1  1  1  1  1  1                  1  1  1  1  1  1  1  1  1
       1  1  1 .                      1  1  1  1  1  1  1  1  1
                              (e)
```

```
      1  1                              1  1  1  1  1
      1  1                              1  1  1  1  1
                                        1  1  1  1  1
                              (f)
```

```
   1  1  1  1                        1  1  1  1  1  1  1
   1  1  1  1                        1  1  1  1  1  1  1
   1  1  1  1                        1  1  1  1  1  1  1
   1  1  1  1                        1  1  1  1  1  1  1
                                     1  1  1  1  1  1  1
                              (g)
```

```
 1               1  1  1  1  1
 1               1  1  1  1  1


        1
                              (h)
```

Fig. 14 Shrinking and reexpansion for elongated part detection. (e) Result of expanding (a). (f) Result of shrinking (e), two steps. (g) Result of expanding (f). (h) Points in (a) but not in (g). Now all the large components are elongated.

various values of k, we can detect elongated parts of S that have various widths ("creeks," "streams," "rivers," etc.) and distinguish them from non-elongated parts ("lakes," "seas," etc.).

The definition of elongatedness just given is noise sensitive; S may vanish under a small amount of shrinking, without being elongated, if there are small holes in S (see Fig. 14d). This is because our elongatedness detection scheme expands and reshrinks the complement $\bar{S}$, so that clusters of small parts of $\bar{S}$ may "fuse" (see Section 9.2.4). We can avoid this problem by working alternately on S and on $\bar{S}$: first shrink $\bar{S}$ and reexpand it by a small amount, to clean up small parts of $\bar{S}$; then shrink S and reexpand it by a larger amount. Small holes in S will disappear under the first procedure, before they can cause S to be incorrectly called elongated when the second procedure is used. This approach is illustrated in Figs. 14e–h.

If S is everywhere elongated, it is often useful to be able to "thin" S into a set of arcs (and curves), without changing the connectedness properties of S. These arcs should consist of points of S whose distances from $\bar{S}$ are all approximately half the width of S. A *thinning* algorithm should reduce any S to such a set of arcs, and it should have no effect on an S that consists only of thin arcs. Thus it should be safe to apply such an algorithm to any given S until no further change takes place, and the result should then be the desired set of thin arcs.

The basic idea of a thinning algorithm is to delete from S simple border points (whose deletion does not locally disconnect S; see Section 9.2.5) that have more than one neighbor in S (so that the ends of thin arcs are never deleted). Note that if S has a very irregular or "hairy" border, its border points may all be either arc ends or nonsimple, so it cannot be properly thinned; an example is

```
    1       1       1

        1   1   1

    1   1   1   1   1

        1   1   1

    1       1       1
```

As in Section 9.2.5, we cannot simultaneously delete from S all simple border points that are not arc ends, since this may cause S to vanish completely, e.g., if S is

```
    1   1   1   1   1   1   1   1 .

    1   1   1   1   1   1   1   1
```

To avoid this, we can, at any given application of the algorithm, delete only border points on certain sides of S. For example, at odd-numbered steps we can delete simple nonend points provided their north or east neighbors are not in S, while at even-numbered steps we can require that the south or west neighbors of the deleted points not be in S. (This alternation between sides also helps to ensure that the curves that remain after thinning are centered in S, as above.) An alternate approach is to check the neighbors of each point on two sides, to determine whether they too will be deleted, and if so, not to delete the given point. The results of thinning using these approaches are generally similar, but not identical, as shown in Fig. 15; this figure also shows the skeleton of the input picture (Section 9.2.3) for comparison with the thinning results.

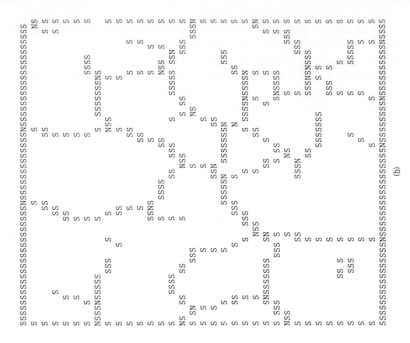

(b)

(a)

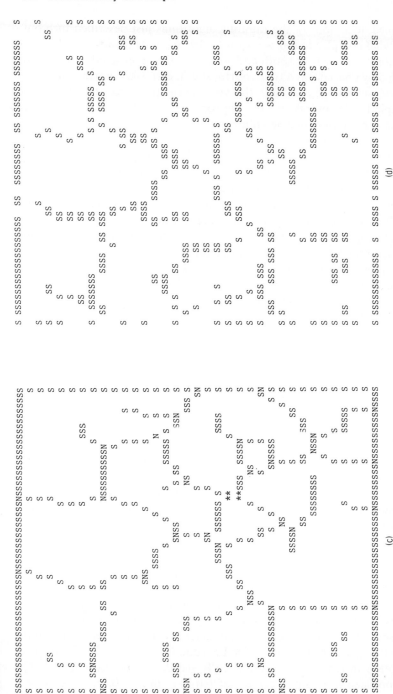
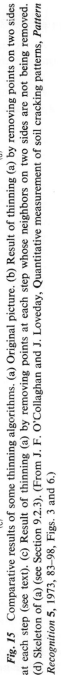

Fig. 15   Comparative results of some thinning algorithms. (a) Original picture. (b) Result of thinning (a) by removing points on two sides at each step (see text). (c) Result of thinning (a) by removing points at each step whose neighbors on two sides are not being removed. (d) Skeleton of (a) (see Section 9.2.3). (From J. F. O'Collaghan and J. Loveday, Quantitative measurement of soil cracking patterns, *Pattern Recognition* **5**, 1973, 83–98, Figs. 3 and 6.)

The detailed definitions of the thinning algorithms just described turn out to be somewhat complicated, if we want to delete points from two sides of S at a time (north and east, or south and west). On the other hand, if we delete points from only one side at a time (north, south, east, west, north, south, ..., at successive steps), we can use the very straightforward algorithm that deletes simple border points (on the appropriate side of the border at each step) that have more than one neighbor in S.

Thinning algorithms were defined for two-valued pictures representing elongated objects S, but they can also be defined for pictures whose gray levels have more than two values, e.g., for an S consisting of nonzero gray levels on a background of zeros. The basic idea is, at each step, to remove only border points having a single nonzero gray level, namely, the lowest level for which simple border points that are not arc ends still remain. The result of applying such an algorithm is a set of curves that tend to lie along ridges of high gray level in the original picture.

Another approach to thinning is to use (thick) curve-following techniques to track the elongated parts of the given S (see the end of Section 9.1.2), and determine a "midline" for each part as it is being tracked. This approach is potentially more powerful than those previously described, since it can take advantage of global shape information obtained about the part being tracked, unlike the methods which thin by deleting individual points on the basis of local criteria.

9.5 BIBLIOGRAPHICAL NOTES

The theory of connectedness in digital pictures was developed by Rosenfeld [30–32]. Variants of the border-following algorithm BF are probably quite old, although detailed descriptions do not seem to be found in the literature before the late 1960's. The algorithm described in Exercise 16 was independently invented by DeLotto [8] and Mason and Clemens [20]. On border-finding algorithms see Minsky and Papert [21, 22]. The run-tracking method of component labeling is described by Butler et al. [7], but its application to counting simply connected objects is old [26]. On the genus see Minsky and Papert [21, 22] and Gray [11].

Distance functions on digital pictures were studied by Rosenfeld and Pfaltz [37, 38]; the gray-weighted generalization and the path constructing algorithm are due to Rutovitz [39]. The skeleton was proposed, for the continuous case, by Blum [4]; see also Blum [5]. For a general treatment of the discrete case, and a method of keeping the skeleton connected, see Mott-Smith [24]. On the gray-weighted skeleton see Levi and Montanari [17]. On the use of the skeleton to represent S economically see Pfaltz and Rosenfeld [27]. Another approach to finding "centers of tension" in an object is described by Kasvand [16]. For applications of propagation and shrinking operations to detect clusters and simplify objects see Moore [23]; Rosenfeld and Pfaltz [38]. An interesting approach to cluster detection using graph-theoretic methods is treated by Zahn [45]. Region-filling by propagation is described by Strong and Rosenfeld [43]. Shrinking operation Φ is due to Golay (see Preston [28]), while operation Ψ was independently invented by Beyer (Minsky and Papert [22]) and by Levialdi [18, 19].

Curve property measurement and curve segmentation using the chain code representation were studied by Freeman and his students beginning in the early 1960's; for a recent review of this work see Freeman [9]. On digital straight line segments see also Rosenfeld [33]. Angle detection using smoothed curvatures is described by Rosenfeld and Johnston [35]; angle detection on the border of a solid object is treated by Belson [3]. On polygonal approximation to curves see Ramer [29], as well as references [30, 31] in Section 8.5. Fourier descriptors of closed curves, based on various equations for the curve, were developed in the 1950's and early 1960's; for a recent review see Zahn and Roskies [46]. On coordinate transformations for the detection of circles and straight lines see Bazin and Benoit [2] and Hough [15].

On the many factors that influence human judgments of shape similarity and shape complexity see the review in Zusne [47]; much of the basic work in this area was done by Attneave in the 1950's. A discussion of "to the left of" can be found in Winston [44]. The use of projections dates back to early character recognition systems that employed scanning slits. Cross sections were also used in much of the early character recognition work. On the use of integral geometry to measure pattern properties see Novikoff [25]; on "shadows" see Glucksman [10]. Segmentation by run tracking is due to Grimsdale et al. [12].

Digital convexity was studied by Sklansky [41] and Hodes [14]; see also Rosenfeld [33]. The parallel method of constructing the convex hull is due to Arcelli and Levialdi [1]. The limitations of P^2/A in the digital case are discussed by Rosenfeld [34]. Perimeter criteria for region merging are discussed by Brice and Fennema [6]; the method of linking regions based on branch point types is taken from Guzman [13]. Shrinking and reexpansion for elongatedness analysis is treated by Rosenfeld et al. [36]. Thinning was described by Sherman [40]; many refinements on the basic algorithm have been developed, some of which are reviewed by Stefanelli and Rosenfeld [42], who also give proofs that such algorithms preserve topology.

REFERENCES

1. C. Arcelli and S. Levialdi, Concavity extraction by parallel processing, *IEEE Trans. Syst. Man Cybernet.* **SMC-1**, 1971, 349–396.

2. M. J. Bazin and J. W. Benoit, Off-line global approach to pattern recognition for bubble chamber pictures, *IEEE Trans. Nucl. Sci.* **NS-12**, Aug. 1965, 291–295.
3. M. Belson, A new boundary method for pictorial pattern recognition, *EASCON '69 Record* (IEEE Publ. 69-C-31-AES), Oct. 1969, 274–279.
4. H. Blum, An associative machine for dealing with the visual field and some of its biological implications, *in* "Biological Prototypes and Synthetic Systems" (E. E. Bernard and M. R. Kare, eds.), pp. 244–260. Plenum Press, New York, 1962.
5. H. Blum, A transformation for extracting new descriptors of shape, *in* "Models for the Perception of Speech and Visual Form" (W. Wathen-Dunn, ed.), pp. 362–380. M.I.T. Press, Cambridge, Massachusetts, 1967.
6. C. R. Brice and C. L. Fennema, Scene analysis using regions, *Artificial Intelligence* **1**, 1970, 205–226.
7. J. W. Butler, M. K. Butler, and A. Stroud, Automatic analysis of chromosomes, "Data Acquisition and Processing in Biology and Medicine," Vol. 3, pp. 261–275. Pergamon, Oxford, 1963.
8. I. DeLotto, Un inseguitore di contorno, *Alta Frequenza* **32**, 1963, 703–705.
9. H. Freeman, Computer processing of line-drawing images, *Comput. Surv.* **6**, 1974, 57–97.
10. H. A. Glucksman, A parapropagation pattern classifier, *IEEE Trans. Electron. Comput.* **EC-14**, 1965, 434–443.
11. S. B. Gray, Local properties of binary images in two dimensions, *IEEE Trans. Comput.* **C-20**, 1971, 551–561.
12. R. L. Grimsdale, F. H. Sumner, C. J. Tunis, and T. Kilburn, A system for the automatic recognition of patterns, *Proc. IEE* **106B**, 1959, 210–221.
13. A. Guzman, Decomposition of a visual scene into three-dimensional bodies, *Proc. Fall Joint Comput. Conf.* pp. 291–304. AFIPS Press, Montvale, New Jersey, 1968.
14. L. Hodes, Discrete approximation of continuous convex blobs, *SIAM J. Appl. Math.* **19**, 1970, 477–485.
15. P. V. C. Hough, Method and means for recognizing complex patterns, U. S. Patent 3069654, 1962.
16. T. Kasvand, Experiments with an on-line picture language, *in* "Frontiers of Pattern Recognition" (S. Watanabe, ed.), pp. 223–264. Academic Press, New York, 1972.
17. G. Levi and U. Montanari, A gray weighted skeleton, *Informat. Contr.* 17, 1970, 62–91.
18. S. Levialdi, Parallel counting of binary patterns, *Electron. Lett.* **6**, 1970, 798–800.
19. S. Levialdi, On shrinking binary picture patterns, *Comm. ACM 15*, 1972, 7–10.
20. S. J. Mason and J. K. Clemens, Character recognition in an experimental reading machine for the blind, *in* "Recognizing Patterns" (P. A. Kolers and M. Eden, eds.), pp. 156–167. MIT Press, Cambridge, Massachusetts, 1968.
21. M. L. Minsky and S. Papert, Linearly unrecognizable patterns, *Proc. Symp. Appl. Math.* **19**, 176–217. Amer. Math. Soc., Providence, Rhode Island, 1967.
22. M. L. Minsky and S. Papert, "Perceptrons, An Introduction to Computational Geometry." MIT Press, Cambridge, Massachusetts, 1969.
23. G. A. Moore, Automatic scanning and computer processes for the quantitative analysis of micrographs and equivalent subjects, *in* "Pictorial Pattern Recognition" (G. C. Cheng *et al.*, eds.), pp. 275–326. Thompson, Washington, D. C., 1968.
24. J. C. Mott-Smith, Medial axis transformations, *in* "Picture Processing and Psychopictorics" (B. S. Lipkin and A. Rosenfeld, eds.), pp. 267–283. Academic Press, New York, 1970.
25. A. B. J. Novikoff, Integral geometry as a tool in pattern perception, *in* "Principles of Self-Organization" (H. von Foerster and G. W. Zopf, eds.), pp. 347–368. Pergamon, New York, 1962.

26. T. C. Nuttall, Apparatus for counting objects, U. S. Patent 2803406, 1957.
27. J. L. Pfaltz and A. Rosenfeld, Computer representation of planar regions by their skeletons, *Comm. ACM* **10**, 1967, 119–122, 125.
28. K. Preston, Jr., The CELLSCAN system, a Leucocyte pattern analyzer, *Proc. Western Joint Comput. Conf.* pp. 173–183. Spartan, New York, 1961.
29. U. Ramer, An iterative procedure for the polygonal approximation of plane curves, *Comput. Graph. Image Proc.* **1**, 1972, 244–256.
30. A. Rosenfeld, Connectivity in digital pictures, *J. ACM* **17**, 1970, 146–160.
31. A. Rosenfeld, Arcs and curves in digital pictures, *J. ACM* **20**, 1973, 81–87.
32. A. Rosenfeld, Adjacency in digital pictures, *Informat. Contr.* **26**, 1974, 24–33.
33. A. Rosenfeld, Digital straight line segments, *IEEE Trans. Comput.* **C-23**, 1974, 1264–1269.
34. A. Rosenfeld, Compact figures in digital pictures, *IEEE Trans. Syst. Man Cybernet.* **SMC-4**, 1974, 211–223.
35. A. Rosenfeld and E. Johnston, Angle detection on digital curves, *IEEE Trans. Comput.* **C-22**, 1973, 875–878.
36. A. Rosenfeld, C. M. Park, and J. P. Strong, Noise cleaning in digital pictures, *EASCON '69 Record* (IEEE Publ. 69-C-31-AES), Oct. 1969, 264–273.
37. A. Rosenfeld and J. L. Pfaltz, Sequential operations in digital picture processing, *J. ACM* **13**, 1966, 471–494.
38. A. Rosenfeld and J. L. Pfaltz, Distance functions on digital pictures, *Pattern Recognit.* **1**, 1968, 33–61.
39. D. Rutovitz, Data structures for operations on digital images, *in* "Pictorial Pattern Recognition" (G. C. Cheng *et al.*, eds.), pp. 105–133. Thompson, Washington, D. C., 1968.
40. H. Sherman, A quasi-topological method for the recognition of line patterns, "Information Processing," pp. 232–237. UNESCO, Paris, 1959.
41. J. Sklansky, Recognition of convex blobs, *Pattern Recognit.* **2**, 1970, 3–10.
42. R. Stefanelli and A. Rosenfeld, Some parallel thinning algorithms for digital pictures, *J. ACM* **18**, 1971, 255–264.
43. J. P. Strong III and A. Rosenfeld, A region coloring technique for scene analysis, *Comm. ACM* **16**, 1973, 237–246.
44. P. H. Winston, Learning structural descriptions from examples, AI Tech. Rep. 231, Artificial Intelligence Lab., MIT, Cambridge, Massachusetts, 1970.
45. C. T. Zahn, Graph-theoretic methods for detecting and describing Gestalt clusters, *IEEE Trans. Comput.* **C-20**, 1971, 68–86.
46. C. T. Zahn and R. Z. Roskies, Fourier descriptors for plane closed curves, *IEEE Trans. Comput.* **C-21**, 1972, 269–281.
47. L. Zusne, "Visual Perception of Form," Academic Press, New York, 1970.

Chapter 10

Description

As indicated at the beginning of Chapter 8, a *description* of a picture (or scene) involves properties of the picture or of its parts, and relationships among the parts. In Chapter 9 we dealt with geometrical properties of picture parts, i.e., with properties that do not depend on the picture's gray levels, but only on the sets of picture points that belong to the given picture parts. The present chapter deals with other types of properties, and with methods of describing pictures in terms of properties and relationships.

An important special case of picture description is *classification*, in which we want to assign the picture to one of a prespecified set of classes; here the description is simply the name of the class to which the picture belongs. For example, we might want to assign pictures of alphabetic characters to the classes "*A*," "*B*," "*C*," etc; bubble chamber pictures to classes representing various types of "events"; pieces of aerial photographs to classes representing terrain types; and so on. Given a set of property values that have been measured for the given picture, methods of *pattern recognition* can be used to decide to which class the picture should be assigned (see the end of Section 8.1.3 for an indication of how this can be done). Pattern recognition will not be covered in this book; for a good introduction to it see, e.g., Duda and Hart [7]. We do, however, cover the types of picture properties that are commonly used for picture classification and description. Picture properties are discussed in Sections 10.1 and 10.2, as well as in Chapter 9.

The standard pattern recognition techniques use sets of property values to characterize a picture. This is a reasonable approach if the picture is homogeneous (e.g., uniformly textured). If the picture is structured, on the other hand, an adequate description should involve picture parts, their properties, and relations among them; it should have the form of a relational structure rather than simply a list of overall property values. Some simple types of relational structures that can be used to describe pictures are discussed in Section 10.3. The difficulty of deriving useful descriptive relational structures from pictures and the use of models to aid in this picture analysis process are treated in Section 10.4.

10.1 PREPROCESSING AND NORMALIZATION

A *picture property* is a function that maps pictures into numbers; the number obtained from a given picture f is the value of the property for that picture. Examples of properties are

(a)   the gray level $f(x_0, y_0)$ of f at some specified point;

(b)   the average gray level of f, i.e., $(1/A)\iint f(x, y)\, dx\, dy$, where A is the area of the region of integration;

(c)   the amount of energy in some band of f's power spectrum, i.e., $\iint_{\mathscr{S}} |F(u,v)|^2\, du\, dv$, where $\mathscr{S}$ is some region in the spatial frequency plane;

(d)   a particular Fourier coefficient of f, say $\iint f(x, y) \cos(mx + ny)\, dx\, dy$;

and so on. Such properties can be computed directly as soon as f is given. Other types of properties, on the other hand, can be computed only after some preliminary operation has been performed on f. This section deals with various types of "preprocessing" operations that are commonly performed on f as preliminaries to property measurement.

Geometrical properties of picture parts can be measured only after the parts in question have been, at least implicitly, extracted from the picture by segmenting it (Chapter 8). Given a picture subset S, examples of properties that we can now measure are

(a)   the number of connected components of S,

(b)   the areas of these components,

(c)   the positions of the centroids of these components,

and so on. In some cases, the segmentation and property measurement can be performed in a single operation. For example, when we match a template f against a picture g, say by cross-correlating them, we can take the highest value of the cross correlation as a property which measures the degree of match between g and f.

Many other transformations of a picture, besides segmentation, can be useful as preprocessing operations to facilitate property measurement. For example, one can sometimes obtain "cleaner" property values by enhancing the picture (e.g., cleaning up noise) before measuring the properties. Many picture properties are most conveniently measured if we first compute some transform of the given picture, e.g., its power spectrum (see Section 10.1.4). Other properties are conveniently measured by first computing some function of a single variable from the given picture or picture subset, e.g., a projection or cross section (Sections 9.4.1 and 9.4.2), an intrinsic equation (Section 9.3.1), or a histogram (Sections 6.2.3 and 9.3.1).

One often wants to define properties whose values are not affected by changes in the lightness, contrast, position, orientation, size, etc. of the given picture or picture subset. In the body of this section, we discuss methods of defining such "invariant" properties. In particular, we can make the property values insensitive to such transformations by "normalizing" or standardizing the picture (or subset) before measuring the properties. Normalization procedures are an important class of preprocessing operations; most of this section deals with such procedures.

10.1.1   Invariant Properties

In many picture classification problems, two pictures that differ only in lightness or contrast, or two picture subsets that differ only in position, orientation, or scale, must belong to the same class. In other words, the classes in such a problem are closed under changes in the parameters of lightness, position, etc.[†] For such problems, one would like to use picture properties that do not depend on these parameters, i.e., *invariant* properties.

If the given picture is represented by an abstract mathematical function, it is easy to formulate what we mean by changes in the parameters just mentioned. Given the picture $f(x, y)$, we can vary its lightness by adding a constant (provided that the result remains nonnegative), yielding $f(x, y) + c$, say; we can vary its contrast by multiplying it by a positive constant, obtaining $cf(x, y)$; and similarly for other simple operations on the gray scale. We can translate, rotate, or rescale f by performing the appropriate coordinate system change to obtain $f(x - a, y - b)$, $f(x \cos \theta + y \sin \theta, -x \sin \theta + y \cos \theta)$, or $f(ax, by)$, and similarly for other geometrical operations. For real pictures, of course, it must be realized that these operations are not always realizable.

[†] It should be realized that this is not true for all picture-classification problems. For example, in character recognition, we may be able to distinguish a comma from an apostrophe only because they occur in different positions, or a p from a d only because they have different orientations.

The size and gray level range of a real picture are finite, so that only finite degrees of variation in lightness, contrast, position, and scale are possible. In a digital picture, moreover, these operations are no longer mathematically simple, as we have seen in Section 6.3. Nevertheless, it is convenient to assume that we can work with these simple mathematical operations when we want to define invariant properties.

It is easy to define simple picture properties that are invariant under changes in lightness or contrast, i.e., under shifting or stretching of the gray scale. Indeed, differences between pairs of gray levels (or of average gray levels) are invariant under gray scale shifting; and ratios of gray levels are invariant under gray scale stretching (or shrinking). If we want invariance under both shifting and stretching, we can use ratios of differences. For example, the expression

$$\frac{f(x_1, y_1) - f(x_2, y_2)}{f(x_3, y_3) - f(x_4, y_4)}$$

does not change in value if we replace $f(x, y)$ by $a \cdot f(x, y) + b$.

The gray level histogram of a picture f is invariant under any one-to-one geometrical operation on f. Hence, statistics or other properties derived from this histogram, such as the average gray level of f, or its maximum gray level, are also invariant under such geometrical operations. More generally, statistical properties of a picture derived by examining points at random will tend to be invariant under these operations.

Many simple geometrical properties of a picture subset S are invariant under translation, rotation, and magnification, as we recall from elementary geometry.[†] For example, the height and width of S (or more generally, its extent in any direction) are invariant under translation. The area, perimeter, and diameter of S (or more generally, the distance between any two of its points) are also invariant under rotation. Ratios of distances and angles between lines are also invariant under magnification, as are the various shape properties of a given S, such as its convexity, dispersedness, and elongatedness. Topological properties of S, such as connectedness or the genus, are invariant under arbitrary "rubber-sheet" distortions of the picture. It should be realized, of course, that these results may be only approximately true for a digital S, once we redigitize it after performing geometrical operations on it.

Many important classes of operations on pictures are closed under composition, i.e., if $\mathcal{O}_1$ and $\mathcal{O}_2$ are operations in the given class, then so is $\mathcal{O}_1 \circ \mathcal{O}_2$. (By definition, applying the composite transformation to a picture f means applying $\mathcal{O}_1$ to f, and then applying $\mathcal{O}_2$ to the resulting picture $\mathcal{O}_1(f)$.) For

---

[†] The various branches of geometry, in fact, are often regarded as the study of properties that remain invariant under various types of geometrical operations.

example, the composite of two gray scale shifts or stretches is a gray scale shift or stretch; and the composite of two translations, rotations, or magnifications is a translation, rotation, or magnification, respectively. The identity operation $\mathscr{I}$, which takes any picture into itself, is a member of each of these classes; and the classes are also closed under inverse, i.e., for each operation $\mathcal{O}$ in the class, there is an operation $\mathcal{O}^{-1}$ in the class such that $\mathcal{O} \circ \mathcal{O}^{-1} = \mathscr{I}$. For example, the inverse of the gray scale shift $f+c$ is the shift $f-c$; the inverse of the stretch cf is f/c; the inverse of translation by (a,b) is translation by $(-a,-b)$; the inverse of rotation by θ is rotation by $-\theta$; and the inverse of magnifying by (a,b) is demagnifying by $(1/a, 1/b)$. Since composition of operations is always associative, i.e., $\mathcal{O}_1 \circ (\mathcal{O}_2 \circ \mathcal{O}_3) = (\mathcal{O}_1 \circ \mathcal{O}_2) \circ \mathcal{O}_3$, it follows that these classes are all *groups* of operations.

Given a group $\mathscr{G}$ of operations on pictures, and given any picture f, it is easy to see that the set of resultant pictures $\mathcal{O}(f)$, for all $\mathcal{O}$ in $\mathscr{G}$, is the same as the set $\mathcal{O}(\mathcal{O}_1(f))$, where $\mathcal{O}_1$ is any particular operation in $\mathscr{G}$. In other words, whichever $\mathcal{O}(f)$ we start with, if we apply every operation in $\mathscr{G}$ to it, we always obtain the same set of all possible $\mathcal{O}(f)$'s. This provides a general, though not always practical, method of defining invariant properties: If we measure some given property for all of the $\mathcal{O}(f)$'s, the resulting set of property values is the same whether we started with f or with any of the $\mathcal{O}(f)$'s. Thus any statistic or other property derived from this set of values, such as its average or its maximum, will be the same whether we started with f or with any $\mathcal{O}_1(f)$. In practice, it may not be possible[†] to construct all of the $\mathcal{O}(f)$'s; but we should be able to obtain approximately invariant properties by using a large, randomly chosen set of $\mathcal{O}(f)$'s.

10.1.2   Gray Scale Normalization

An important approach to defining invariant properties of a given picture f is to *normalize* f, i.e., to develop a method of standardizing f, such that whether we start with f or with any one of the pictures $\mathcal{O}(f)$, we obtain the same standardized version. Any property of the standardized version of f can then be regarded as an invariant property of f.

Let us first consider how we might normalize f with respect to shifting of the gray scale. Let μ be the average gray level of f, and let μ_0 be any gray level value. If we shift the gray scale of f by the amount $\mu_0 - \mu$, we obtain a new picture f_0 whose average gray level is μ_0. Suppose now that in place of f,

---

[†] One case in which we are, in effect, constructing all of the $\mathcal{O}(f)$'s is when we cross correlate a template with a picture, since this amounts to matching the template with all possible translated versions of the picture. The maximum (or average, etc.) of the resulting match values is thus a translation-invariant property of the picture.

we had started with some $f+c$ that differs from f by a gray scale shift. The average gray level of $f+c$ is $\mu+c$; and if we shift the gray scale of $f+c$ by the amount $\mu_0-(\mu+c)$, we again obtain the same picture f_0. In other words, whether we start with f or with any gray scale-shifted version of f, we can obtain a standard f_0 by shifting the gray scale of the given picture by the amount $\mu_0-\mu$, where μ is the average gray level of the given picture.

This rather trivial example can be generalized as follows: Given a set of operations ($\mathcal{O}$'s), let $\mathcal{P}(f)$ be a property of f such that, for some numerical value w, there is exactly one $\mathcal{O}$ such that $\mathcal{P}(\mathcal{O}(f)) = w$. Then we can use $\mathcal{O}(f)$ as a normalized version of f. In the example just given, $\mathcal{P}$ is average gray level, w is μ_0, and there is in fact only one gray scale-shifted version of f that has a given average gray level μ_0. The same approach also works for gray scale stretching and shrinking; here again we can use average gray level, since there is only one gray scale-stretched (or shrunk) version of f that has a given average gray level.

This simple approach breaks down if the set of possible picture operations is too large; for example, if both shifting and stretching of the gray scale are allowed, there is no longer a unique $\mathcal{O}(f)$ that has a given average gray level. We can, however, guarantee uniqueness even in this case by imposing two requirements on our normalized picture, e.g., that it have both a given average gray level μ and a given standard deviation of gray levels σ. (In fact, we can stretch the gray scale to achieve the desired σ, and then shift to achieve the desired μ.) In general, given a family of operations involving k parameters, we can attempt to achieve uniqueness by imposing k requirements on the normalized picture.

These examples suggest that if we want to normalize with respect to a still wider variety of gray scale operations, a possible approach would be to construct a normalized picture that has a given gray level histogram (e.g., a flat histogram), since this imposes the greatest possible number of requirements on the normalized picture's (discrete) gray scale. In fact, this method of normalization is commonly used to standardize a picture's gray scale, even when the set of possible gray scale operations is not known.

It should be realized that for digital pictures, these methods of normalization will usually not work exactly, since the normalized pictures that we construct may have to be requantized. It should, however, be possible to achieve at least an approximate normalization using this approach even in the digital case.

10.1.3 Geometrical Normalization

Just as we did for gray scale normalization, we can also achieve normalization with respect to geometrical operations, such as translation, rotation, and magnification, by operating on the given picture so that it has specified property values. Some standard methods of doing this will be described in this section.

If we regard the gray level at each point (x, y) of the given picture f as the "mass" of (x, y), we can define the *centroid* (or "center of gravity") of f, as well as the moments (of inertia) of f about specified points or lines. The centroid of f is the point $(\bar{x}, \bar{y})$ whose coordinates are given by

$$\bar{x} = \int\int xf(x, y)\, dx\, dy \Big/ \int\int f(x, y)\, dx\, dy$$

and

$$\bar{y} = \int\int yf(x, y)\, dx\, dy \Big/ \int\int f(x, y)\, dx\, dy$$

For digital pictures, the integrals become sums. As a simple example, we can compute the centroid of

$$
\begin{array}{ccc}
2 & 1 & 1 \\
3 & 1 & 0 \\
3 & 2 & 1
\end{array}
$$

as follows (where the origin is at the leftmost point of the bottom row):

$$\sum\sum f(x, y) = 14; \qquad \sum\sum xf(x, y) = 8; \qquad \sum\sum yf(x, y) = 12$$
$$\bar{x} = 4/7, \qquad \bar{y} = 6/7$$

For picture subsets, the computation is even simpler, since the only possible values for f are 0 and 1 (here $f = \varphi_S$ is the "overlay" representing the picture subset S).

It is easily verified that if f is translated, say by (a, b), then the position of f's centroid is translated by the same amount. Thus there is a unique translation of f that has its centroid in any given position. It follows, as in Section 10.1.2, that if we want to normalize f with respect to translation, we can simply shift f so that its centroid is in some standard position, say at the origin. If f is a digital picture, and this shift is not by an integer amount, the resulting normalized f will have to be redigitized; but in any case, this method should allow us to achieve at least approximate normalization. We shall suppose in the following paragraphs that the centroid of f is at the origin.

The *moment of inertia* of f about the origin is

$$m_f = \int\int (x^2 + y^2) f(x, y)\, dx\, dy$$

It is easily verified that if f is magnified, so that it becomes $f(cx, cy)$, say, then m_f is multiplied by $1/c^4$. Thus there is a unique positive scale factor c

that yields a given value of m_f. It follows as in the preceding paragraph that we can normalize f with respect to magnification by magnifying it so that m_f has some given value. Here again, in the digital case, this normalization can only be approximate, since the magnified f must be redigitized.

The moment of inertia of f about the line $y = x \tan \theta$ through the origin is

$$m_\theta = \int\int (x \sin \theta - y \cos \theta)^2 f(x, y) \, dx \, dy$$

It is easily verified that if f is rotated about the origin, say by the angle φ, then $m_{\theta + \varphi}$ for the rotated f is the same as m_θ for the original f. Let θ_0 be the angle for which m_{θ_0} is as small as possible. There may be several such angles (e.g., all the m_θ's are evidently equal if f is a circle centered at the origin); if there is a unique one, then the line $y = x \tan \theta_0$ is called the *principal axis of inertia* of f. As above, we can normalize f with respect to rotation around the origin by rotating it so that its principal axis of inertia has some standard orientation, say vertical. As usual, this normalization is only approximate in the digital case.

We can compute θ_0 explicitly as follows: Let

$$m_{20} = \int\int x^2 f(x, y) \, dx \, dy, \qquad m_{11} = \int\int xy f(x, y) \, dx \, dy,$$

$$m_{02} = \int\int y^2 f(x, y) \, dx \, dy$$

(where the integrals become sums in the digital case); these m_{ij}'s are just the second moments of the picture f (see Section 10.2.2). Then we have

$$m_\theta = m_{20} \sin^2 \theta - 2m_{11} \sin \theta \cos \theta + m_{02} \cos^2 \theta$$

To find the θ_0 that minimizes m_θ, we differentiate m_θ with respect to θ, and set the result to zero, obtaining

$$2m_{20} \sin \theta_0 \cos \theta_0 - 2m_{11}(\cos^2 \theta_0 - \sin^2 \theta_0) - 2m_{02} \cos \theta_0 \sin \theta_0 = 0$$

or

$$m_{20} \sin 2\theta_0 - 2m_{11} \cos 2\theta_0 - m_{02} \sin 2\theta_0 = 0$$

which implies

$$\tan 2\theta_0 = 2m_{11}/(m_{20} - m_{02})$$

Since $\tan 2\theta_0 = 2 \tan \theta_0/(1 - \tan^2 \theta_0)$, we can obtain $\tan \theta_0$ as a root of the quadratic equation

$$\tan^2 \theta_0 + \frac{m_{20} - m_{02}}{m_{11}} \tan \theta_0 - 1 = 0$$

Incidentally, this last equation is readily equivalent to

$$(m_{11} \tan \theta + m_{20})^2 - (m_{20} + m_{02})(m_{11} \tan \theta + m_{20}) + (m_{20} m_{02} - m_{11}) = 0$$

which implies that $m_{11} \tan \theta + m_{20}$ is an eigenvalue of the matrix

$$\begin{pmatrix} m_{20} & m_{11} \\ m_{11} & m_{02} \end{pmatrix}$$

It can be shown, in fact, that the principal axis θ_0 is in the direction of the eigenvector corresponding to the larger eigenvalue of this matrix.

Exercise 1. Let $f(x, y) = 1$ if $(x^2/a^2) + (y^2/b^2) \leqslant 1$, and $f(x, y) = 0$ otherwise, so that f is the characteristic function of an ellipse. Prove that the centroid of f is at the origin, and that the principal axis of f coincides with the major axis of the ellipse. ▮

Exercise 2. What is the moment of inertia of f about an arbitrary line $ax + by + c = 0$? Show that any line for which this moment is a minimum must pass through the centroid of f. ▮

The principal axis of inertia of f can be regarded as a line that "best fits" f. More generally, one can find higher-order curves that "best fit" f in various senses. For example, given a general quadratic curve

$$q(x, y) \equiv ax^2 + bxy + cy^2 + ux + vy + w = 0$$

we can attempt to find the values of the coefficients a, b, c, u, v, w such that

$$\int\int (q(x, y))^2 f(x, y)$$

is a minimum. The curve having these coefficients would be a sort of "quadratic principal axis" for f. Given f's best-fitting quadratic curve q_0, one can attempt to perform "shape normalization" on f by transforming coordinates so that q_0 becomes some standard type of curve—for example, if q_0 is an ellipse, we could transform to make it a circle.

If f is two-valued, say $f = \varphi_S$, where S is a picture subset, another approach to normalizing f can be used. First, we can translate S so that (say) its leftmost and uppermost points are at the left and top edges of the picture, respectively. We can also rescale S so that its width and height have specified values; this normalizes it with respect to magnification.

To normalize S with respect to rotation, we can find (say) a rectangle circumscribed around S that has smallest possible area, and then rotate S so that the long side of this rectangle has some standard orientation, say vertical. Note that, like the principal axis of inertia, this rectangle may not be unique. Note also that in a digital picture, these methods of normalization are only approximate, since we may have to redigitize after rescaling or rotating.

The circumscribed rectangle of S that has smallest possible area is a rectangle that "best fits" S. We can also use circumscribed shapes other than rectangles, e.g., arbitrary quadrilaterals. Given some best-fitting quadrilateral Q of S, or in fact any quadrilateral that can be defined in terms of S, one can perform "shape normalization" on S by transforming coordinates so that Q acquires some standard shape, e.g., square. An example in which quadrilateral fitting is appropriate is the case of slanted alphanumeric characters (e.g., italics); here, if we can find a "shearing" transformation of the form $x' = x + \lambda y$, $y' = y$ that takes the best-fitting quadrilateral into a rectangle, we may be able to transform the slanted characters into upright characters.

The principal-axis and best-fitting rectangle methods of rotational normalization give analogous, but not identical, results. Examples of the results obtained by these methods, using a thresholded chromosome picture as input, are shown in Fig. 1.

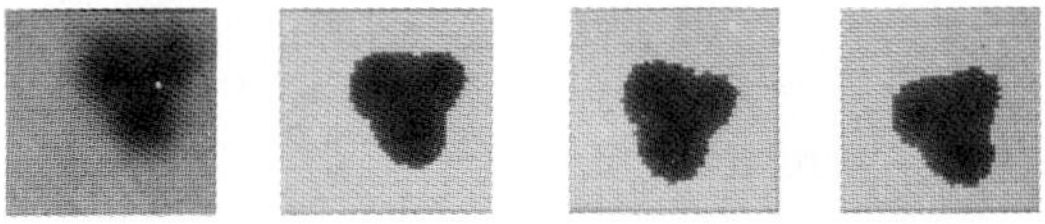

(a)   (b)   (c)   (d)

Rotated by (degrees)	Rectangle dimensions		
	Width	Height	Area
0	28	28	784
10	27	29	783
20	27	29	783
30	30	30	900
40	29	30	870
50	27	31	837
60	27	31	837
70	28	31	868
80	28	30	840

(e)

Fig. 1 Geometrical normalization. (a) Chromosome. (b) Semithresholded version of (a). (c) Result of rotating (b) about its centroid to make its principal axis vertical. (d) Result of rotating (b) to make a smallest circumscribed rectangle upright. (e) Dimensions of circumscribed rectangles of (b) for various rotations.

10.1.4 Transform Normalization

Another approach to normalizing a picture f with respect to some set of operations is to create a transform $\mathcal{T}(f)$ of f that is invariant under the given operations, i.e., $\mathcal{T}(\mathcal{O}(f)) = \mathcal{T}(f)$, where $\mathcal{O}$ is any one of the operations. Several examples of this approach will be described in this section.

The autocorrelation R_f (Section 2.1.4) and the power spectrum $|F|^2$ of a picture f remain the same when f is shifted. Thus these transforms of f are invariant under translation of f, and any properties derived from them can be regarded as translation-invariant properties of f. Note that these transforms are also invariant under rotation of f by 180°.

It should be pointed out that two pictures can have the same autocorrelation (and hence the same power spectrum, since this is just the Fourier transform of the autocorrelation) even if they do not differ by a translation or a 180° rotation. As a simple example, consider the pictures

0	0	0	0	0	0	0		0	0	0	0	0	0	0
0	0	3	10	3	0	0	and	0	0	1	6	9	0	0
0	0	0	0	0	0	0		0	0	0	0	0	0	0

which both have autocorrelation

0	0	0	0	0	0	0
0	9	60	118	60	9	0
0	0	0	0	0	0	0

Exercise 3. Show that if two pictures each have at most two points with nonzero gray level, and they have the same autocorrelation, then they can only differ by a translation and/or a 180° rotation.[†] ∎

As we shall see in Section 10.2, simple properties of the autocorrelation and power spectrum of f, such as their values at specified points, or their integrals over specified regions, provide useful descriptors of the texture of f.

[†] It can be shown that if two pictures have the same second-order autocorrelation, defined by

$$\int \int f(x, y) f(x+a, y+b) f(x+c, y+d) \, dx \, dy$$

then they differ only by a translation. On the question of when the autocorrelation determines the original function see R. L. Adler and A. G. Konheim, A note on translation invariants, *Proc. Amer. Math. Soc.* **13**, 1962, 425–429; J. A. McLaughlin and J. Raviv, Nth-order autocorrelation in pattern recognition, *Informat. Contr.* **12**, 1968, 121–142; D. Chazan and B. Weiss, Higher order autocorrelation functions as translation invariants, *ibid.* **16**, 1970, 378–383.

Cross sections of R_f or $|F|^2$ have also been found useful as a source of properties. In particular, note that the values of R_f along a circle C of radius r, centered at the origin, measure the degree to which f matches itself when "nutated" around itself in an orbit of radius r. If f is two-valued, say $f = \varphi_S$, then the value of R_f at a point (a, b) measures the probability that $(x + a, y + b)$ will be in S, given that (x, y) is in S, i.e., if we drop a line segment L of length $\sqrt{a^2 + b^2}$ and slope $\tan^{-1}(b/a)$ on the picture in a random position, then $R_f(a, b)$ measures the probability that if one end of L is in S, so is the other end. Thus the integral of R_f along C measures the probability that if we drop a line segment L of length r on the picture in a random position and orientation, and one end of L is in S, so is the other end. As these last examples show, integral-geometric methods can be used to estimate simple properties of the autocorrelation.

Translation invariance can also be obtained by using the one-dimensional power spectrum of a projection of the given picture f. Indeed, let f_θ be the projection of f on a line in direction θ (see Section 9.4.1). Thus f_θ is invariant under translation of f in the direction perpendicular to θ. Moreover, the power spectrum $|F_\theta|^2$ of f_θ is also invariant under translation of f in direction θ, hence in any direction.[†]

Analogous methods can be used, at least in principle, to obtain a transform of f that is invariant under rotation or magnification of f. For example, suppose that we use a polar coordinate system (r, θ) for f, and let the polar coordinate "autocorrelation" be defined by

$$f_0(\lambda, \varphi) \equiv \int_0^{2\pi} \int_0^\infty f(r, \theta) f(\lambda r, \theta + \varphi) r \, dr \, d\theta$$

This is evidently invariant under rotation of f about the origin. Moreover, it changes only by a constant factor if f is magnified; thus a ratio of its values at two points (say) is invariant under magnification of f. If f_0 is computed not for f itself, but for R_f, we also have invariance under translation of f. It should be realized, of course, that this polar coordinate method is not readily applicable if f is a digital picture. A polar coordinate digital picture would not be easy to store compactly in a computer, and its construction would involve unequally spaced sampling of the original picture.

Some simpler methods of obtaining rotation invariance can be defined, but they too depend on a polar coordinate representation of f. For example, the integral

$$\bar{f}(r) \equiv \int_0^{2\pi} f(r, \theta + \varphi) \, d\varphi$$

[†] Indeed, F_θ is equal to the cross section of F (the Fourier transform of f) along the line through the origin with slope $\theta + 90°$; see Sections 6.1.1 and 7.1.2.

is evidently invariant under rotation of f about the origin. Indeed, $\bar{f}(r)$ is just what f would look like if it were rotated very rapidly about the origin; this clearly does not depend on the initial orientation of f.

As a generalization of this idea[†] let $\mathcal{G}$ be a group of picture operations (see the end of Section 10.1.1), and let each operation $\mathcal{O}$ in $\mathcal{G}$ be linear, i.e., for all pictures f_1, f_2 we have $\mathcal{O}(f_1 + f_2) = \mathcal{O}(f_1) + \mathcal{O}(f_2)$. (These conditions evidently hold if we let $\mathcal{G}$ be the rotations about the origin.) Let us apply every $\mathcal{O}$ in $\mathcal{G}$ to f, and take the sum (or average) of all the resulting $\mathcal{O}(f)$'s, say $\bar{f} = \mathcal{O}_1(f) + \mathcal{O}_2(f) + \cdots$. As indicated at the end of Section 10.1.1, for any $\mathcal{O}$ in $\mathcal{G}$, the set of pictures $\mathcal{O}_1(\mathcal{O}(f)), \mathcal{O}_2(\mathcal{O}(f)), \ldots$, is the same as the set $\mathcal{O}_1(f), \mathcal{O}_2(f), \ldots,$; hence if we start with $\mathcal{O}(f)$ instead of f, we obtain the same $\bar{f}$.

Another polar coordinate-based approach to rotation invariance makes use of the fact that $f(r, \theta)$ is a periodic function of θ with period 2π. We can thus expand this function in a Fourier series; and the power spectrum of this Fourier series is invariant under rotation of f about the origin. In Section 9.3.3 we used a similar method to obtain a normalized representation of a curve: We represented the curve by an intrinsic equation, e.g., slope as a function of arc length; this is a periodic function with period equal to the circumference of the curve, and it is also invariant under translation of the curve. We then expanded this function in a Fourier series, and took its power spectrum, to obtain rotation invariance. In either case, we can obtain scale invariance by taking ratios of Fourier coefficients. It should be pointed out that the intrinsic equation itself can be made rotation-invariant by measuring the slopes relative to an intrinsically defined direction, e.g., the mean slope or the most frequently occurring slope, rather than relative to the x-axis. Analogous methods can be used to obtain normalized curve representations based on other types of intrinsic equations. In fact, suppose we are given the parametric equations of a curve, say $x = f(t)$, $y = g(t)$ where $0 \leqslant t \leqslant 1$; then we can let $z = x + jy$, and expand this complex-valued function of f in a Fourier series as above.

Still another polar coordinate approach to rotation and magnification invariance is as follows: The power spectrum $|F|^2$ rotates when f is rotated, and demagnifies when f is magnified (see Section 2.1.4). Suppose that we scan $|F|^2$ along a spiral centered at the origin, where the turns of the spiral are spaced logarithmically; and that we sample $|F|^2$ at evenly spaced angular positions around each turn of the spiral. Let us now put the samples into a rectangular array A whose rows correspond to turns of the spiral. Rotating f then cyclically shifts A horizontally, while magnifying f shifts A vertically,

[†] See W. Pitts and W. S. McCulloch, How we know universals—the perception of auditory and visual forms, *Bull. Math. Biophys.* **9**, 1947, 127–147.

since the demagnification becomes a shift on the logarithmic radial scale. (This description is not accurate near the top or bottom of A, since we have only used finitely many turns of the spiral; but if we cover a wide enough frequency range, the extreme high and low frequencies can be ignored.) These shifts can be eliminated by taking the (cyclic) autocorrelation or power spectrum of A.

10.2 PROPERTIES

As indicated at the beginning of Section 10.1, a property is a function that maps pictures into numbers. Properties can be real valued (e.g., Fourier coefficients) or integer valued (e.g., numbers of connected components). Two-valued properties, that are either true or false for a given picture, are an important special case; such properties are sometimes called *predicates*. Some of these properties have clear-cut mathematical definitions (e.g., "is connected"), while others (e.g., "is coarsely textured" or "is large") are basically fuzzy, and can be made two-valued only by applying arbitrary thresholds to numerical-valued properties (such as coarseness and area).

Geometrical properties of picture subsets that do not depend on the gray levels of the points in the given subset were discussed in detail in Chapter 9. In this section we deal with properties that do depend on gray level—textural properties and moments, for example. Such properties can be measured over an entire picture or over any subset of a picture. For simplicity, we will usually assume that they are being measured over the entire picture.

Most of the picture properties that are used for picture classification and description are chosen on heuristic grounds, i.e., because they are easy to measure and appear to be relevant to the desired goals. In most cases, there is no mathematical theory that can be used to determine optimal properties of given types for given purposes.[†] (Theoretical guidance is available only in certain simple cases, e.g., if we are dealing with pictures g obtained by adding noise to an ideal prototype f, and we want to compute the degree of match between g and f by convolving some template h with g, we have seen in Section 8.3.2 that the best h to use, under certain assumptions, is f itself.) Ideally, such a theory would be based on models for the classes of pictures in question; for an interesting approach along these lines see Grenander [15, 16].

[†] Given a set of properties, mathematical "feature selection" techniques can be used to find optimal subsets of these properties for a given classification task; this approach belongs to pattern recognition theory, and will not be treated here (see, e.g., Duda and Hart [7]). Many early "learning machine" pattern recognition systems began with large sets of properties, often generated randomly, and attempted to modify these properties (or their contributions to the classification decision) so as to improve classification performance.

Another factor becomes important if we want not merely to classify pictures, but to describe them in the same way that people do. The properties used in such a description must be compatible with those used by the human visual system; otherwise, meaningful dialog about the pictures between humans and machines would be difficult or impossible.

Section 10.2.1 considers properties that can be measured by examining only small parts of a picture; such properties are called *local*. We also discuss properties whose values depend on sets of local property values, e.g., textural properties. Section 10.2.2 discusses linear properties, which are always of the form $\iint hf$ for some "template" h. Section 10.2.3 introduces some approaches to measuring the complexity of picture properties.

10.2.1   Local and Textural Properties

A picture property is called *local* if its value depends only on some small portion of the given picture. For example, the value of a local operation (e.g., the Laplacian) at any point is a local property, since it depends only on the part of the picture in the immediate neighborhood of that point. The extreme case of a local property is a "point property," whose value depends only on a single point of the picture.

Exercise 4.   If $f_1, ..., f_n$ are distinct pictures, show that there exist points $(x_1, y_1), ..., (x_{n-1}, y_{n-1})$ such that no two of the n pictures have the same values at all $n-1$ of the points. Thus any set of n distinct pictures can be distinguished using at most $n-1$ point properties. (See A. Glovazky, Determination of redundancies in a set of patterns, *IEEE Trans. Informat. Theory* **IT-2**, 1956, 151–153; A. Gill, Minimum-scan pattern recognition, *ibid*. **IT-5**, 1959, 52–57; B. H. Mayoh, Optimum classification of objects, Algorithm 83, *Comm. ACM* **5**, 1963, 167–168.)   ∎

More formally, let S be a subset of the plane, and let $\mathscr{P}$ be a picture property.[†] Suppose that, for all pictures $f(x, y)$, the property value $\mathscr{P}(f)$ depends only on the restriction of f to S—in other words, in computing $\mathscr{P}(f)$, we can ignore the values of f at all points outside S. If the diameter of S is small, $\mathscr{P}$ is called local. In the digital case, we can always find a *smallest* set S of points such that $\mathscr{P}(f)$ depends only on the restriction of f to S, since our point sets are all finite; this smallest S is called the *set of support* of $\mathscr{P}$.

Individual local properties are useful primarily in cases where the pictures in question can be normalized with respect to translation, rotation, and magnification, since otherwise the value of a local property in a given position

---

[†] There should be no confusion with the use of $\mathscr{P}$ to denote the probability of an event in Chapters 2 and 5.

will not always represent information about the same part of the picture. Examples of useful individual local properties include values of the auto-correlation or power spectrum of a picture, which give translation-invariant information about the picture's texture; and streak-detector values in given parts of a geometrically normalized character, which give information about the presence of strokes in the character.

Sets of local properties and counts of local property values can be useful even without normalization. For example, if the overlay φ_S contains only 2×2 patterns of the forms

$$\begin{matrix} 1 & 1 \\ 1 & 1 \end{matrix} , \quad \begin{matrix} 1 & 1 \\ 0 & 0 \end{matrix} , \quad \begin{matrix} 1 & 0 \\ 0 & 0 \end{matrix} , \quad \begin{matrix} 0 & 0 \\ 0 & 0 \end{matrix}$$

and their rotations by multiples of 90°, it is not hard to see that S must con-sist of a set of solid rectangles of 1's on a background of 0's (how else could $\begin{smallmatrix} 1 & 1 \\ 1 & 0 \end{smallmatrix}$ patterns and their rotations be entirely absent?). Moreover, if the patterns

$$\begin{matrix} 1 & 0 \\ 0 & 0 \end{matrix} , \quad \begin{matrix} 0 & 1 \\ 0 & 0 \end{matrix} , \quad \begin{matrix} 0 & 0 \\ 0 & 1 \end{matrix} , \quad \text{and} \quad \begin{matrix} 0 & 0 \\ 1 & 0 \end{matrix}$$

are present exactly once each, S must be a single rectangle.

Statistics of local property values provide useful information about the *texture* of a picture. For example, in a "busy" picture, the average value of the gradient or of the Laplacian should be high, whereas in a smooth picture it should be low. In a highly directional picture, say one containing many streaks in direction θ, the average value of the directional difference in direc-tion θ should be low, while that in direction $\theta + (\pi/2)$ should be high.

A wide variety of local properties can be used in texture analysis. One can use the values of difference operators, as just mentioned; or more generally, one can use operators that have high values where specified local patterns are present in the picture, e.g., lines, corners, etc.

One can use not only means, but also other statistics of local properties as textural properties. Useful statistics include the standard deviation, or the variance, which measure the spread of the property values around the mean; the third moment, which measures asymmetry about the mean; the median, and various other p-tiles—or, equivalently, the numbers of picture points at which the given property takes on various ranges of values.

Figure 2 shows histograms of the values of x- and y-difference operators for the five pictures used in Fig. 9 of Chapter 8. As expected, the busier pictures contain higher numbers of points that have high difference values.

When using statistics of local properties as texture measures, it is important to realize that the values of such statistics can be sensitive to nontextural factors as well. For example, the average value of the gradient depends not

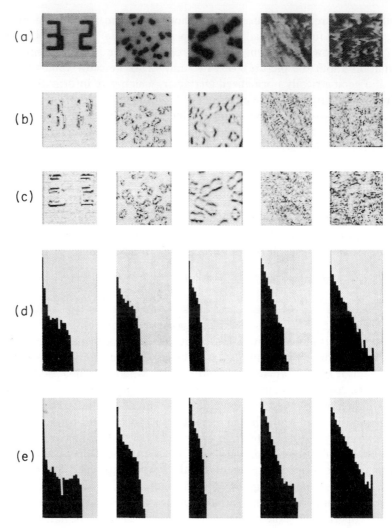

Fig. 2 Histograms of $|\Delta_x|$ and $|\Delta_y|$, scaled logarithmically, for five pictures. (a) Input picture f. (b) $|\Delta_x f|$. (c) $|\Delta_y f|$. (d) Histogram of (b). (e) Histogram of (c).

only on busyness, but also on contrast; if we stretch or shrink a picture's gray scale, this average goes up or down, even though the degree of busyness of the picture does not change. It is often desirable to perform gray scale normalization (Section 10.1.2) on a picture before measuring its textural properties.

Statistics of the picture's gray levels are not generally useful as textural properties. (The gray level variance, for example, is the same for any two-valued picture that has equal numbers of 1's and 0's, whether the 1's are all in one half of the picture, in alternate rows, in a checkerboard pattern, or randomly distributed!) Statistics of *joint* gray level occurrences, on the other hand, representing the frequencies with which various pairs of gray levels occur at neighboring points of the picture, do provide useful textural information. For example, in a smooth picture, the neighbors of a point with gray level z are likely to have gray levels close to z, whereas in a busy picture this will not be true. Such statistics can be made directionally sensitive by using only neighbors in specified directions.

If the given picture has k gray levels, the numbers of occurrences of neighboring points having various pairs of gray levels can be stored in a $k \times k$ matrix, for any specified displacement (a, b), or set of such displacements, between the neighboring points. The (i, j) element of this matrix is the number of times that a point having gray level z_i occurs in position (a, b) relative to a point having gray level z_j. For example, in the four-level picture

$$
\begin{array}{cccc}
2 & 2 & 3 & 3 \\
1 & 3 & 3 & 2 \\
0 & 0 & 3 & 1 \\
1 & 0 & 0 & 1
\end{array}
$$

the matrix for $(a, b) = (1, 0)$ is

	0	1	2	3
0	2	1	0	0
1	1	0	0	1
2	0	0	1	1
3	1	1	1	2

Such matrices for the five pictures used in Fig. 2 are shown in Figs. 3a–e. For simplicity, the 64 gray levels in these pictures have been divided into eight ranges (0–7, 8–15, ..., 56–63), and the (i, j) element of the matrix is the number of times that a point whose gray level is in the ith range occurs in position $\pm(4, 0)$, $\pm(0, 4)$, $\pm(4, 4)$, or $\pm(4, -4)$ relative to a point whose level is in the jth range.

It can be seen from Figs. 3a–e that in a smooth picture, the high values in the matrices are concentrated near the main diagonals, since the neighbors of a point tend to have the same gray level as the point itself; whereas in a busy picture, the high values should tend to spread further away from the diagonal. This suggests that the moment of inertia of such a matrix about its main diagonal could be used as a measure of busyness. Various other statistical properties derived from these matrices have also been found useful in texture analysis. The results depend, of course, on the size of the displacement(s) (a, b) that one uses, as seen from Fig. 3f. As mentioned earlier, directional sensitivity is achieved by using matrices derived from displacements (a, b) in different directions (see Figs. 3g and h).

A class of properties that have commonly been used for texture analysis is

(a)	1294	1236	48	17	21	14	44	73
	1236	1684	86	36	54	50	98	167
	48	86	30	13	14	13	15	33
	17	36	13	11	6	7	5	8
	21	54	14	6	12	5	4	20
	14	50	13	7	5	8	9	17
	44	98	15	5	4	9	29	33
	73	167	33	8	20	17	33	89

(b)	23	47	2	0	1	1	0	0
	47	1002	425	74	48	43	81	28
	2	425	1722	384	152	145	300	116
	0	74	384	192	47	38	93	41
	1	48	152	47	17	20	40	17
	1	43	145	38	20	18	35	11
	0	81	300	93	40	35	52	16
	0	28	116	41	17	11	16	8

(c)	0	0	0	0	0	0	0	0
	0	73	188	16	8	8	7	1
	0	188	2223	441	171	166	257	57
	0	16	441	178	81	90	219	75
	0	8	171	81	36	44	113	52
	0	8	166	90	44	41	116	55
	0	7	257	219	113	116	251	126
	0	1	57	75	52	55	126	62

(d)	1016	329	210	104	80	47	11	0
	329	276	245	128	123	84	8	0
	210	245	249	166	161	114	5	0
	104	128	166	147	178	186	6	0
	80	123	161	178	250	285	21	0
	47	84	114	186	285	443	35	0
	11	8	5	6	21	35	12	0
	0	0	0	0	0	0	0	0

(e)	70	56	57	81	156	187	13	1
	56	37	49	81	138	162	7	1
	57	49	74	89	166	220	10	1
	81	81	89	114	258	309	15	1
	156	138	166	258	567	683	26	1
	187	162	220	309	683	982	28	2
	13	7	10	15	26	28	4	1
	1	1	1	1	1	2	1	0

Fig. 3 (a)–(e)

(f)

284	136	95	65	44	35	6	0
136	110	122	95	64	33	3	1
95	122	129	163	141	67	4	0
65	95	163	223	313	158	10	1
44	64	141	313	891	677	20	2
35	33	67	158	677	1720	58	2
6	3	4	10	20	58	15	2
0	1	0	1	2	2	2	2

(g)

1128	317	156	60	41	18	9	0
317	318	242	120	100	67	6	0
156	242	284	185	154	104	3	0
60	120	185	176	186	163	2	0
41	100	154	186	254	303	17	0
18	67	104	163	303	442	38	0
9	6	3	2	17	38	16	0
0	0	0	0	0	0	0	0

(h)

738	310	275	185	138	70	19	0
310	208	221	141	162	116	13	0
275	221	208	129	136	139	8	0
185	141	129	102	141	192	5	0
138	162	136	141	216	249	21	0
70	116	139	192	249	342	22	0
19	13	8	5	21	22	2	0
0	0	0	0	0	0	0	0

Fig. 3 Gray level co-occurrences for the five pictures in Fig. 2a. For simplicity, gray levels have been combined into 8 groups $(0-7, 8-15, ..., 56-63)$. (a)–(e) Matrix element (i, j) is proportional to the number of times that a gray level in the ith group occurs at distance 4 from a gray level in the jth group in any one of the four principal directions (horizontal, vertical, and the two diagonals). (f) Same for the last of the five pictures, using distance 1. (g)–(h) Same for the next-to-last picture, using distance 4 in the two diagonal directions only.

the gray level transition probabilities. The probability p_{ij} measures the frequency with which a point having gray level z_j occurs immediately following a point having level z_i on a row of the picture. This frequency is just the (i, j) element of the matrix corresponding to displacement $(1, 0)$.

Other types of picture statistics can also be used for texture analysis. For example, one could use the frequencies of occurrence of runs of constant gray level (or gray level in a given range) having various lengths and directions. (Long runs will tend to occur in a given direction if the gray level transition probabilities p_{ij} in that direction are low for $i \neq j$ and high for $i = j$.) Other statistical properties of cross sections of the picture in various directions could be used in place of run length statistics. The values of such properties for a given picture can be estimated using the random sampling methods of integral geometry.

Still another approach to texture analysis is to use simple properties of the picture's autocorrelation or power spectrum as texture measures. For example, in a busy picture the value of the autocorrelation should fall off very rapidly from its peak at zero displacement, while in a smooth picture the falloff should be more gradual. Directional biases in the picture should show up as directional differences in the rate of falloff. Similarly, the power spectrum of a busy picture should have relatively high power at high spatial frequencies, compared to the spectrum of a smooth picture; while directional biases in the picture should give rise to directional biases in the spectrum. Periodic patterns in a picture should produce very high values in the autocorrelation (for a displacement corresponding to the pattern period) and in the power spectrum (at the fundamental spatial frequency of the pattern).

The power spectra of the five pictures used in Figs. 2 and 3 were shown in Fig. 11a of Chapter 8; they are shown again, with their values scaled logarithmically, in Fig. 4a. The broader spread of the power for the busier pictures, and its directional biases for the pictures that had such biases, are readily apparent. Figure 4c shows, in bar graph form, the results of integrating the spectra with respect to θ (in polar coordinates), i.e., it plots the average

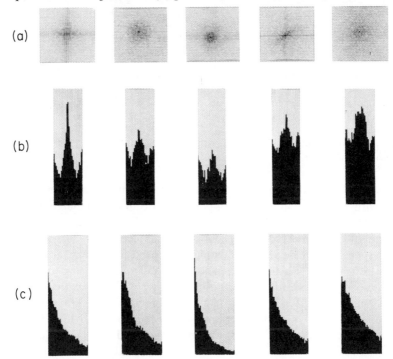

Fig. 4   Power spectra for the five pictures in Fig. 2a. (a) Log power spectra, $\log(|F|^2 + 1)$; for $|F|^2$ itself see Fig. 11a of Chapter 6. (b) $\int |F|^2 \, dr$. (c) $\int |F|^2 \, d\theta$.

power on a circle centered at the origin, as a function of the radius r of the circle. Figure 4b similarly shows the results of integration with respect to r, i.e., the average power along a line through the origin, as a function of the slope θ of the line. Average values of the power in an angular sector emanating from the origin, or in a circular ring centered at the origin, are useful textural properties for many purposes.

10.2.2   Linear Properties

We saw in Section 2.1.2 that any shift-invariant linear operation is a convolution. Thus if $\mathcal{O}$ is an operation that takes pictures into pictures, and $\mathcal{O}$ is linear and shift invariant, then there exists a function h such that

$$\mathcal{O}(f) = \int\int_{-\infty}^{\infty} h(x-\alpha, y-\beta) f(\alpha, \beta) \, d\alpha \, d\beta$$

for all pictures f, i.e., $\mathcal{O}(f)$ is the convolution of h with f.

It can be shown that, similarly, any linear property is a "template match" property. In other words, if $\mathcal{P}$ is an operation that takes pictures into numbers, and $\mathcal{P}$ is linear, then there exists a function h such that

$$\mathcal{P}(f) = \int\int_{-\infty}^{\infty} h(\alpha, \beta) f(\alpha, \beta) \, d\alpha \, d\beta$$

for all pictures f. (We recall (Section 8.3) that by the Cauchy–Schwarz inequality, $\int\int hf$ is a maximum, for a given $\int\int f^2$, when $f = ch$ for some constant c; this justifies regarding $\int\int hf$ as a measure of the match between the template h and the picture f.) More precisely, one can prove the following theorem (in which all integrals are understood to be taken over the entire plane):

Theorem.[†]   Let Φ be the set of functions such that $\int\int f^2 < \infty$ for all f in Φ. Let $\mathcal{P}$ be any real-valued, linear function defined on Φ, and suppose that there exists a real number M such that, for all f in Φ,

$$\mathcal{P}(f) \leqslant M \left(\int\int f^2 \right)^{1/2}$$

Then there exists an h in Φ such that, for all f in Φ,

$$\mathcal{P}(f) = \int\int hf$$

[†] This theorem is known as the *Riesz representation theorem*, and can be found in books on Hilbert space theory.

Proof: If $\mathscr{P}(f) = 0$ for all f, this is trivial (just take $h = 0$). Otherwise, it can be shown that there exists an h_0 in Φ such that

$$\mathscr{P}(h_0) \Big/ \left(\int \int h_0^2 \right)^{1/2} = \max \mathscr{P}(f) \Big/ \left(\int \int f^2 \right)^{1/2} \neq 0$$

Thus $M = \mathscr{P}(h_0)/(\int \int h_0^2)^{1/2}$ still gives an upper bound on $\mathscr{P}$. We shall now show that if g in Φ is such that $\mathscr{P}(g) = 0$, then $\int \int h_0 g = 0$. Indeed, we have, for any real λ,

$$M^2 \int \int h_0^2 = (\mathscr{P}(h_0))^2 = (\mathscr{P}(h_0 - \lambda g))^2$$

$$\leqslant M^2 \int \int (h_0 - \lambda g)^2$$

$$= M^2 \left[\int \int h_0^2 - 2\lambda \int \int h_0 g + \lambda^2 \int \int g^2 \right]$$

so that $0 \leqslant -2\lambda \int \int h_0 g + \lambda^2 \int \int g^2$. If $\int \int g^2 \neq 0$, we can take $\lambda = \int \int h_0 g / \int \int g^2$ to obtain $0 \leqslant -(\int \int h_0 g)^2$, which implies $\int \int h_0 g = 0$; while if $\int \int g^2 = 0$, we have $g = 0$, so that $\int \int h_0 g = 0$ in any case.

Now for any f in Φ we can write

$$f = [f - (\mathscr{P}(f)/\mathscr{P}(h_0)) h_0] + (\mathscr{P}(f)/\mathscr{P}(h_0)) h_0$$

If we call the bracketed term g, we see that $\mathscr{P}(g) = 0$, so that by the preceding paragraph, $\int \int h_0 g = 0$. Thus for any real γ we have

$$\int \int (\gamma h_0) f = \int \int (\gamma h_0) g + (\mathscr{P}(f)/\mathscr{P}(h_0)) \int \int \gamma h_0^2$$

$$= \gamma \int \int h_0 g + \mathscr{P}(f) \left[\gamma \int \int h_0^2 / \mathscr{P}(h_0) \right]$$

$$= \mathscr{P}(f) \left[\gamma \int \int h_0^2 \Big/ M \left(\int \int h_0^2 \right)^{1/2} \right]$$

Taking $\gamma = M/(\int \int h_0^2)^{1/2}$ thus gives $\int \int (\gamma h_0) f = \mathscr{P}(f)$, so that $h = \gamma h_0$ satisfies the theorem. ∎

A wide variety of templates (h's) can be used to compute linear picture properties. If the pictures in question can be normalized with respect to position, orientation, and scale, it is possible to use entire pictures as templates. For example, when attempting to identify characters from a specific type font, one can use entire characters as templates, and classify an unknown character as being the same as the one whose template it best matches. Rather than normalizing with respect to translation, one can cross correlate each template with the picture, as shown in Section 8.3, and find the position having the greatest degree of match.

More generally, when dealing with a class of pictures that can be constructed from a set of basic subpictures, e.g., characters from a set of strokes, one can use the strokes as templates. Still more generally, one can initially use a large set of arbitrary patterns as templates, and employ mathematical feature selection techniques to find subsets of these patterns which contribute the most to the classification decision. Many early pattern recognition systems actually employed *random* templates (or, in some cases, nonlinear properties with randomly chosen sets of support) to define initial property sets.

Another important class of linear properties makes use of templates that are mathematically simple functions, e.g., polynomials, sinusoids, Walsh functions, etc. (When the template function is a sinusoid, the resulting property value is a Fourier coefficient of the given picture.) If we use a set of mutually orthogonal templates, such as sinusoids, Walsh functions, etc., we obtain a set of uncorrelated properties.

If we take the template function $h(x, y)$ to be the monomial $x^i y^j$, we obtain a property known as the (i, j)th *moment* of the given picture, which we shall denote by m_{ij}. As we saw in Section 10.1.3, the coordinates of the centroid of the picture are

$$\bar{x} = m_{10}/m_{00}, \qquad \bar{y} = m_{01}/m_{00}$$

If we use a new coordinate system in which $(\bar{x}, \bar{y})$ is the origin, the moments computed in this system are called *central moments*. Such moments are evidently translation-independent properties of the original picture. Rotation- and scale-invariant properties can also be defined in terms of moments. For example, the moment of inertia of the picture about the origin, $m_{20} + m_{02} = \iint (x^2 + y^2) f(x, y) \, dx \, dy$, is readily invariant under rotation about the origin; and a ratio of two moments that have the same value of $i+j$, e.g., m_{10}/m_{01}, is invariant under magnification.

Exercise 5. Prove that the $(1,0)$ and $(0,1)$ *central* moments are always zero. ∎

If we substitute $-x$ for x and $-y$ for y in the definition of m_{ij}, we obtain

$$m_{ij} = \int \int x^i y^j f(x, y) \, dx \, dy = \int \int (-x)^i (-y)^j f(-x, -y) \, dx \, dy$$

It follows that if $f(x, y)$ is symmetric about the origin, i.e., $f(-x, -y) = f(x, y)$, we have $m_{ij} = (-1)^{i+j} m_{ij}$ for all i and j. Thus if $i+j$ is odd, m_{ij} must be zero. Similarly, if we only substitute $-y$ for y, we obtain

$$m_{ij} = \int \int x^i (-y)^j f(x, -y) \, dx \, dy$$

Thus if $f(x, y)$ is symmetric about the x-axis, i.e., $f(x, -y) = f(x, y)$, we must have $m_{ij} = (-1)^j m_{ij}$, so that $m_{ij} = 0$ if j is odd; and analogously, if $f(x, y)$ is symmetric about the y-axis, m_{ij} must be zero if i is odd. Moments for which i, j, or $i+j$ is odd can thus be used as measures of asymmetry about the y-axis, the x-axis, and the origin, respectively. Moments also provide a basis for normalizing a picture, e.g., by transforming it to make specified moments, or combinations of moments, zero; see also Section 10.1.3.

Exercise 6.   Prove that if $f(x, y)$ is symmetric about the line $y = x$ (i.e., $f(y, x) = f(x, y)$), then $m_{ij} = m_{ji}$ for all i, j.   ∎

Exercise 7.   Compute the $(2, 0)$, $(1, 1)$, and $(0, 2)$ central moments for the pictures

$$\begin{matrix} & 1 & \\ 1 & 1 & 1 \end{matrix} \quad , \quad \begin{matrix} & 1 & \\ 1 & 1 & 1 \end{matrix} \quad , \quad \text{and} \quad \begin{matrix} 1 & & \\ 1 & 1 & 1 \end{matrix}$$

where all values not shown are zero.   ∎

10.2.3   Property Complexity

As we saw at the beginning of Section 10.2.1, some picture properties depend only on parts of the given picture; if these parts are small, we call the properties "local." We also saw some examples of how picture properties could be defined in terms of sets of local properties. In this section we explore these ideas further and show how they can be used to define measures of the complexity of picture properties. In what follows, the number of points in the set S is denoted by $|S|$.

To start with a simple example, suppose that we have a two-valued picture φ_S whose set of points having value 1 is S. If the points of φ_S are $p_1, \ldots, p_N$, then we can express various two-valued properties of φ_S as Boolean functions of these N Boolean variables. For example, the property "$|S| = 1$" can be expressed as the function

$$\bigvee_{i=1}^{N} (\bar{p}_1 \wedge \cdots \wedge \bar{p}_{i-1} \wedge p_i \wedge \bar{p}_{i+1} \wedge \cdots \wedge \bar{p}_N)$$

(where $\bigvee$ denotes logical OR, $\wedge$ denotes logical AND, and the overbars denote negation), since this function has value 1 if and only if exactly one of the p_i's has value 1.

We can now define the *length* of a property $\mathscr{P}$ as the number of variables in the shortest Boolean function expression for $\mathscr{P}$. For example, the length of the property "$|S| = 1$," if we represent it by the Boolean function just given, is N^2. It can, in fact, be shown that the length of this particular property cannot be a linear function of N. On the other hand, the property

"$|S| \geqslant 1$" can be expressed as

$$\bigvee_{i=1}^{N} p_i$$

so that its length is N. If a property has a small set of support (i.e., it depends only on a small part of the picture), its length will also be small; for example, the property "p_j is in S" has length 1, since it has the Boolean function expression p_j. Evidently, a property has length 0 if and only if it is a constant, i.e., it is true (or false) for every S.

It is evident that if the properties $\mathscr{P}$ and $\mathscr{Q}$ have lengths m and n, respectively, then logical combinations of $\mathscr{P}$ and $\mathscr{Q}$ such as $\mathscr{P} \vee \mathscr{Q}$, $\mathscr{P} \wedge \mathscr{Q}$, etc., have lengths at most $m+n$. It can be shown that properties such as "S is convex" and "S is connected" cannot have lengths that are linear functions of N. Interestingly, no familiar property has ever been shown to have length that is more than a quadratic function of N.

A more interesting concept of property complexity is obtained by considering two-valued properties that are obtained by thresholding the values of other properties. Specifically, let $\mathscr{S}$ be any set of real-valued properties $\mathscr{P}_i$, and let the property $\mathscr{Q}$ be defined by

$$\mathscr{Q}(f) = \begin{cases} 1 & \text{if } \sum_{\mathscr{S}} a_i \mathscr{P}_i(f) \geqslant t \\ 0 & \text{otherwise} \end{cases}$$

where the coefficients a_i and the threshold t are real numbers. We then say that $\mathscr{Q}$ is a "linear threshold property" with respect to $\mathscr{S}$. We can now define the *order* of $\mathscr{Q}$ to be k if there exists an $\mathscr{S}$ such that

(a) $\mathscr{Q}$ is linear threshold with respect to $\mathscr{S}$;
(b) the set of support of every $\mathscr{P}_i$ in $\mathscr{S}$ has at most k points,
and no such $\mathscr{S}$ exists for any smaller value of k.

Evidently, a property has order 0 if and only if it is a constant. There are many nontrivial properties of order 1; for example, for any set of picture points $T = \{q_1, ..., q_M\}$, the property "$T \subseteq S$" is of order 1. (Such a property is called a *mask*.) Indeed, let $\mathscr{P}_i$ be the property "q_i is in S," so that $\mathscr{P}_i$ has a one-point set of support. Then the $T \subseteq S$ property is equivalent to

$$\sum_{i=1}^{M} \mathscr{P}_i \geqslant M$$

so that it is linear threshold with respect to the set of $\mathscr{P}_i$'s. Note that the $T \subseteq S$ property depends on the position of the set T, and changes if the picture is translated. It can be shown that the only translation-invariant linear threshold properties of order 1 are those of the form "$|S| \geqslant k$" (or $\leqslant$, or $<$, or $>$).

Any two-valued property $\mathscr{P}$ is linear threshold with respect to the set of all masks. Indeed, $\mathscr{P}$ is a Boolean function of the Boolean variables $p_1, \ldots, p_N$, and so can be expressed in normal form as

$$\bigvee_{i=1}^{K} (p_{i1} \wedge \cdots \wedge p_{iN})$$

where each p_{ij} is either p_j or $\bar{p}_j$. For any given two-valued picture, only one of these K conjunctions can be 1; hence their disjunction is the same as their sum. Moreover, each conjunction is equivalent to a product, if we replace $\bar{p}_j$'s by $(1-p_j)$s. In the resulting sum of products, we can multiply out and group like terms to obtain a linear combination of products of p_i's with integer coefficients; but a product of p_i's is just a mask, since it is 1 if and only if all its p_i's are in S.

As we saw in Section 9.4.3, S is convex if and only if, whenever it contains two points, it also contains every point on the line segment joining them. For any three points p, q, r, with r on the line segment from p to q, let $\mathscr{P}_{p,q,r}$ be the property "p and q are in S but r is not in S." Then clearly "S is convex" is equivalent to

$$\sum -\mathscr{P}_{p,q,r} \geqslant 0$$

so that the convexity property is linear threshold with respect to the set of properties $\mathscr{P}_{p,q,r}$, making it of order (at most) 3.

As we saw in Section 9.1.3, the genus of S can be computed by counting the numbers of occurrences of various 2×2 (or smaller) patterns in φ_S. It follows readily that a property such as "S has genus $\geqslant 0$" is of order at most 4. In fact, it can be shown that functions of the genus are the only properties of bounded order (i.e., order that does not grow with the picture size) that are invariant under connectedness-preserving transformations of the picture. It follows that a property such as "S is connected" does not have bounded order, since it is not a function of the genus.

Many other simple picture properties do not have bounded order; "$|S|$ is odd" is an example. In fact, simple logical combinations of low-order properties need not have low order; for example, for any k, there exist properties $\mathscr{P}$ and $\mathscr{Q}$ of order 1 such that $\mathscr{P} \wedge \mathscr{Q}$ and $\mathscr{P} \vee \mathscr{Q}$ have order $>k$. Moreover, even if a property is linear threshold with respect to some given set of properties $\mathscr{S}$, the coefficients in the linear combination may span an impractically large range. For example, when we express "$|S|$ is odd" as a linear threshold function with respect to the set of all masks, it turns out that the ratio of the largest to the smallest coefficient grows exponentially with the picture size.

There are many other ways of defining property complexity. For example, we can say that $\mathscr{Q}$ has *diameter* d if $\mathscr{Q}$ is linear threshold with respect to a set of properties whose sets of support all have diameters $\leqslant d$, where d is as small as possible. Clearly a property has order 0 or 1 if and only if it has diameter 0. Here again, however, many familiar properties, such as "S is connected," turn out to have diameters that grow with the picture size.

The complexity of a property $\mathscr{P}$ can also be defined in terms of a grammar G or automaton A whose language is just the set of pictures that have property $\mathscr{P}$; the complexity of $\mathscr{P}$ can be measured by the number of symbols required to write out the shortest possible definition of such a G or A. The formal approach to describing "picture languages" in terms of automata or grammars will not be pursued further here; for selected references to the rapidly growing literature on this topic see Section 10.5.

10.3 RELATIONAL STRUCTURES

As pointed out at the beginning of Chapter 8 and of this chapter, a picture description can be thought of as a *relational structure* that involves parts of the picture, properties of these parts, and relationships among them. Given such a structure, it becomes relatively straightforward to answer questions about the picture, or to generate statements about it, provided that these questions or statements involve only the given parts, properties, and relations. In particular, it is easy to determine whether the picture contains specified arrangements of parts, i.e., whether the structure contains specified substructures.

In simple cases, one can derive a useful relational structure from a picture by segmenting the picture into parts, measuring properties of the parts, and establishing relationships between them, as in this and the two preceding chapters. Unfortunately, this is often not a straightforward process; we shall discuss this point further in Section 10.4.

Relational structures that describe pictures can involve entities at many levels, ranging from single points to complex objects or configurations of objects. It is often convenient to organize these structures hierarchically: The picture is composed of regions, or of objects on a background; objects are composed of regions, bounded by edges or lines; regions, edges, and lines are composed of points. At each of these levels, various specialized types of relational structures will be appropriate; these structures can be thought of as picture description notations or "languages." In Section 10.3.1 we discuss the types of entities, properties, and relations that can be used in describing pictures that consist of simple, discrete objects. Section 10.3.2 contains a similar discussion of the description of objects that can be represented by simple line drawings. Section 10.3.3 briefly treats the uses of such languages in formulating models for classes of pictures (or objects).

10.3.1   Description at the Picture Level

A picture can be described, in a general way, in terms of objects that are present in it, and properties of and relations among these objects. This superficial level of description usually involves a limited set of object types, properties, and relationships, which often have simple names. It should be realized that this simplicity is deceptive; it may be very difficult to determine, by analyzing the picture, just which objects are present, and just which properties or relations hold for these objects (see, e.g., the discussion of "to the left of" at the beginning of Section 9.4).

A partial list of properties that might be used in descriptions at this level is as follows:

(a)   brightness: black, gray, white; light, dark; uniform, shaded;
(b)   color: red, orange, etc.;
(c)   texture: smooth, grainy, mottled, striped;
(d)   size: length, area, volume; height, width, depth; large, small; tall, short; wide, narrow;
(e)   orientation: horizontal, vertical, oblique;
(f)   shape: solid, hollow; compact, jagged, elongated, etc.

Note that some of these are two-valued predicates (although they may have very fuzzy definitions!), e.g., black, smooth, large; while others are numerical valued, e.g., length. It should be realized that not all of these properties are easy to measure, or even to define, e.g., shading, textural properties, and three-dimensional attributes such as volume, depth, and solidity.

The relationships that might be used at this level include those involving comparison of properties, e.g., "darker, smoother, longer, ... than"; "the same color, size, ... as." In addition, there are spatial relationships such as

(a)   containment: part of, inside, containing, surrounding;
(b)   adjacency: touching, next to, on top of, etc.;
(c)   direction: above, behind, to the left of, etc.;
(d)   distance: near, etc.

There are also ternary (and higher-order) spatial relations, such as "between" and "beyond," as well as those involving relative values of binary relations, e.g., "nearer than." Still other types of relations involve mechanics as well as geometry, e.g., "supports" and "leans on." The relations mentioned here are all two-valued, but one can also allow quantitative relationships based on measurement of angles, distances, etc.

As a very simple example of a description at this level, consider how we might describe the "arch" shown in Fig. 5a:

(1)   The objects *A*, *B* and *C* are rectangles with horizontal and vertical sides.

(2)   *A* and *B* have the same height and width, and their positions have the same *y*-coordinates.

(3)   *A* and *B* do not touch.

(4)   *A* and *B* support *C*.

Note that we could have given many other pieces of information about Fig. 5a, e.g., the colors and sizes of the objects, or the exact position of *C* relative to *A* and *B*. Some of the omitted information would have been redundant, for example, the fact that *A* and *B* have the same area, or that *C* is on top of *A* and *B*. Indeed, given a flexible descriptive language, most of the statements about a given picture that can be made in that language will be redundant. Of course, it may sometimes be advantageous not to discard redundant information, since keeping it saves the trouble of deriving it every time it is needed.

Exercise 8.   What constraints must be satisfied by the sizes and positions of rectangles *A*, *B*, and *C* in Fig. 5a in order that *C* be stably supported by *A* and *B*? You may assume that these rectangles all have the same uniform density (i.e., mass per unit area).   ▮

The list of properties and relations given previously is by no means exhaustive. If we tried, for example, to construct a description of a face as an arrangement of two eyes, a nose, and a mouth, we would quickly discover that we need a better vocabulary for describing the shapes of objects, and symmetries in the positions of objects.

Descriptions of the type discussed so far can be conveniently represented by labeled graphs in which:

(a)   each node corresponds to an object, and is labeled with a list of property names and property values that hold for that object;

(b)   each arc between a pair of objects is labeled with a list of names and values of the relations that hold between that pair of objects.

Note that the arcs must be directed, since the relations may not be symmetric (e.g., we must be able to distinguish "*A* is to the left of *B*" from "*B* is to the left of *A*"); thus the graph should more properly be called a "digraph." Note also that there can be more than one arc between a given pair of nodes; such a graph is more properly called a "multigraph." The graph corresponding to the description of Fig. 5a given above [statements (1)–(4)] is shown in Fig. 5b. It should be pointed out that ternary and higher-order relations ("between," etc.) cannot be naturally represented in such a graph structure.

The graph structure just described is strongly object oriented; it is easy to

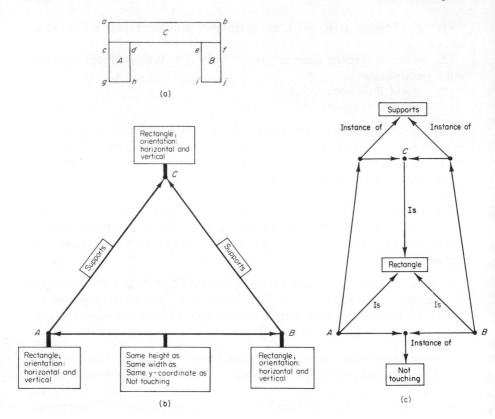

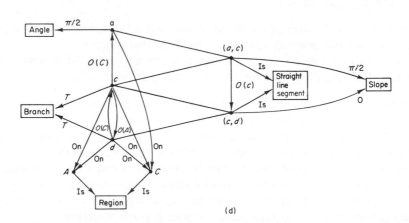

Fig. 5 (a)–(d)

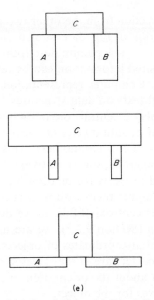

(e)

Fig. 5 A simple picture and its descriptions. (a) "Arch." (b) Partial description of (a) in terms of properties of and relations between objects. (c) Partial description of (a) treating properties and relations as entities in their own right. (d) Partial description of (a) in terms of vertices, arcs, and regions where $O(c)$ is clockwise order at vertex c and $O(C)$ is clockwise order on border of region C. (e) Some examples of arches that fit the description in (b).

extract from the graph the properties of a given object A, or the relations into which A enters, but it is harder to determine which objects enjoy a given property $\mathscr{P}$, or enter into a given relation $\mathscr{R}$, without searching through the entire graph. Alternative types of graph structures can be devised in which objects, properties, and relations are all treated alike. For example, we can use a structure of the following type:

(a)   Nodes correspond to objects, to properties, or to relations.

(b)   If object A has property $\mathscr{P}$, we join node A to node $\mathscr{P}$ by an arc, and label the arc with the value of $\mathscr{P}$ for A.

(c)   If objects A and B are in relation $\mathscr{R}$, we create an "instance node" $\mathscr{R}_i$, joined to $\mathscr{R}$ by an arc labeled "instance of." We then join A and B to $\mathscr{R}_i$, and label the arc(s) with the value of $\mathscr{R}$ for (A, B).

(d)   Property or relation nodes can themselves be joined by arcs if they are themselves related, e.g., we might join "supports" to "on top of," since the former is a special case of the latter.

This type of graph can also be used to represent certain ternary relations, e.g., to represent "C is between A and B," one could join A, B, and C to an instance node $\mathscr{R}_i$ of the "between" relation, but have the arc between C and

$\mathcal{R}_i$ go in the opposite direction from the other two arcs. A graph of this type representing our description of Fig. 5a is shown, in part, in Fig. 5c.

We shall not discuss here how picture description graphs can be stored in a computer, nor how desired information can be retrieved from these graphs using search algorithms. Computer representation and searching of graphs are discussed in standard texts on data structures.

Spatial relationships play a central role in picture descriptions. It is often desirable to specify them quantitatively, i.e., to give the spatial coordinates of objects (perhaps relative to one another). This can, of course, be done using a picture description graph, since (relative) coordinates are just a special kind of numerical-valued property (or relationship). If we want to give exact coordinates, we must compute them with respect to specific reference points on the objects (e.g., their centroids). [When we describe a picture subset S by specifying its skeleton (Section 9.2.3), we are using a special type of description in terms of spatial coordinates of objects: S is a union of "disks" of specified radii whose centers are at the skeleton points.] Similarly, to give quantitative information about the orientation of an object, we must introduce specific reference axes for the object.

Another way of representing quantitative spatial relationships is to introduce special "attachment points" on the objects. We can then represent the adjacency of objects A and B by specifying that a particular attachment point of A coincides with a particular attachment point of B. To represent the relative positions of nonadjacent objects A and B in this way, one can introduce an "invisible" object C of appropriate size and orientation that links A to B; thus adjacency of A to C and of C to B guarantees the desired relative positions of A and B. We can obtain arbitrarily detailed descriptions in this way by taking the objects to be single picture points which attach to their four (or eight) neighbors, so as to build up arbitrary connected regions as arrays of points.

10.3.2 Description at the Object Level

In Section 10.3.1 we discussed the description of a picture in terms of relations among objects; we now consider how one might describe the objects themselves. In many cases, objects are composed of small numbers of regions, each of which has brightness, color, or texture that is uniform, or perhaps varies at a uniform rate. We can abstractly represent such a complex of regions as a line drawing, where the lines represent the edges between the regions (or between regions and background); for each region in the drawing, we must also specify its brightness, color, or texture. In this section we discuss methods of describing line drawings by relational structures.

As indicated in Section 9.3.2, a curve can be segmented into arcs at its branch points, at its "angles" (or curvature maxima), and at its inflections (or curvature zero crossings). If the curves in a line drawing are segmented in this way, one can construct a relational structure description of the drawing that might include the following types of information:

(a)   properties of the arcs, such as straightness, (average) curvature, (average) slope, arc length, chord length, etc;

(b)   properties of the "vertices" at which arcs begin and end—in particular, whether they are endpoints, angles (and if so, whether acute, obtuse, etc.), inflections, or branch points of various types (see, e.g., Section 9.4.4);

(c)   incidence relations between arcs and vertices—which arcs emanate from which vertices, and which vertices lie on which arcs;

(d)   order relations—in particular, the order in which the arcs are encountered as one moves around a given vertex (say clockwise);

(e)   border relations between regions and the sequences of arcs that bound them, distinguishing between outer borders and hole borders (see Section 9.1.1). (When a region border is followed in a specified sense, say clockwise, then if it is the outer border, the region is always on its right, while if it is a hole border, the region is on its left. Thus, with respect to a given sense of border following, it becomes possible to identify angles as convex or concave.)

As a simple example of this level of description, let us once again consider the "arch" in Fig. 5a, but this time as a line drawing rather than as three rectangular objects. At this level, a description could incorporate the following information:

(1)   Vertices a, b, g, h, i, j are right angles; vertices c, d, e, f are "T"-type branch points.

(2)   There are arcs joining the pairs of vertices (a, b), (g, h), (i, j), (c, d), (d, e), (e, f), (a, c), (b, f), (c, g), (d, h), (e, i), and (f, j). They are all straight line segments; the first six are horizontal, the last six vertical.

(3)   The clockwise order of the line segments at vertex c is (c, a), (c, d), (c, g) [and similarly for the other three branch points].

(4)   In clockwise order, the cyclic sequence of vertices (a, b, f, e, d, c) is on the outer boundary of a region; all its angles are convex (and similarly for other sequences that form region boundaries).

In this type of description, the arcs, vertices and regions can be represented by the nodes of a labeled graph, and their incidence relations by arcs of the graph. Order relations at vertices and around region borders can be incorporated by "threading" directed cycles through the graph. Part of such a graph, for Fig. 5a, is shown in Fig. 5d. One could also create a graph representation in which the properties and relations, as well as the entities, correspond to nodes (compare Figs. 5b and 5c); the details are left as an exercise to the reader. It should be mentioned that descriptions at this level are very similar to the types of descriptions of line drawings that are used in computer graphics.

The list of properties and relations for line drawing description given previously is by no means complete—for example, it omitted relations involving comparisons of property values; relations of adjacency between regions; size and shape properties of regions; etc. Even our abbreviated list allows us to formulate highly redundant descriptions.

Exercise 9. Write a description of Fig. 5a using the line segments as objects and their endpoints as attachment points (see the end of Section 10.3.1). ▌

Exercise 10. Give descriptions of a set of ideal capital letters $(A, B, ...)$ using the concepts in this section; you can ignore serifs. ▌

A single region has a boundary that is a simple closed curve (or a set of such boundaries, if the region has holes). We can segment such a curve at its angles and inflections, and describe it in terms of the resulting parts as above; here the graph of incidence relations is always a cycle, with vertices and arcs alternating. It we want a more detailed description, we can represent each arc by its chain code (Section 1.2.3), i.e., by a string of integers representing slopes of successive unit segments that approximate the arc. An example of a description at the chain code level is our characterization of digital straight line segments in Section 9.3.1.

10.3.3 Descriptions as Models

In some cases, the descriptions that we formulate using the methods of Sections 10.3.1 and 10.3.2 will completely determine the picture (or object) in question, in the sense that it could be reconstructed exactly from the description. More often, however, we want to create a partial description that captures the relevant aspects of the picture, and leaves other aspects unspecified. Such a description constitutes a *model* for a class of pictures, where the given picture is supposed to be representative of the class.

As a simple example, consider the description of the arch (Fig. 5a) that we gave in Section 10.3.1. Here we never specified

(a)   the brightnesses, colors, or textures of the rectangles;

(2)   their sizes (except that A and B are the same size);

(3)   their positions (except that A and B have collinear bases, and that they both support C).

Thus a wide variety of arches fit our definition; some examples are shown in Fig. 5e. Note that part (3) of the arch description in Section 10.3.1 was a negative condition (A and B do not touch); such conditions are often important in modeling classes of pictures.

We can obtain models that are invariant under various classes of operations by omitting quantitative information from a description, or by giving this information in relative rather than absolute form. For example, by not specifying object brightness information, we obtain a model that is invariant under arbitrary transformations of the gray scale (with the understanding that the objects always differ from the background). By not specifying the positions of objects in an absolute sense, but only relative to one another, we obtain a translation-invariant model; and similarly, by specifying orientations or sizes in a relative manner, we obtain rotation or magnification invariance.

It is much harder to formulate descriptions that generalize over nontrivial classes of geometrical distortions. How, for example, could we formulate a model for a reasonably broad class of acceptable A's (not just for a single ideal A shape, as in Exercise 10), and still ensure that anything grossly ill-formed will be excluded? The class of acceptable A's is very hard to define; the allowable tolerances on the stroke lengths and orientations are highly interdependent. The following exercises provide a variety of other examples of varying degrees of difficulty; the reader should not expect to be able to do all of them!

Exercise 11.   Formulate models, in terms of arcs and vertices, for the following classes of objects:

(a)   quadrilaterals,
(b)   simple quadrilaterals (whose sides do not cross one another), and
(c)   convex quadrilaterals.   ∎

Exercise 12.   Formulate models, in terms of chain codes, for the following classes of arcs:

(a)   simple arcs (that do not cross themselves),
(b)   closed curves, and
(c)   closed curves that bound elongated regions.   ∎

Exercise 13.   Formulate models, in terms of skeletons, for the following classes of objects:

(a)   unions of nonoverlapping disks,
(b)   connected regions, and
(c)   simply connected regions.   ∎

It should be mentioned that formal mathematical models have been developed for a variety of picture description languages. These models are analogous to the grammar and automaton models used in the theory of formal languages. In particular, formal models exist for classes of languages such as

   (a)   arbitrary relational structures, represented by labeled graphs ("webs");
   (b)   arrangements of entities in specified relative positions;
   (c)   complexes of entities joined at specified attachment points;
   (d)   arrays of points;
   (e)   strings and cycles (e.g., chain codes).

Such models are potentially important for the development of a theory of the computational complexity of picture description processes (see the end of Section 10.2.3). They are also of some use in modeling real-world classes of pictures; but this approach has had only limited success to date. Because of their specialized mathematical nature, these classes of formal models will not be discussed further here.

10.4 PICTURE ANALYSIS

As indicated at the beginning of Section 10.3, deriving a useful descriptive relational structure from a given picture or scene is often quite difficult. This task is usually called *picture analysis* or *scene analysis* (the latter term is generally used only in connection with "three-dimensional" scenes, in which perspective and occlusions of objects by other objects play significant roles). In all but the most trivial cases, one cannot simply segment the picture into parts, and create a relational structure from this information; indeed, it is rarely possible to specify in simple terms how to extract the desired parts, or even how to define the desired properties and relations.

Section 10.4.1 briefly discusses how models for the class of pictures being analyzed can be used to aid in the process of analysis. Section 10.4.2 illustrates the use of models in picture analysis by discussing an example, involving the computer verification of handwritten signatures, in some detail. Section 10.4.3 comments on the difficulty of constructing explicit models for nontrivial classes of pictures, and on some consequences of this limitation.

10.4.1 The Need for Models

In attempting to analyze a picture, one should certainly make use of whatever prior knowledge is available about the class of pictures to which the given picture belongs. Such knowledge can be regarded as a model, perhaps a very informal one, for the class of pictures in question. The model should be used to guide the choice of picture processing, segmentation, and property measurement operations that will be performed on the given picture. These operations may have to vary from place to place in the picture, to reflect the fact that the picture (as described by the model) consists of distinctive parts. The model should also be used to evaluate the results of these operations, so that the operations can be modified, or further operations performed, as necessary.

Models can be used to guide the picture analysis process on many different levels. The following are a few examples that indicate the range of possibilities:

(a)   At one extreme, we may know that the picture is one of a specified finite set of possibilities, e.g., alphanumeric characters. In this case we may only need to match the picture with a finite set of templates in order to determine which one is actually present.

(b)   More generally, we may know that the picture consists of specified parts in specified spatial relationships, e.g., a human face, containing eyes, nose, mouth, etc.; a chest x ray, consisting of heart, lungs, ribs; a specific printed circuit board, containing components and connectors; and so on. In such cases we can look for the parts, and as we find them, this guides us in where to look for other parts. Of course, the parts themselves may not be trivial to recognize; models for the parts are needed too!

(c)   Still more generally, we may know that certain parts must be present, but not where they are located, e.g., chromosomes in a mitotic cell, or blocks on a tabletop. Here success in finding some of the parts is of less help in finding the others. In some cases, however, detection of one part may impose constraints on the other parts that should be present, e.g., detecting a bubble chamber "event" implies that certain tracks should exist nearby; detecting a bridge in an aerial photograph implies that certain rivers or roads should exist nearby; and so on.

(d)   Further, we may be quite uncertain about the forms that the parts themselves may take, e.g., letters in a handwritten word. The presence of the parts in particular combinations imposes constraints on these forms, although it is often very difficult to formulate these constraints.

As these examples indicate, picture analysis can be a very difficult task. Successful picture analysis systems have, however, been developed that handle problems at all of the levels of difficulty listed above. In the next section we outline the steps in one such system, involving the computer analysis of handwritten signatures, and show how knowledge about the problem domain is used, in an explicit or implicit manner, at each step.

10.4.2   An Example: Handwritten Signature Verification

As an illustration of the use of models to direct the analysis of a picture, let us consider the problem of computer verification of handwritten signatures. The work described here is part of the Ph.D. dissertation of R. N. Nagel (Department of Computer Science, University of Maryland, 1976).

Computer analysis of handwriting is a relatively unexplored research area. Most of the work done in this area has made use of real-time input (i.e., the handwriting is input to the computer by means of a light pen or tablet), so that the time sequence of strokes in the handwriting is known. Very little has been done on the analysis of handwriting that is input to the computer by scanning and digitizing of source documents.

Automatic reading of unknown handwriting is a very difficult problem. Handwriting varies greatly, even for a single individual. It is hard to segment unknown handwritten words into letters or strokes. This becomes much easier, although it is still quite nontrivial, when the words are known, as in the case of the handwritten signature of a known person.

The task of signature verification is to determine whether a given signature is genuine or a forgery. We consider here only the problem of detecting free-hand forgeries, where the forger made no attempt to simulate or trace a genuine signature. In the latter case, the forged signature has the proper shape, but differs from a genuine signature in the quality of the strokes. Free-hand forgeries, on the other hand, differ from a genuine signature with respect to the values of various size and shape features.

The modeling of a handwritten signature is a highly nontrivial task, since even genuine signatures can be quite variable in size and shape; template matching is not a viable approach to signature verification. (In fact, if two signatures match exactly, this is considered good evidence that at least one of them is a simulation or tracing forgery.) Certain features of a signature are, however, believed to be relatively invariant for a given writer; these include ratios of small letter height to signature width, of tall letter height to small letter height, and of distances between letters. To measure these features, it is necessary to segment the signature at least partially—in particular, to determine the small and tall letter heights, and the positions of the letters.

Our signature analysis program begins with a picture containing a digitized signature. The first task is to extract the signature from the background. As we saw in Sections 8.1.1 and 8.1.2, Figs. 1 and 6, this can be done reasonably well by thresholding the picture, where the threshold is selected by examination of the picture's histogram. This step makes use of some basic assumptions about normal writing materials, namely, that the writing contrasts strongly with the paper, and occupies only a small fraction of the picture area, so that the histogram should have two well separated peaks of very different sizes. (Note that we need not assume here that the writing is darker than the paper.)

We next determine the position of the signature in the picture, and the locations of the central, upper, and lower letter zones, by examining the x- and y-projections of the thresholded picture. (We assume here that the signature is approximately horizontal. A badly slanted signature could be detected at this stage by examining projections in oblique directions, and finding the one having least extent; this projection should be perpendicular to the direction of writing. This scheme is based on our knowledge that the signature is elongated in the direction of writing, which is true for all but the shortest names. A more refined scheme could be devised that would make use of cross sections.) These projections, for the signature in Fig. 1a of Chapter 8, are shown in Fig. 6.

An initial guess as to the horizontal position of the signature can be made

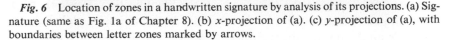

Fig. 6 Location of zones in a handwritten signature by analysis of its projections. (a) Signature (same as Fig. 1a of Chapter 8). (b) x-projection of (a). (c) y-projection of (a), with boundaries between letter zones marked by arrows.

based on the extent of the x-projection. This can be improved later when the first and last letters have been detected. The central letter zone should correspond to the main peak on the y-projection. The boundaries of this zone can be found by looking for positions flanking the peak where (e.g.) a smoothed second difference of the projection changes sign. They are marked with arrows in Fig. 6c. Knowledge of the spelling of the signature is an important factor here, since the shape of the y-projection depends critically on the numbers of letters in the central, upper, and lower zones.

Our next step is to detect the upper- and lower-zone letters. We do this by using the raster tracking method of connected component labeling described in Section 9.1.3. Each connected component in the upper zone receives a unique label, and the points at which it enters the central zone are noted; these points are the best indicators of the sequence of the upper-zone letters, since the letters may have considerable horizontal extent, and often even overlap one another, in the upper zone itself. The lower-zone components are treated similarly. We also store the locations of all uppermost and lowermost runs of signature points, in each zone; these should correspond to "tops" and "bottoms" of letters. If there are too many connected components, the thresholding of the signature can be reconsidered.

We must now try to identify the upper- and lower-zone letters. To guide this process, we make use of a table that contains the following information:

(1) the expected positions of these letters (where they enter the middle zone) relative to the ends of the signature (we shall refer to these positions below as "landmark" positions);

(2) the middle-zone letters that should occur between each consecutive pair of such letters.

Such a table, for the signature of Fig. 6, is shown in Fig. 7a.

Given a consecutive pair of detected landmarks (where letters enter the central zone from above or below), we examine the central-zone tops and bottoms between them and attempt to verify that the proper number of central-zone letters is present. This examination makes use of our prior knowledge about the expected numbers, sizes, and relative positions of the tops and bottoms in these letters. If the distance between the landmarks and the number of intermediate letters are both close to the expected values in the table, we tentatively identify the landmarks and proceed to the next pair. (The actual closeness criteria used are somewhat complex and will not be described in detail here.) A table of such tentative identifications is shown in Fig. 7b.

If a landmark cannot be identified, the previous landmark can be re-examined (by changing the tolerances on the identification decision). Also, when all pairs of consecutive landmarks have been processed, global checks are made on the results: the sequences of upper- and lower-zone letters must be compatible; letters that enter both zones must have been found in consistent positions; and the numbers of expected and detected landmarks must coincide. If these checks are not satisfied, the identifications must be reconsidered. In particular, if too many or too few landmarks were found, the locations of the upper and lower zones must be reconsidered.

When a final identification table is obtained, the letter heights, etc., can be measured. This information can then be used to determine whether the signature is genuine or forged. A table of these measurements for the signature of Fig. 6 is shown in Fig. 7c.

10.4.3   Some Difficulties with Models

As the signature verification example in the last section shows, it is possible to perform highly complex picture analysis tasks by making effective use of prior knowledge about the class of pictures in question. In the signature example, this knowledge was not expressed in the form of an explicit relational structure; it was implicitly embedded in the signature analysis programs

(a)

Upper zone letter		Ideal distance to next letter or end	Letters in this interval having a middle-zone component
	G	.225	o, r, d
	d	.1324	o, n
	J	.0176	—
	V	.157	a, n, d
	d	.0762	e, r
	B	.1798	r, u, g
Lower zone letter	(Beginning)	.3733	o, r, d, o, n
	J	.549	a,n,d,e, r, r, u
	G	0	—

(b) Location (col. nos.)

From	To	Identification	Measured distance to next letter or end
25	28	G	.2095
82	85	d	.1107
114	114	J	
130	131	J	.0158
136	137	V	
140	140	V	
154	155	V	.1423
192	195	d	.0751
215	227	B	.1976
		(Beginning)	.3636
120	120	J	
129	130	J	.5731
276	277	g	0

(c)

Upper- or lower-zone letter	Its height (in rows)	Height of middle-zone letter just preceding it	Ratio
d	13	4	3.25
J	21	5	4.2
d	21	6	3.5
B	35	5	7.0
J	19	5.5	3.45
g	44	10	4.4

Average middle-zone letter height: 5.83
Signature width: 253
Number of letters: 17
Average letter width: 14.9
Ratio of height to width: .39

Fig. 7 Steps in the analysis of the signature shown in Fig. 6. (a) Expected distances between upper- and lower-zone letters (fractions of signature length), and middle-zone letters that should occur between them. (b) Upper- and lower-zone letter identifications assigned to detected "landmarks" (see text). (c) Measurement of letter-height ratios.

themselves. The use of programs or procedures as a representation for knowledge has proved to be a powerful tool in the development of picture and scene analysis systems.

The signature example is illustrative of a wide class of problems in which it does not appear to be possible to formulate our knowledge about the problem domain as an explicit structural model. Indeed, the picture parts, properties, and relations that would enter into such a model may themselves be difficult or impossible to define in explicit terms. In particular

(1) It is hard to specify in advance how a noisy picture, perhaps composed of textured regions, should be segmented, and to describe the allowable variations that characterize a desired class of textures.

(2) Similarly, it is hard to specify in advance how a complex shape should be segmented, and to describe the allowable distortions and "geometrical noise" that characterize a desired class of shapes.

(3) Relational structures built up from noisy pictures will themselves be noisy; many combinations of possible segmentations, etc., have to be considered, and comparison with models becomes a very complicated process.

(4) The allowable variations, in either case, will probably depend on the pictorial *context* of the texture or shape, e.g., on what other textures or shapes are present in the given picture.

(5) In addition, the allowable variations may depend on the *meaning* of the picture, i.e., on information about the real-world situation that is depicted; this information may not itself be present in the picture in any form, but it can influence our interpretation of what is present. If a picture analysis system is to make effective use of real-world knowledge, it must also be able to make inferences from given information, since it will usually be impractical to provide all possible needed information in an explicit form. Thus such a system will have to possess powerful problem-solving capabilities.

In cases where we cannot define models explicitly, picture analysis can no longer be regarded as a process of deriving a relational structure from a picture and only then verifying it by matching it with a model. Instead, we use the implicit, program-embedded model to derive a relational structure from the picture, and this process itself constitutes an implicit verification that the structure "matches," i.e., is consistent with our knowledge about the given class of pictures.

If we were able to formulate explicit models in a wider variety of cases, it might also be possible to develop systematic methods of designing picture analysis systems for given classes of pictures, based on models for the classes. As it is, the design of such systems involves considerable trial and error, and it is often difficult to explain why the system succeeded or failed in a given case.

There is an interesting analogy between the problem of describing complex pictures and the problem of understanding utterances in natural language. Explicit models for classes of pictures can be regarded as analogous to grammars; as indicated at the end of Section 10.3.3, a variety of picture-model formalisms analogous to grammars have in fact been defined. It is difficult to formulate explicit picture models when there is a high degree of variability, or when noise is present; and the same is true about grammars. (Imagine having to formulate a grammar for digitized speech waveforms of utterances!) Nontrivial picture models have to be context dependent; similarly, grammars for nontrivial subsets of a natural language are usually (at least) context-sensitive. Using a model to guide the analysis of a picture requires considerable backtracking, and the same is true when one attempts to parse a natural-language sentence using a context-sensitive grammar. Finally, useful picture models must incorporate real-world knowledge, which may not refer directly to the structure of the picture; and similarly, the comprehension of natural language requires semantic, as well as syntactic, information. The achievement of language comprehension and "picture comprehension" by computers will constitute major steps toward the development of intelligent machines.

10.5 BIBLIOGRAPHICAL NOTES

The use of invariant properties dates back to the earliest days of pictorial pattern recognition. Geometrical normalization techniques have been used for applications as varied as character recognition, target recognition, and chromosome karyotyping. In particular, on the use of best-fitting conic sections see Paton [42]; on the use of circumscribed quadrilaterals see Nagy and Tuong [38]. Transform normalization techniques are also quite classical; see especially Horwitz and Shelton [22], Kain [25], and Doyle [6]. In particular, on the use of integral-geometric techniques to measure properties of the autocorrelation see Tenery [48] and Moore [37]. Geometrical properties of perspective transformations and invariants of such transformations are not treated in this book; see Duda and Hart [7, Chapters 10–11].

Local properties were used, in early work on character recognition, to classify geometrically normalized patterns. A review of textural properties that have been used for pictorial pattern classification may be found in Hawkins [20]. The approach based on co-occurring pairs of gray levels is treated by Haralick et al. [18]. Picture and subpicture templates have been used for many years to define "template match" properties for character recognition and other applications. On moments see Giuliano et al. [14, 32], Hu [23, 24], Alt [1], and Shimbel [47]. On the length, order, and diameter of picture properties see Minsky and Papert [35, 36], and Hodes [21].

Relational structures that describe or define pictures, particularly line drawings, have been extensively studied in connection with computer graphics systems; for a recent review see Williams [49], as well as the collection of papers by Nake and Rosenfeld [39]. Some early examples of relational structure models for objects and line drawings are due to Grimsdale *et al.* [17] and Sherman [46]. On models for pictures or scenes in terms of objects see Evans [10], Kirsch [27], and Lipkin *et al.* [30]. On models for line drawings see also Marill *et al.* [31]; models for curves based on chain coding are discussed by Feder [11]. Generalized relational structures that describe classes of pictures are treated by Winston [51]. On relational structure matching see Barrow and Popplestone [3] and Barrow *et al.* [4].

Specific classes of pictures or objects for which models have been formulated include printed and written alphabetic characters (Eden [8–9]; Narasimhan [41]); chromosomes (Ledley [29]); polyhedra (Roberts [43] and many others); nuclear spark and bubble chamber pictures (Shaw [44, 45]; Feder [12]); and mathematical notation (Anderson [2]). Some of these models are formulated in terms of curve segments and their shapes; others, in terms of entities joined at attaching points; still others, in terms of entities placed in specified relative positions.

For other early discussions of the "syntactic" approach to picture analysis see Minsky [34], Ledley [28], and Narasimhan [40]. For reviews, see Miller and Shaw [33] and Fu [13], and the collections of papers by Kaneff [26] and Nake and Rosenfeld [39]. References dealing with formal models (grammars, automata) will not be given here; see the review papers cited in Section 1.3 for bibliographies, as well as for references to more recent applications.

When models are used to direct the analysis of a picture, they can be represented as data structures, or they can be embodied in the picture analysis procedures themselves. For some recent examples of these two approaches see Harlow [19] and Winograd [50]; the latter deals with natural language understanding of descriptions of simple scenes. Several high-level programming languages have been developed that facilitate the incorporation of knowledge into procedures; for a recent review of these see Bobrow and Raphael [5].

REFERENCES

1. F. L. Alt, Digital pattern recognition by moments, *J.ACM* **9**, 1962, 240–258. Also *in* "Optical Character Recognition" (G. L. Fischer *et al.*, eds.), pp. 153–179. Spartan, Baltimore, Maryland, 1962.
2. R. H. Anderson, Syntax-directed recognition of handprinted two-dimensional mathematics, *in* "Interactive Systems for Experimental Applied Mathematics" (M. Klerer and J. Reinfelds, eds.), pp. 436–459. Academic Press, New York, 1968.
3. H. G. Barrow and R. J. Popplestone, Relational descriptions in picture processing, *in*

"Machine Intelligence" (B. Meltzer and D. Michie, eds.), Vol. 6, pp. 377–396. University Press, Edinburgh, Scotland, 1971.

4. H. G. Barrow, A. P. Ambler, and R. M. Burstall, Some techniques for recognizing structure in pictures, *in* "Frontiers of Pattern Recognition" (S. Watanabe, ed.), pp. 1–29. Academic Press, New York, 1972.

5. D. G. Bobrow and B. Raphael, New programming languages for artificial intelligence research, *Comput. Surv.* **6**, 1974, 153–174.

6. W. Doyle, Operations useful for similarity-invariant pattern recognition, *J.ACM* **9**, 1962, 259–267.

7. R. O. Duda and P. E. Hart, "Pattern Classification and Scene Analysis," Wiley, New York, 1973.

8. M. Eden, On the formalization of handwriting, *Proc. Symp. Appl. Math.* **12**, 83–88. Amer. Math. Soc., Providence, Rhode Island, 1961.

9. M. Eden, Handwriting and pattern recognition, *IRE Trans. Informat. Theory* **IT-8**, 1962, 160–166.

10. T. G. Evans, A heuristic program to solve geometric-analogy problems, *Proc. Spring Joint Comput. Conf.* pp. 327–338. Spartan, New York, 1964.

11. J. Feder, Languages of encoded line patterns, *Informat. Contr.* **13**, 1968, 230–244.

12. J. Feder, Plex languages, *Informat. Sci.* **3**, 1971, 225–241.

13. K. S. Fu, "Syntactic Methods in Pattern Recognition." Academic Press, New York, 1974.

14. V. E. Giuliano. P. E. Jones, G. E. Kimball, R. F. Meyer, and B. A. Stein, Automatic pattern recognition by a Gestalt method, *Informat. Contr.* **4**, 1961, 332–345.

15. U. Grenander, Toward a theory of patterns, *Symp. Probability Methods Anal.* pp. 79–111. Springer, New York, 1967.

16. U. Grenander, A unified approach to pattern analysis, *Advan. Comput.* **10**, 1970, 175–216.

17. R. L. Grimsdale, F. H. Sumner, G. J. Tunis, and T. Kilburn, A system for the automatic recognition of patterns, *Proc. IEE* **106B**, 1959, 210–221.

18. R. M. Haralick, K. Shanmugam, and I. Dinstein, Textural features for image classification, *IEEE Trans. Sys. Man Cybernet.* **SMC-3**, 1973, 610–621.

19. C. A. Harlow, Image analysis and graphs, *Comput. Graph. Image Proc.* **2**, 1973, 60–82.

20. J. K. Hawkins, Textural properties for pattern recognition, *in* "Picture Processing and Psychopictorics" (B. S. Lipkin and A. Rosenfeld, eds.), pp. 347–370. Academic Press, New York, 1970.

21. L. Hodes, The logical complexity of geometric properties in the plane, *J.ACM* **17**, 1970, 339–347.

22. L. P. Horwitz and G. L. Shelton, Jr., Pattern recognition using autocorrelation, *Proc. IRE* **49**, 1961, 175–185.

23. M. K. Hu, Pattern recognition by moment invariants, *Proc. IRE* **49**, 1961, 1428.

24. M. K. Hu, Visual pattern recognition by moment invariants, *IRE Trans. Informat. Theory* **IT-8**, 1962, 179–187.

25. R. Y. Kain, Autocorrelation pattern recognition, *Proc. IRE* **49**, 1961, 1085–1086.

26. S. Kaneff, "Picture Language Machines," Academic Press, London, 1970.

27. R. A. Kirsch, Computer interpretation of English text and picture patterns, *IEEE Trans. Electron. Comput.* **EC-13**, 1964, 363–376.

28. R. S. Ledley, "Programming and Utilizing Digital Computers," pp. 364–367. McGraw-Hill, New York, 1962.

29. R. S. Ledley, High-speed automatic analysis of biomedical pictures, *Science* **146**, 1964, 216–223.

30. L. E. Lipkin, W. C. Watt, and R. A. Kirsch, The analysis synthesis, and description of biological images, *Ann. N. Y. Acad. Sci.* **128**, 1966, 986–1012.

31. T. Marill, A. K. Hartley, T. G. Evans, B. H. Bloom, D. M. R. Park, T. P. Hart, and D. L. Darley, CYCLOPS-1: a second-generation recognition system, *Proc. Fall Joint Comput. Conf.* pp. 27–33. Spartan, New York, 1963.

32. R. F. Meyer, V. E. Giuliano, and P. E. Jones, Analytic approximation and translational invariance in character recognition, *in* "Optical Character Recognition" (G. L. Fischer *et al.*, eds.), pp. 181–195. Spartan, Baltimore, Maryland, 1962.

33. W. F. Miller and A. C. Shaw, Linguistic methods in picture processing—a survey, *Proc. Fall Joint Comput. Conf.*, pp. 279–290. Thompson, Washington, D. C., 1968.

34. M. L. Minsky, Steps toward artificial intelligence, *Proc. IRE* **49**, 1961, 8–30.

35. M. L. Minsky and S. Papert, Linearly unrecognizable patterns, *Proc. Symp. Appl. Math.* **19**, 176–217. Amer. Math. Soc., Providence, Rhode Island, 1967.

36. M. L. Minsky and S. Papert, "Perceptrons: An Introduction to Computational Geometry." MIT Press, Cambridge, Massachusetts, 1969. See also the review by H. D. Block, *Informat. Contr.* **17**, 1970, 510–522.

37. D. J. H. Moore, An approach to the analysis and extraction of pattern features using integral geometry, *IEEE Trans. Syst. Man Cybernet.* **SMC-2**, 1972, 97–102.

38. G. Nagy and N. Tuong, Normalization techniques for hand-printed numerals, *Comm. ACM* **13**, 1970, 475–481.

39. F. Nake and A. Rosenfeld (eds.), "Graphic Languages." North-Holland Publ., Amsterdam, 1972.

40. R. Narasimhan, Labelling schemata and syntactic descriptions of pictures, *Informat. Contr.* **7**, 1964, 151–179.

41. R. Narasimhan, Syntax-directed interpretation of classes of pictures, *Comm. ACM* **9**, 1966, 166–173.

42. K. Paton, Conic sections in chromosome analysis, *Pattern Recognit.* **2**, 1970, 39–51.

43. L. G. Roberts, Machine perception of three-dimensional solids, *in* "Optical and Electro-Optical Information Processing" (J. T. Tippett *et al.*, eds.) pp. 159–197. MIT Press, Cambridge, Massachusetts, 1965.

44. A. C. Shaw, A formal picture description scheme as a basis for picture processing systems, *Informat. Contr.* **14**, 1969, 9–52.

45. A. C. Shaw, Parsing of graph-representable pictures, *J.ACM* **17**, 1970, 453–481.

46. H. Sherman, A quasi-topological method for the recognition of line patterns, "Information Processing," pp. 232–237. UNESCO, Paris, 1959.

47. A. Shimbel, A logical program for the simulation of visual pattern recognition, *in* "Principles of Self-Organization" (H. von Foerster and G. W. Zopf, Jr., eds.), pp. 521–526. Pergamon, Oxford, 1962.

48. G. R. Tenery, A pattern recognition function of integral geometry, *IEEE Trans. Military Electron.* **MIL-7**, 1963, 196–199.

49. R. Williams, A survey of data structures for computer graphics systems, *Comput. Surv.* **3**, 1971, 1–21.

50. T. Winograd, "Understanding Natural Language." Academic Press, New York, 1972. Also appeared as *Cognitive Psychol.* **3**, 1972, 1–191.

51. P. H. Winston, Learning structural descriptions from examples, AI Tech. Rep. 231, Artificial Intelligence Lab., MIT, Cambridge, Massachusetts, 1970.

Index